PREALGEBRA

SECOND EDITION

K. Elayn Martin-Gay
University of New Orleans

Prentice Hall
Upper Saddle River
New Jersey 07458

Library of Congress Cataloging-in-Publication Data

Martin-Gay, K. Elayn
Prealgebra / K. Elayn Martin-Gay.—2nd ed.
p. cm.
Includes index.
ISBN 0-13-242470-3
1. Arithmetic. I. Title.
QA107.M293 1997
513'.1—dc21 97-3525
CIP

Sponsoring Editors, Melissa Acuña and Ann Marie Jones
Editor-in-Chief, Jerome Grant
Editorial Director, Tim Bozik
Production Editors, Barbara Mack and Robert C. Walters, PMI
Managing Editor, Linda Behrens
Assistant Vice President of Production and Manufacturing, David W. Riccardi
Executive Managing Editor, Kathleen Schiaparelli
Development Editor, Emily J. Keaton
Marketing Manager, Jolene Howard
Marketing Assistant, Jennifer Pan
Creative Director, Paula Maylahn
Art Director, Amy Rosen
Assistant to Art Director, Rod Hernandez
Art Manager, Gus Vibal
Interior Design, Geri Davis, The Davis Group, Inc.
Cover Design, Bruce Kenselaar
Photo Editor, Lori Morris-Nantz
Photo Research, Beaura Katherine Ringrose
Manufacturing Buyer, Alan Fischer
Manufacturing Manager, Trudy Pisciotti
Supplements Editor/Editorial Assistant, April Thrower
Cover Photo, Art Wolfe / Tony Stone Images

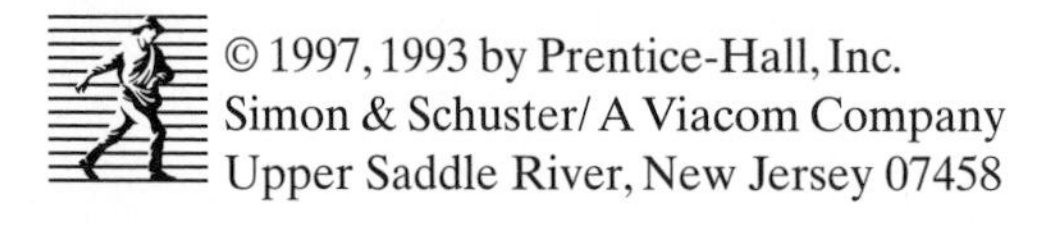

Printed in the United States of America

10 9 8 7 6 5

ISBN 0-13-242470-3

Prentice-Hall International (UK) Limited, *London*
Prentice-Hall of Australia Pty. Limited, *Sydney*
Prentice-Hall Canada Inc., *Toronto*
Prentice-Hall Hispanoamericana, S.A., *Mexico City*
Prentice-Hall of India Private Limited, *New Delhi*
Prentice-Hall of Japan, Inc., *Tokyo*
Simon & Schuster Asia Pte. Ltd., *Singapore*
Editora Prentice-Hall do Brasil, Ltda., *Rio de Janeiro*

To Jewett B. Gay, and to the memory of her husband, Jack Gay

CONTENTS

PREFACE

ABOUT THE BOOK

This book was written to help students make the transition from arithmetic to algebra. To reach this goal, I introduce algebraic concepts early and repeat them as I treat traditional arithmetic topics, thus laying the groundwork for the next algebra course your students will take. A second goal was to show students the relevancy of the mathematics in everyday life and in the workplace.

In preparing this second edition, I considered the comments and suggestions of colleagues throughout the country and of the many users of the first edition. The numerous features that contributed to the success of the first edition have been retained. This updated revision includes new mathematical content and increased attention to geometric concepts, data interpretation, problem solving, and real-life applications. I have carefully chosen pedagogical features to help students understand and retain concepts. The key content and pedagogical features are described on this and the following pages.

In addition, the supplements to this text have been enhanced and the range of supplements increased, offering a complete integrated teaching and learning package for maximum support and effectiveness.

KEY PEDAGOGICAL FEATURES IN THE SECOND EDITION

Readability and Connections Many reviewers of this edition as well as users of the previous edition have commented favorably on the readability and clear, organized presentation. I have tried to make the writing style as clear as possible while still retaining the mathematical integrity of the content. As new topics are presented, efforts have been made to relate the new ideas to those that the students may already know. Constant reinforcement and connections within problem-solving strategies, geometric concepts, pattern recognition, and situations from everyday life can help students gradually master both new and old information.

Accessible Real-World Applications Many new practical applications are found throughout the book in worked-out examples and exercise sets. The applications were carefully chosen to be accessible, to help motivate students, and to strengthen their understanding of mathematics in the real world. They help show connections to a wide range of areas such as consumer applications, biology, environmental issues, allied health, business, entertainment, history, art, finance, sports, and physical science, as well as to important related mathematical areas such as geometry. Many involve current and interesting real-life data. Sources for data include newspapers, magazines, government publications, and reference books.

Unique Exercise Sets Each exercise set is divided into two parts. Both parts contain graded problems. The first part is carefully keyed to worked examples in the text. A student can gain confidence and then move on to the remaining exercises, which are not keyed to examples. There are ample exercises throughout the book, including end-of-chapter reviews, tests, and cumulative reviews. In addition, each exercise set contains one or more of the following features.

Mental Mathematics These problems are found at the beginning of many exercise sets. They are mental warmups that reinforce concepts found in the accompanying section and increase students' confidence before they tackle an exercise set. By relying on their own mental skills, students increase not only their confidence in themselves, but also their number sense and estimation ability.

Review Exercises Formerly called Skill Review, these exercises are found at the end of each section after Chapter 1. These problems are keyed to earlier sections and review concepts learned earlier in the text that are needed in the next section or in the next chapter. These exercises show the links between earlier topics and later material.

Conceptual and Writing Exercises These exercises, now found in almost every exercise set, are keyed with the icon ❏. They require students to show an understanding of a concept learned in the corresponding section. This is accomplished by asking students questions that require them to use two or more concepts together. Some of these exercises require students to stop, think, and explain in their own words the concept(s) used in the exercises they have just completed. Guidelines recommended by the American Mathematical Association of Two-Year Colleges (AMATYC Crossroads guidelines) and other professional groups suggest incorporating writing in mathematics courses to reinforce concepts.

Practice Problems Throughout the text, each worked example has a parallel problem called a Practice Problem, found in the margin. Practice Problems invite students to be actively involved in the learning process before beginning the section exercise set. Practice Problems *immediately reinforce* a concept after it is developed.

Reminder Reminders, formerly Helpful Hint boxes, contain practical advice on problem solving. Reminders appear in the context of material in the chapter and give students extra help in understanding and working problems. They are highlighted in a box for quick reference.

Visual Reinforcement of Concepts The text contains a wealth of graphics, models, and illustrations to visually clarify and reinforce concepts. These include new bar charts, line graphs, application illustrations, calculator screens, and geometric figures.

Scientific Calculator Explorations and Exercises Scientific Calculator Explorations contain examples and exercises placed appropriately throughout the text to instruct students on the proper use of the calculator and reinforce concepts.

Additional exercises building on the skills developed in the Explorations may be found in exercise sets throughout the text, and are marked with an icon . The inside back cover of the text includes a brief description of selected keys on a scientific calculator for reference as desired.

Group Activities Each chapter opens with a photograph and description of a real-life situation. At the close of the chapter, students can apply the mathematical and critical thinking skills they have learned to make decisions and answer the questions in the Group Activity, which is related to the chapter-opening situation. The Group Activity is a multi-part, often hands-on, problem. These new situations, designed for student involvement and interaction, allow for a variety of teaching and learning styles.

Answers and suggestions specific to the Group Activities are available in the Annotated Instructor's Edition.

Chapter Highlights Found at the end of each chapter, the new Chapter Highlights contain key definitions, concepts, *and examples* to help students better understand and retain what they have learned.

Chapter Review and Test The end of each chapter contains a review of topics introduced in the chapter. These review problems are keyed to sections. The chapter test is not keyed to sections.

Cumulative Review Each chapter after the first contains a cumulative review. Each problem contained in the cumulative review is actually an earlier worked example in the text that is referenced in the back of the book along with the answer. Students who need to see a complete worked-out solution, with explanation, can do so by turning to the appropriate example in the text.

Functional Use of Color and Design Elements of the text are highlighted with color or design to make it easier for students to read and study.

Videotape and Software Icons At the beginning of each section, videotape and software icons are displayed. The icons help remind students that these learning aids are available should they wish to use them in reviewing concepts and skills at their own pace. These learning aids have direct correlation with the text and emphasize the text's methods of solution.

NEW KEY CONTENT FEATURES IN THE SECOND EDITION

Greater Integration of Geometry There is increased emphasis and coverage of geometric concepts. This was accomplished by including a greater integration of geometry throughout the text such as focusing more on finding the perimeter and area of composite figures.

For instance, perimeter is now introduced earlier, in Section 1.2. Then further perimeter concepts are introduced in Section 3.6 and used throughout as appropriate. Similarly, the material on area is introduced in Section 1.5 and used throughout, with special emphasis in Section 5.9. Circumference of a circle, volume, polygons, the Pythagorean theorem, and other geometric concepts are introduced and integrated as appropriate.

More Intuitive Introduction to Graphing and Descriptive Statistics Reading tables and bar and line graphs is gradually introduced (beginning in Section 1.1), and then integrated throughout and reviewed and expanded in Section 6.1. Throughout this review, an emphasis is slowly placed on the concept of paired data. This leads naturally to the idea of an ordered pair of numbers and the rectangular coordinate system, introduced in Section 6.2. Chapter 6 also includes a new section on averages, medians, and modes.

Polynomials I have added a new chapter on polynomials. This chapter allows for increased flexibility in the amount and use of algebra topics needed for students to succeed in this course and to prepare them for future courses.

Problem-Solving Approach This is introduced with a new six-step process that is integrated throughout the text. The six steps are UNDERSTAND, ASSIGN, TRANSLATE, SOLVE, CHECK, and STATE. The repeated use of these steps in a variety of examples shows their wide applicability. Reinforcing the steps can increase students' comfort level and confidence in tackling problems.

A special emphasis and strong commitment are given to contemporary and practical applications of algebra. Real data were drawn from a variety of sources, including magazines, newspapers, government publications, and reference books.

Increased Opportunities to Use Technology Optional calculator explorations and exercises are integrated appropriately throughout the text.

New Examples Additional detailed step-by-step examples were added where needed. Many of these reflect real-life situations. Examples are used in two ways—numbered, as formal examples, to check or increase understanding of a topic, and unnumbered, to introduce a topic or informally discuss the topic.

New Exercises A significant amount of time was spent on the exercise sets. New exercises and additional examples help address a wide range of student learning styles and abilities. New kinds of exercises, strategically placed, include group activities, conceptual and writing exercises, real data applications, optional calculator exercises, and reading tables, charts, and graphs. In addition, the mental math, computational, and word problems were refined and enhanced. There is now a total of approximately 5,000 exercises.

ACKNOWLEDGMENTS

First, as usual, I would like to thank my husband, Clayton, for his constant encouragement. I would also like to thank my children, Eric and Bryan, for continuing to eat my cooking and going on adventures with me.

I would also like to thank my extended family for their invaluable help and wonderful sense of humor. Their contributions are too numerous to list. They are Rod and Karen Pasch; Peter, Michael, Christopher, Matthew, and Jessica Callac; Stuart, Earline, Melissa, and Mandy Martin; Mark, Sabrina, and Madison Martin; Leo and Barbara Miller; and Jewett Gay.

A special thank you to all users of the first edition of this text who made suggestions for improvements that were incorporated into the second edition. I would also like to thank the following reviewers of this text:

Rebecca C. Benson-Beaver, *Valencia Community College*
Karen Sue Cain, *Eastern Kentucky University*
Celeste Carter, *Richland College*
Carol Chapman, *Orange Coast College*
Camille Cochrane, *Shelton State Community College*
Margarita Fresquez, *Palo Alto College*
Terry Y. Fung, *Kean College of New Jersey*
Mary Ellen Gallegos, *Sante Fe Community College*
Mark Greenhalgh, *Fullerton College*
Teresa Hasenauer, *Indian River Community College*
Ruth Ann Henke, *Southern Illinois University at Edwardsville*
John Heublein, *Kansas State University—Salina*
Maryann E. Justinger, *Erie Community College—South*
Robert Kaiden, *Lorain County Community College*
Stephan Kinholt, *Green River Community College*
Helen Kirk, *Palo Alto College*
Patricia Lanz, *Erie Community College—South Campus*
Michael Montano, *Riverside Community College*
Sam Tinsley, *Richland College*

Special thanks to Cheryl Roberts for contributing to the overall accuracy of the book. Phyllis Barnidge and Penny Korn did an excellent job of providing and checking answers and solutions. Emily Keaton was invaluable for her many suggestions during the development of the second edition. I very much appreciated the writers and accuracy checkers of the supplements to accompany this text. Last, but by no means least, a special thanks to the staff at Prentice Hall for their support and assistance: Melissa Acuña, Ann Marie Jones, Barbara Mack, Robert Walters, Linda Behrens, Alan Fischer, Amy Rosen, Gus Vibal, Paula Maylahn, April Thrower, Evan Girard, Jolene Howard, Jennifer Pan, Gary June, Jerome Grant, and Tim Bozik.

K. Elayn Martin-Gay

ABOUT THE AUTHOR

K. Elayn Martin-Gay has taught mathematics at the University of New Orleans for over 18 years and has received numerous teaching awards, including the local University Alumni Association's Award for Excellence in Teaching.

Over the years, Elayn has developed a videotaped lecture series to help her students better understand algebra. This highly successful video material is the basis for the five-book series *Prealgebra, Beginning Algebra, Intermediate Algebra, Introductory and Intermediate Algebra*, a combined approach, and *Intermediate Algebra: A Graphing Approach.*

SUPPLEMENTS FOR THE STUDENT

PRINTED SUPPLEMENTS

STUDENT SOLUTIONS MANUAL (ISBN 0–13–258237–6)

Detailed step-by-step solutions to odd-numbered text and review exercises.
Solutions to all chapter practice tests.
Solution methods reflect those emphasized in the text.
Includes study skills and note-taking suggestions.
Ask your bookstore about ordering.

STUDENT STUDY GUIDE (ISBN 0–13–248299–5)

Additional step-by-step worked out examples and exercises.
Practice tests.
Two practice final examinations.
Solution methods reflect those emphasized in the text.
Includes study skills and note-taking suggestions.
Ask your bookstore about ordering.

THE NEW YORK TIMES SUPPLEMENT

A free newspaper from Prentice Hall and *The New York Times.*
Interesting and current articles on mathematics.
Invites discussion and writing about mathematics.
Created new each year.

HOW TO STUDY MATH (ISBN 0–13–020884–1)

MEDIA SUPPLEMENTS

VIDEOTAPE SERIES

(Sample Video ISBN 0–13–258146–9);
(Video Series ISBN 0–13–258187–6)
Specifically keyed to the textbook by section.
Presentation and step-by-step examples by the textbook author using the terminology and solution methods emphasized in the text.
Comprehensive coverage.

MATHPRO TUTORIAL SOFTWARE

(IBM Network-User ISBN 0–13–258203–1);
(IBM Single-User ISBN 0–13–268822–0);
(Mac Network-User ISBN 0–13–258211–2);
(Mac Single-User ISBN 0–13–840653–7)
Text-specific tutorial exercises.
Interactive feedback.
Graded and recorded Practice Problems.
New user interface, glossary, and expressions editor for ease of use and flexibility.
Network version available.

SUPPLEMENTS FOR THE INSTRUCTOR

PRINTED SUPPLEMENTS

ANNOTATED INSTRUCTOR'S EDITION (ISBN 0–13–269317–8)

Answers to exercises on the same text page.
Instructor's answers also include answers and pedagogical suggestions for group activities.

INSTRUCTOR'S SOLUTIONS MANUAL (ISBN 0–13–258195–7)

Solutions to even-numbered exercises and chapter tests.
Graphics computer-generated for clarity.
Notes to the Instructor.

TEST ITEM FILE (ISBN 0–13–258153–1)

Six forms (A,B,C,D,E, and F) of Chapter Tests.
- Three forms contain multiple-choice items.
- Three forms contain free-response items.

Two forms of Cumulative Review Tests.
- Every two chapters.

Final Exams.
- Four forms with free-response scrambled items.
- Four forms with multiple-choice scrambled items.

Answers to all items.

MEDIA SUPPLEMENTS

TESTPRO2 COMPUTERIZED TESTING

(Sample Disk IBM, ISBN 0–13–258104–3);
(Sample Disk Mac, ISBN 0–13–258112–4);
(IBM, ISBN 0–13–258161–2);
(Mac, ISBN 0–13–258179–5)
Comprehensive text-specific testing.
Generates test questions and drill worksheets from algorithms keyed to the text learning objectives.
Edit or add your own questions.
Compatible with Scantron or other possible scanners.

USING THE INTERNET AND A WEB BROWSER

Using the Internet and a Web Browser, such as Netscape, can add to your mathematical resources. The following is a list of some of the sites that may be worth your or your students' visit.

Prentice Hall Home Page	http://www.prenhall.com
The Mathematical Association of America	http://www.maa.org
The American Mathematical Society	http://www.ams.org
The National Council of Teachers of Mathematics	http://www.nctm.org
The Census Bureau	http://www.census.gov

INTERNET GUIDE (ISBN 0–13–268616–3)

This guide provides a brief history of the Internet, discusses the uses of the World Wide Web, and describes how to find your way within the Internet and how to reach others on it. Contact your local Prentice Hall representative regarding the Internet Guide.

Photo Credits

p.xvii Tony Freeman/PhotoEdit

p. xx Jim Williamson

CHAPTER 1 CO Laima Druskis/Simon & Schuster/PH College **p. 53** Teri Stratford/Simon & Schuster/PH College **p. 78** Air France

CHAPTER 2 CO Tony Freeman/PhotoEdit

CHAPTER 3 CO Rosemary Weller/Tony Stone Images

CHAPTER 4 CO Marabella's

CHAPTER 5 CO D. C. Lowe/FPG International

CHAPTER 6 CO Irene Springer/Simon & Schuster/PH College

CHAPTER 7 CO Dick Luria/FPG International **p. 480** Niesenbahan/Switzerland Tourism

CHAPTER 8 CO David Young-Wolff/PhotoEdit

CHAPTER 9 CO Matura/Gamma-Liaison, Inc.

How to Use the Text: A Guide for Students

Prealgebra, Second Edition has been designed as just one of the tools in a fully integrated learning package to help you develop prealgebra skills. Our goal is to encourage your success and mastery of the mathematical concepts introduced in this text. Take a few moments now to see how this text will help you excel.

CHAPTER

2

Integers

Investigating Integers

Popular board games often involve instructions for a player to move his or her playing piece forward several spaces each turn. Occasionally a player might receive an instruction to move backward. These instructions to move forward or backward a certain number of spaces can be represented by positive and negative integers.

In the chapter Group Activity on page xxx, you will have the opportunity to play a game based on integers.

The photo application at the opening of every chapter and **applications throughout** offer real-world scenarios that connect mathematics to your life. In addition, at the end of the chapter, a group activity or discovery-based project further shows the chapter's applicability.

Page 101

Group Activity

Investigating Integers

Materials:
- Colored thumbtacks
- coin
- six-sided die
- tape

Try the following activity. The object is to have the largest absolute value at the end of the game.

1. Attach this page to a piece of cardboard with tape. Each group member should choose a different colored thumbtack as his or her playing piece. Insert each thumbtack at the starting place 0 on the number line at the left.

	Signed Number
Turn 1:	
Turn 2:	
Turn 3:	
Turn 4:	
Turn 5:	
Total	

3. Continue taking turns until each group member has taken five turns. Verify your final position on the number line by finding the total of the integers in the table. Do your total and final position agree?

There is an opportunity to **explore** an exercise that relates to the chapter-opening photo as a group activity or discovery-based project.

Page 143

STUDY FOR SUCCESS

As you study, **make connections**–this text's organization can help you. There are features in this text designed to help you relate material you are learning to previously mastered material. Math topics are tied to real life as often as possible. Read the learning objectives at the beginning of every section.

Save time by having a plan. Follow the **six-step process,** and you will find yourself successfully solving a wide range of problems.

PROBLEM-SOLVING STEPS

1. UNDERSTAND the problem. During this step, don't work with variables, but simply become comfortable with the problem. Some ways of accomplishing this are to
   * Read and reread the problem.
   * Construct a drawing.
2. ASSIGN a variable to an unknown in the problem. Use this variable to represent any other unknown quantities.
3. TRANSLATE the problem into an equation.
4. SOLVE the equation.
5. CHECK the proposed solution in the stated problem.
6. STATE your conclusion.

Page 192

REMINDER When simplifying expressions with exponents, notice that parentheses make an important difference.

$(-3)^2$ and -3^2 do not mean the same thing.
$(-3)^2$ means $(-3)(-3) = 9$.
-3^2 means the opposite of $3 \cdot 3$, or -9.
Only with parentheses is the -3 squared.

"Reminders" contain practical advice and provide extra help in understanding and working problems.

Page 135

CHAPTER 9 HIGHLIGHTS

DEFINITIONS AND CONCEPTS	EXAMPLES
SECTION 9.1 ADDING AND SUBTRACTING POLYNOMIALS	
A **polynomial** is a monomial or a sum or difference of monomials.	Polynomials $5x^2 - 6x + 2,\ -\frac{9}{10}y,\ 7$
To add polynomials, combine like terms.	Add $(7z^2 - 6z + 2) + (5z^2 - 4z + 5)$. $(7z^2 - 6z + 2) + (5z^2 - 4z + 5)$ $= 7z^2 + 5z^2 - 6z - 4z + 2 + 5$ Group like terms. $= 12z^2 - 10z + 7$ Combine like terms.

The new **Chapter Highlights** contain key definitions, concepts, and examples to help students better understand and retain what they have learned.

Page 654

After reading the example, try the **Practice Problem** to help you better understand the concept.

PRACTICE PROBLEM 3
How much fencing is needed to enclose a square field 50 yards on a side?

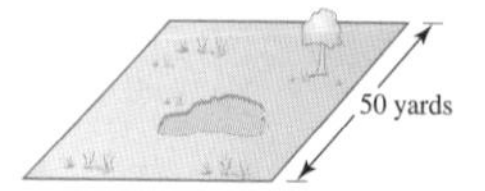

Answers:
2. 64 centimeters
3. 200 yards

EXAMPLE 3
Find the perimeter of a square table top if each side is 5 feet long.

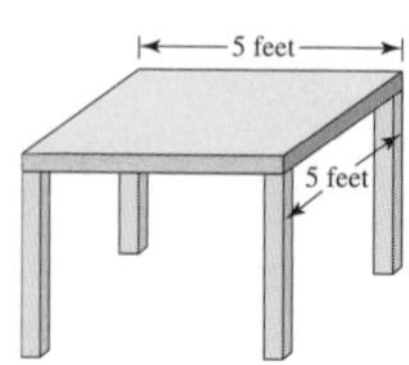

Solution:

The formula for the perimeter of a square is $P = 4s$. Use this formula and replace s by 5 feet.

Page 200

Check Your Understanding. Expand It. Explore!

Good exercise sets are essential to the make-up of a solid prealgebra textbook. The exercises you will find in this textbook are designed to help you understand skills and concepts as well as challenge and motivate you. Note, too, the Highlights, Test, Review and Cumulative Review found at the end of each chapter.

Mental Math

1. Find John Steven's net pay if his monthly salary is $1635 and his total deductions are $635.
2. Find Joann Bosticis's commission on total sales of $27,650 if she is paid 10% of total sales.
3. Find Tim Franklin's gross wages if he worked 40 hours a week and is paid $10.00 per hour.
4. Mr. Carleson makes a monthly salary of $2000.00. Find his annual gross pay.

Confidence building **Mental Math** problems are in many sections.

Page 598

Solve. See Examples 1–5.

13. A polygon has sides of length 5 feet, 3 feet, 2 feet, 7 feet, and 4 feet. Find its perimeter.
14. A triangle has sides of length 8 inches, 12 inches, and 10 inches. Find its perimeter.
15. Baseboard is to be installed in a square room that measures 15 feet on one side. Find how much baseboard is needed.
16. Find how much fencing is needed to enclose a rectangular rose garden 85 feet by 15 feet.
17. If a football field is 53 yards wide and 120 yards long, what is the perimeter?
18. A stop sign has eight equal sides of length 12 inches. Find its perimeter.

19. A metal strip is being installed around a workbench that is 8 feet long and 3 feet wide. Find how much stripping is needed.
20. Find how much fencing is needed to enclose a rectangular garden 70 feet by 21 feet.
21. If the stripping in Exercise 19 costs $3 per foot, find the total cost of the stripping.
22. If the fencing in Exercise 20 costs $2 per foot, find the total cost of the fencing.

Build your confidence with the beginning exercises; the first part of each exercise set is keyed to already worked examples. Then try the remaining exercises

Conceptual and Writing Exercises bring together two or more concepts and often require "in your own words" written explanation.

Page 206

Calculator Explorations

Compound Interest Factor

A compound interest factor may be found by using your calculator and evaluating the formula

$$\textbf{compound interest factor} = \left(1 + \frac{r}{n}\right)^{nt}$$

where r is the interest rate, t is the time in years, and n is the number of times compounded per year. For example, we stated earlier that the compound interest factor for 10 years at 8% compounded semiannually is 2.19112. Let's find this factor by evaluating the compound interest factor formula when $r = 8\%$ or 0.08, $t = 10$, and $n = 2$ (compounded semiannually means 2 times per year). Thus,

$$\text{compound interest factor} = \left(1 + \frac{0.08}{2}\right)^{2\cdot 10} \quad \text{or} \quad \left(1 + \frac{0.08}{2}\right)^{20}$$

To evaluate, press the keys

The display will read 2.1911231. Rounded to 5 decimal places, this is 2.19112.

Find the compound interest factor. Use the table in Appendix D to check your answer.

1. 5 years, 9%, compounded quarterly 1.56051
2. 15 years, 14%, compounded daily 8.16288
3. 20 years, 11%, compounded annually 8.06231

Calculator explorations and exercises are woven into the appropriate sections to reinforce concepts and motivate **discovery-based learning.**

Page 589

Review Exercises

Evaluate each expression using the given replacement numbers. See Section 2.5.

67. x^3 when $x = -3$
68. y^3 when $y = -5$
69. $2y$ when $y = -7$
70. $-5a$ when $a = -4$
71. $3z - y$ when $z = 2$ and $y = 6$
72. $7a - b$ when $a = 1$ and $b = -5$
73. $a^2 + 2b + 3$ when $a = 4$ and $b = 5$
74. $yx - z^2$ when $y = 6, x = 6,$ and $z = 6$

Review Exercises review concepts learned earlier that are needed in the next section or next chapter.

Page 237

STUDENT RESOURCES

1.2

ADDING WHOLE NUMBERS AND PERIMETER

OBJECTIVES

1. Add whole numbers.
2. Identify properties of addition.
3. Find the perimeter of a polygon.
4. Solve problems by adding whole numbers.

1 If one computer in an office has a 2-megabyte memory and a second computer has a 4-megabyte memory, the total memory in the two computers can be found by adding 2 and 4.

$$2 \text{ megabytes} + 4 \text{ megabytes} = 6 \text{ megabytes}$$

The **sum** is 6 megabytes of memory. Each of the numbers 2 and 4 is called an **addend.**

Page 11

Text-specific videos hosted by the award-winning teacher and author of ***Prelgebra, Second Edition*** cover each objective in every chapter section as a supplementary review.

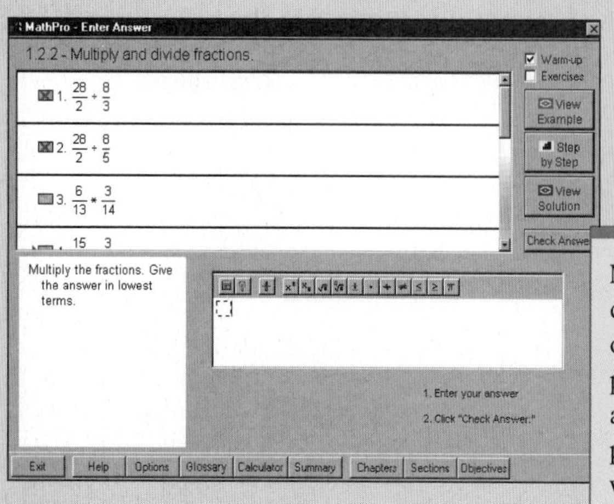

MathPro Tutorial Software, developed around the content of ***Prelgebra, Second Edition,*** provides interactive warm-up and graded algorithmic practice problems with step-by-step worked solutions.

ALSO AVAILABLE:

The New York Times/ Themes of the Times

Newspaper-format supplement—*ask your professor about this exciting free supplement!*

CHAPTER

1

Whole Numbers and Introduction to Algebra

Analyzing the Earning Power of a College Degree

Enrollments at colleges and universities are increasing. However, tuition costs are growing at rates faster than the rate of inflation. Some economists believe that this trend in increasing college enrollments despite escalating tuition costs is due to the increased earning power of a person with a college degree.

In the chapter Group Activity on page 87, you will have the opportunity to analyze how average annual wages have changed for holders of high school diplomas and holders of bachelor's degrees.

Mathematics is an important tool for everyday life. Knowing basic mathematical skills can help simplify many tasks. For example, we use operations on whole numbers to create a monthly budget. Of course, we need to know the cost of expenses such as rent, utilities, food, car payments, car maintenance, and so on.

This chapter covers basic operations on whole numbers. Knowledge of these operations provides a good foundation on which to build further mathematical skills.

TAPE PA 1.1

1.1

PLACE VALUE AND READING TABLES

OBJECTIVES

1. Determine the place value of a digit in a whole number.
2. Write a whole number in words and in standard form.
3. Write the expanded form of a whole number.
4. Graph whole numbers on a number line.
5. Compare two whole numbers.
6. Read tables.

The **digits** 0, 1, 2, 3, 4, 5, 6, 7, 8, and 9 can be used to write numbers. For example, the **whole numbers** are 0, 1, 2, 3, 4, 5, 6, 7, 8, 9, 10, 11, ... The three dots (. . .) after the 11 is called an **ellipsis** and means that this list continues indefinitely. That is, there is no largest whole number. The smallest whole number is 0.

1 The position of each digit in a number determines its **place value.** A place-value chart is shown next with the whole number 48,337,000 entered.

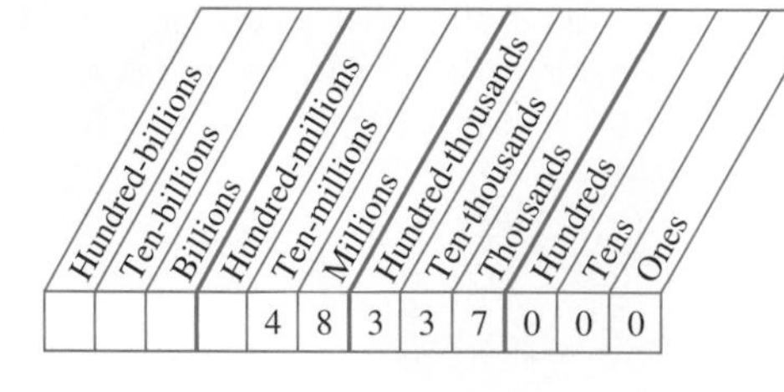

The two 3's in 48,337,000 represent different amounts because of their different placement. The place value of the 3 to the left is hundred-thousands. The place value of the 3 to the right is ten-thousands.

PRACTICE PROBLEM 1

Find the place value of the digit 7 in each whole number.

a. 72,589,620 **b.** 67,890

c. 50,722

Answers:

1a. ten-millions

b. thousands **c.** hundreds

EXAMPLE 1

Find the place value of the digit 4 in each whole number.

a. 48,761 **b.** 249 **c.** 524,007,656

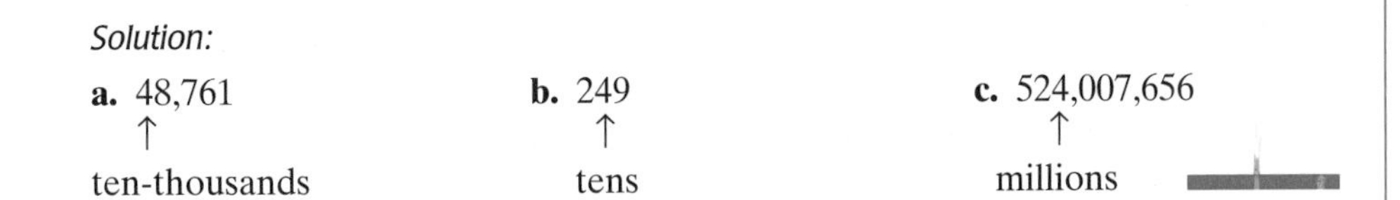

A whole number such as 1,083,664,500 is written in **standard form.** Notice that commas separate the digits into groups of threes. Each group of three digits is called a **period,** so the number 1,083,664,500 has four periods. The names of the first four periods are shown.

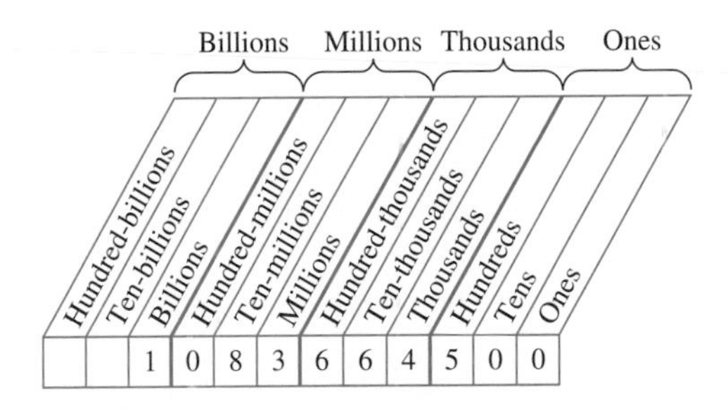

2 To write a whole number in words, write the number in each period followed by the name of the period. (The ones period is usually not written.) This same procedure can be used to read a whole number.

In words, 1,083,664,500 is written or read as "one **billion,** eighty-three **million,** six hundred sixty-four **thousand,** five hundred."

> REMINDER The name of the ones period is not used when reading and writing whole numbers. For example, 9,265 is read as "nine **thousand,** two hundred sixty-five."
>
> Also, the word "and" is not used when reading and writing whole numbers. It is used when reading and writing mixed numbers and some decimal values as shown later in this text.

EXAMPLE 2

Write 52,447 in words.

Solution:

fifty-two	**thousand,**	four hundred forty-seven
number in period	name of period	number in period

To write a whole number in standard form, write the number in each period followed by a comma.

PRACTICE PROBLEM 2

Write 321,670,200 in words.

Answer:

2. three hundred twenty-one million, six hundred seventy thousand, two hundred

PRACTICE PROBLEM 3

Write twenty-six thousand seventy-one in standard form.

EXAMPLE 3

Write two million, five hundred sixty-four thousand, three hundred fifty in standard form.

Solution:

2,564,350

A comma is not written after the ones period.

> REMINDER A comma may or may not be inserted in a four-digit number. For example, both 9,386 and 9386 are acceptable ways of writing nine thousand, three hundred eighty-six.

3 The place value of a digit can be used to write a number in **expanded form.** The **expanded form** of a number shows each digit of the number with its place value. For example, 5672 is written in expanded form as

5 thousands		6 hundreds		7 tens		2 ones
↑ ↑		↑ ↑		↑ ↑		↑ ↑
digit place value	+	digit place value	+	digit place value	+	digit place value

5672 = **5000** + **600** + **70** + **2**

PRACTICE PROBLEM 4

Write 1,047,608 in expanded form.

EXAMPLE 4

Write 706,449 in expanded form.

Solution:

700,000 + 6,000 + 400 + 40 + 9

4 We can picture whole numbers as equally spaced points on a line, called the **number line.** The whole numbers are written in "counting" order on the number line beginning with 0. An arrow at the right end of the number line means that the whole numbers continue indefinitely.

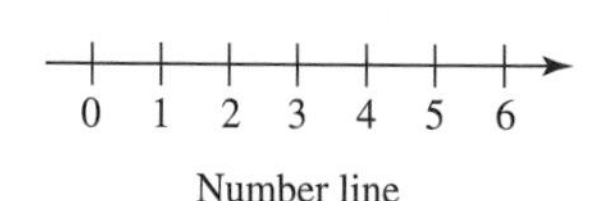

Number line

A whole number is **graphed** by placing a dot on the number line at the point corresponding to the number. For example, 5 is graphed on the number line as shown.

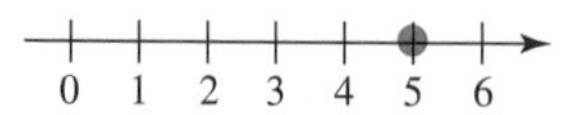

5 For any two numbers graphed on a number line, the number to the **right** is the **greater number,** and the number to the **left** is the **smaller number.** For example, 3 is **to the left** of 6, so 3 **is less than** 6. Also, 6 is **to the right** of 3, so 6 **is greater than** 3.

Answers:

3. 26,071

4. 1,000,000 + 40,000 + 7000 + 600 + 8

We use the symbol $<$ to mean "is less than" and the symbol $>$ to mean "is greater than." For example, the true statement **"3 is less than 6"** can be written in symbols as

$$3 < 6$$

Also, **"6 is greater than 3"** is written as

$$6 > 3$$

The statements $3 < 6$ and $6 > 3$ are both **true statements.** The statement $5 < 2$ is a **false statement** because 5 is greater than 2, not less than.

EXAMPLE 5

Insert $<$ or $>$ to make a true statement.

a. 5 50 **b.** 101 0 **c.** 29 27

Solution:

a. $5 < 50$ **b.** $101 > 0$ **c.** $29 > 27$

PRACTICE PROBLEM 5

Insert $<$ or $>$ to make a true statement.

a. 0 19 **b.** 18 32

c. 107 103

> REMINDER One way to remember the meaning of the symbols $<$ and $>$ is to think of them as arrowheads "pointing" toward the smaller number. For example, $5 < 11$ and $11 > 5$ are both true statements.

6 **Tables** are often used to organize facts. The table below shows the countries that have won the most medals during the Olympic winter games. (Although the medals are won by athletes from the various countries, for simplicity, we will state that countries have won the medals.)

MOST MEDALS
WINTER GAMES (1924–94)

	GOLD	SILVER	BRONZE	TOTAL
USSR[1]	99	71	71	241
Norway	73	77	64	214
U.S.	53	55	39	147
Austria	36	48	44	128
Germany[2]	45	43	37	125
Finland	36	45	42	123
GDR[3]	39	36	35	110
Sweden	39	26	34	99
Switzerland	27	29	29	85
Italy	25	21	21	67

[1]Includes former USSR to 1992, Russia 1994.
[2]Includes West Germany 1968–88.
[3]GDR (East Germany) 1968–88.
Source: The Guinness Book of Records, 1996

Answers:
5a. $<$ **b.** $<$ **c.** $>$

For example, by reading from left to right along the line marked U.S., we find that the United States has won 53 gold, 55 silver, and 39 bronze medals for the years 1924–1994.

PRACTICE PROBLEM 6

Use the Winter Games table to answer the following questions.

a. How many bronze medals has Austria won during the winter games of the Olympics?

b. Which countries shown have won more than 70 gold medals?

EXAMPLE 6

Use the Winter Games table on page 5 to answer the following.

a. How many silver medals has Sweden won during the winter games of the Olympics?

b. Which country shown has won fewer gold medals than Switzerland?

Solution:

a. Read from left to right across the line marked Sweden until the "Silver" column is reached. We find that Sweden has won 26 silver medals.

b. Switzerland has won 27 gold medals while Italy has won 25, so Italy has won fewer gold medals than Switzerland.

Answers:

6a. 44 bronze medals

b. USSR and Norway

EXERCISE SET 1.1

Determine the place value of the digit 5 in each whole number. See Example 1.

1. 352 **2.** 5890 **3.** 5207

4. 6527 **5.** 62,500,000 **6.** 79,050,000

7. 5,070,099 **8.** 51,682,700 **9.** 68,507,280,699

10. 1,350,000,000

Write each whole number in words. See Example 2.

11. 5420 **12.** 3165

13. 26,990 **14.** 42,009

15. 1,620,000 **16.** 3,204,000

17. 53,520,170 **18.** 47,033,107

19. 97,060,550,000 **20.** 77,790,000,000

Write each number in the sentence in words. See Example 2.

21. At this writing, the population of Libya is 5,248,401. (*Source: U.N. Statistical Yearbook*)

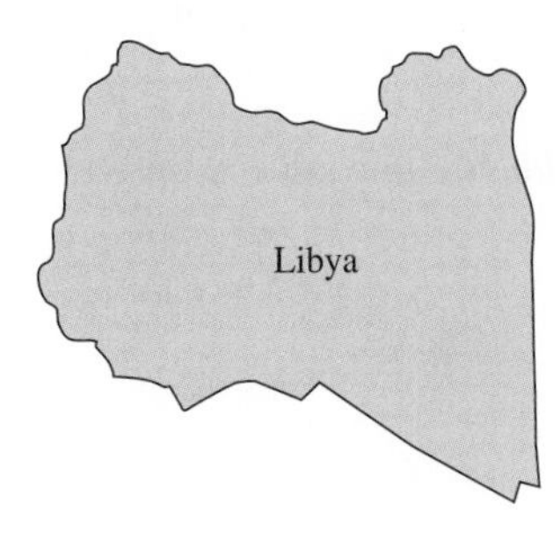

22. Liz Harold has the number 16,820,409 showing on her calculator display.

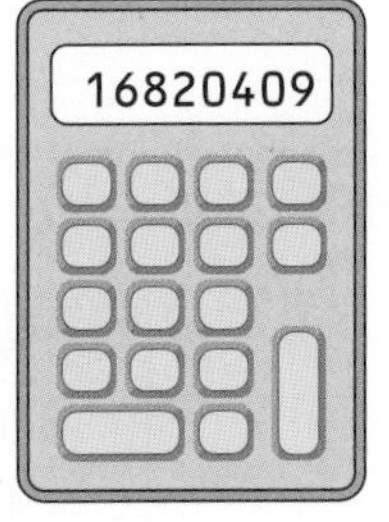

23. In 1994, zinc mines in the United States mined 570,162 metric tons of zinc. (*Source:* U.S. Bureau of Mines)

24. The highest point in Montana is at Granite Peak, at an elevation of 12,799 feet. (*Source:* U.S. Geological Survey)

Write each whole number in standard form. See Example 3.

25. Six thousand, five hundred eight **26.** Three thousand, three hundred seventy

ANSWERS

1. ________
2. ________
3. ________
4. ________
5. ________
6. ________
7. ________
8. ________
9. ________
10. ________
11. ________
12. ________
13. ________
14. ________
15. ________
16. ________
17. ________
18. ________
19. ________
20. ________
21. ________
22. ________
23. ________
24. ________
25. ________
26. ________

ANSWERS

27. ______
28. ______
29. ______
30. ______
31. ______
32. ______
33. ______
34. ______
35. ______
36. ______
37. ______
38. ______
39. ______
40. ______
41. ______
42. ______
43. ______
44. ______
45. ______
46. ______
47. ______
48. ______
49. ______
50. ______
51. ______
52. ______
53. ______
54. ______
55. ______
56. ______
57. ______
58. ______
59. ______
60. ______
61. ______

27. Twenty-nine thousand, nine hundred

28. Forty-two thousand, six

29. Six million, five hundred four thousand, nineteen

30. Ten million, thirty-seven thousand, sixteen

31. Three million, fourteen

32. Seven million, twelve

33. Fifteen billion, forty thousand, sixteen

34. Eight hundred nine billion, sixty-two million, four thousand, five hundred fifty-seven

Write the whole number in each sentence in standard form. See Example 3.

35. The world's tallest self-supporting structure is the CN Tower in Toronto, Canada. It is one thousand, eight hundred twenty-one feet tall. (*Source: World Almanac,* 1996)

36. The average distance between Earth and the sun is more than 93 million miles.

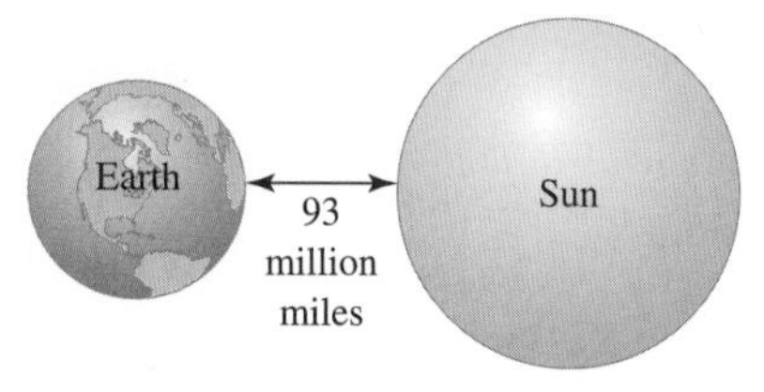

37. Grain storage facilities in Toledo, Ohio, have a capacity of sixty-three million, one hundred thousand bushels. (*Source:* Chicago Board of Trade Market Information Department)

38. Bobby Hurley of Duke University holds the NCAA Men's Division I record for most career assists. He is credited with one thousand, seventy-six assists from 1990 to 1993.

Write each whole number in expanded form. See Example 4.

39. 406 **40.** 789 **41.** 5290 **42.** 6040

43. 62,407 **44.** 20,215 **45.** 30,680 **46.** 99,032

47. 39,680,000 **48.** 47,703,029

Insert < or > to make a true statement. See Example 5.

49. 3 8 **50.** 7 10 **51.** 9 0 **52.** 0 12

53. 6 2 **54.** 15 14 **55.** 22 0 **56.** 39 47

The table below shows the five longest rivers in the world. Use this table to answer Exercises 57 through 61. See Example 6.

RIVER	MILES
Chang jiang-Yangtze (China)	3964
Amazon (Brazil)	4000
Tenisei-Angara (USSR)	3442
Mississippi-Missouri (U.S.)	3740
Nile (Egypt)	4145

57. Write the length of the Amazon River in words.

58. Write the length of the Tenisei-Angara River in words.

59. Write the length of the Nile River in expanded form.

60. Write the length of the Mississippi-Missouri River in expanded form.

61. Which river is the longest in the world?

The table below shows the countries that have won the most medals during the Olympic Summer Games. Use this table to answer Exercises 62 through 65. See Example 6.

MOST MEDALS
SUMMER GAMES (1896–94)*

	GOLD	SILVER	BRONZE	TOTAL
U.S.	789	603	518	1,910
USSR[1]	442	361	333	1,136
Germany[2]	186	227	236	649
Great Britain	177	224	218	619
France	161	175	191	527
Sweden	133	149	171	453
GDR[3]	154	131	126	411
Italy	153	126	131	410
Hungary	136	124	144	404
Finland	98	77	112	287

*Excludes medals won in Official Art competitions in 1912–48.
[1]Includes Czarist Russia to 1912, CIS 1992, Russia 1994.
[2]Germany 1896–1964 and 1992, West Germany 1968–88.
[3]GDR (East Germany) 1968–88.

Source: The Guinness Book of Records, 1996

62. How many silver medals has France won?

63. How many bronze medals has Great Britain won?

64. Which country shown has won fewer total medals than Hungary?

65. Which countries shown have won more than 400 gold medals?

66. Write the largest four-digit number that can be made from the digits 3, 6, 7, and 2 if each digit must be used once.

— — — —

67. Write the largest five-digit number that can be made using the digits 4, 5, and 3 if each digit must be used at least once.

— —, — — —

68. If a number is given in words, describe the process used to write this number in standard form.

69. If a number is written in standard form, describe the process used to write this number in expanded form.

70. The Pro-Football Hall of Fame was established on September 7, 1963 in this town. Use the information and the diagram below to find the name of the town.

*Alliance is East of Massillon.

*Dover is between Canton and New Philadelphia.

*Massillon is not next to Alliance.

*Canton is North of Dover.

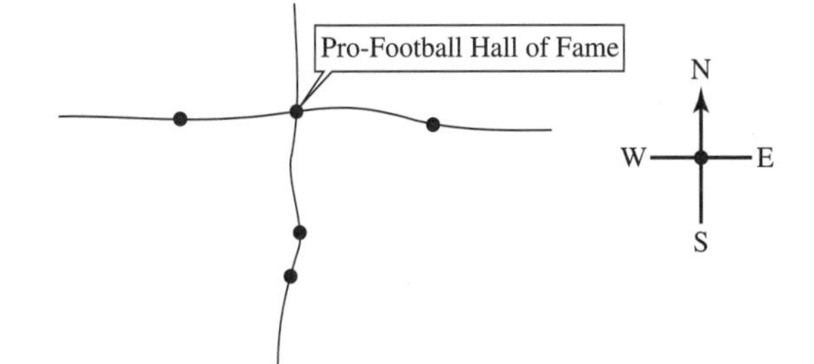

ANSWERS

62. ______

63. ______

64. ______

65. ______

66. ______

67. ______

68. ______

69. ______

70. ______

TAPE PA 1.2

1.2

ADDING WHOLE NUMBERS AND PERIMETER

OBJECTIVES

1. Add whole numbers.
2. Identify properties of addition.
3. Find the perimeter of a polygon.
4. Solve problems by adding whole numbers.

1 If one computer in an office has a 2-megabyte memory and a second computer has a 4-megabyte memory, the total memory in the two computers can be found by adding 2 and 4.

$$2 \text{ megabytes} + 4 \text{ megabytes} = 6 \text{ megabytes}$$

The **sum** is 6 megabytes of memory. Each of the numbers 2 and 4 is called an **addend.**

2	+	4	=	6
↑		↑		↑
addend		addend		sum

Addition of whole numbers can be visualized on a number line using arrows. On the number line, a whole number can be represented anywhere by an arrow of appropriate length pointing to the right.

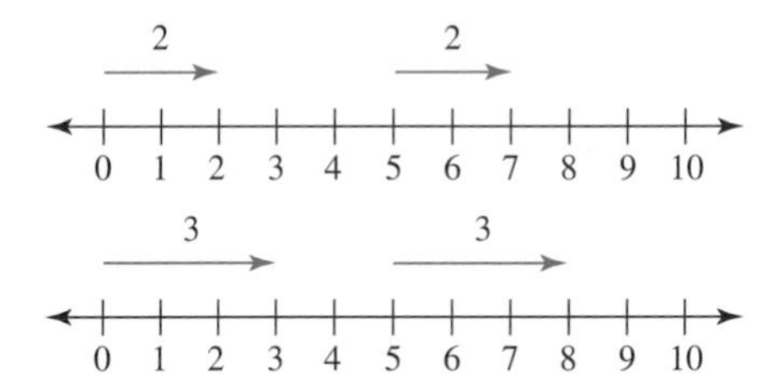

Both arrows represent 2. They both point to the right and they are both 2 units long.

Both arrows represent 3. They both point to the right and they are both 3 units long.

To add 3 and 2 on a number line, start at 0 on the number line and draw an arrow representing 3, the first number. From the tip of this arrow, draw another arrow representing 2, the second number. The tip of the second arrow ends at their sum, 5.

$3 + 2 = 5$

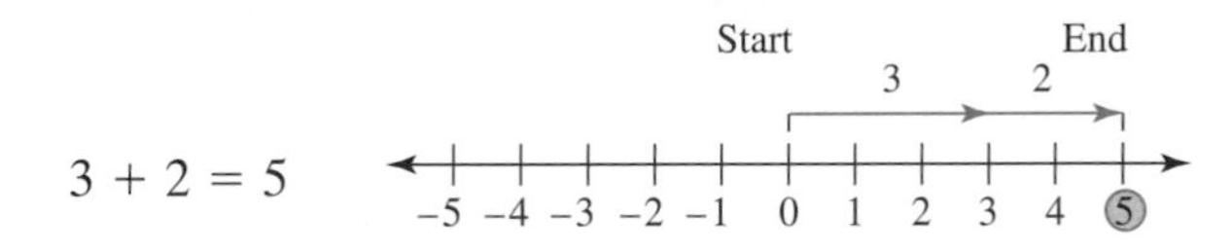

Adding on a number line can be tedious, especially with larger numbers. Instead, we memorize sums of one-digit whole numbers and then use place value to add larger numbers. (For your review, Appendix A contains a summary of all the possible ways that the numbers 0 through 9 can be added two at a time.)

To add whole numbers, add the digits in the ones place, then the tens place, then the hundreds place, and so on. For example, add

2236 + 160.

$$\begin{array}{r} 2\ 2\ 3\ 6 \\ +\ 1\ 6\ 0 \\ \hline 2\ 3\ 9\ 6 \end{array}$$

Line up numbers vertically so that the place values correspond. Then add digits in corresponding place values, starting with the ones place.

(6: sum of ones; 9: sum of tens; 3: sum of hundreds; 2: sum of thousands)

PRACTICE PROBLEM 1
Find the sum: 7235 + 542.

EXAMPLE 1

Find the sum: 23 + 136.

Solution:

$$\begin{array}{r} 23 \\ +\ 136 \\ \hline 159 \end{array}$$

When the sum of digits in corresponding place values is more than 9, "carrying" is necessary. For example, add 365 + 89. First, add the ones-place digits.

$$\begin{array}{r} {}^{1} \\ 3\ 6\ 5 \\ +\ \ 8\ 9 \\ \hline 4 \end{array}$$

5 ones + 9 ones = **14 ones** or **1 ten** + **4 ones.** Write the 4 ones in the ones place and carry the 1 ten to the tens place.

Next, add the tens-place digits.

$$\begin{array}{r} 1\ 1\ \\ 3\ 6\ 5 \\ +\ \ 8\ 9 \\ \hline 5\ 4 \end{array}$$

1 ten + 6 tens + 8 tens = **15 tens** or **1 hundred** + **5 tens.** Write the 5 tens in the tens place and carry the 1 hundred to the hundreds place.

Next, add the hundreds-place digits.

$$\begin{array}{r} 11 \\ 365 \\ +\ 89 \\ \hline 454 \end{array}$$

1 hundred + 3 hundreds = 4 hundreds.

Write 4 in the hundreds place.

PRACTICE PROBLEM 2
Add: 27,364 + 92,977.

EXAMPLE 2

Add: 34,285 + 149,761.

Solution:

$$\begin{array}{r} 11\ 1 \\ 34{,}285 \\ +\ 149{,}761 \\ \hline 184{,}046 \end{array}$$

Answers:
1. 7777
2. 120,341

2 Before we continue adding whole numbers, let's review some properties of addition that you may have already discovered.

The first property of addition that we will review is the **addition property of 0.** This property reminds us that the sum of 0 and any number is that same number.

ADDITION PROPERTY OF 0

The sum of 0 and any number is that number. For example,

$$7 + 0 = 7$$
$$0 + 7 = 7$$

Next, notice that we can add any two whole numbers in any order. For example,

$$4 + 5 = 5 + 4$$

or

$$9 = 9$$

We call this special property of addition the **commutative property of addition.**

COMMUTATIVE PROPERTY OF ADDITION

Changing the **order** of two addends does not change their sum. For example,

$$2 + 3 = 3 + 2$$

EXAMPLE 3

Rewrite each sum using the commutative property of addition.

a. $1 + 5$ **b.** $23 + 47$ **c.** $542 + 112$

Solution:

a. $1 + 5 = 5 + 1$ **b.** $23 + 47 = 47 + 23$

c. $542 + 112 = 112 + 542$

PRACTICE PROBLEM 3

Rewrite each sum using the commutative property of addition.

a. $6 + 2$ **b.** $39 + 85$

c. $465 + 92$

Another property that can help us when adding numbers is the **associative property of addition.** This property states that, when adding numbers, the grouping of the numbers can be changed without changing the sum. We use parentheses below to group numbers. They indicate what numbers to add first. For example,

$$2 + \underbrace{(1 + 5)} = \underbrace{(2 + 1)} + 5$$

or

$$2 + 6 = 3 + 5$$

or

$$8 = 8$$

Answers:

3a. $6 + 2 = 2 + 6$
b. $39 + 85 = 85 + 39$
c. $465 + 92 = 92 + 465$

ASSOCIATIVE PROPERTY OF ADDITION

Changing the **grouping** of addends does not change their sum. For example,

$$3 + (5 + 7) = (3 + 5) + 7$$

EXAMPLE 4

Rewrite each sum using the associative property of addition.

a. $9 + (4 + 6)$ **b.** $10 + (74 + 55)$ **c.** $(5 + 100) + 842$

Solution:

a. $9 + (4 + 6) = (9 + 4) + 6$ **b.** $10 + (74 + 55) = (10 + 74) + 55$

c. $(5 + 100) + 842 = 5 + (100 + 842)$

PRACTICE PROBLEM 4

Rewrite each sum using the associative property of addition.

a. $3 + (5 + 8)$

b. $22 + (15 + 18)$

c. $(74 + 67) + 204$

Together, these properties tell us that we can add whole numbers using any order and grouping that we want.

When adding numbers, it is often helpful to look for two or three numbers whose sum is 10, 20, and so on.

EXAMPLE 5

Add: $13 + 2 + 7 + 8 + 9$.

Solution:

$$13 + 2 + 7 + 8 + 9 = 39$$

$$20 + 10 + 9$$

$$39$$

PRACTICE PROBLEM 5

Add: $11 + 7 + 8 + 9 + 13$.

EXAMPLE 6

Find the sum: $1647 + 246 + 32 + 85$.

Solution:

```
  122
 1647
  246
   32
+  85
-----
 2010
```

PRACTICE PROBLEM 6

Find the sum:
$19 + 5042 + 638 + 526$.

3 A special application of addition is finding the perimeter of a polygon. Geometric figures such as triangles, squares, and rectangles are called **polygons**.

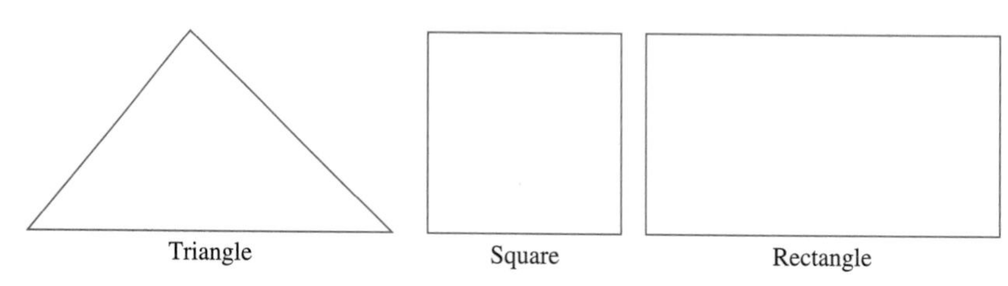

Answers:

4a. $3 + (5 + 8) = (3 + 5) + 8$

b. $22 + (15 + 18) = (22 + 15) + 18$

c. $(74 + 67) + 204 = 74 + (67 + 204)$

5. 48

6. 6225

A polygon can be described as a flat figure formed by line segments connected at their ends. (For more review, see Appendix E.)

The perimeter of a polygon is the **distance around** the polygon. This means that the perimeter of a polygon is the sum of the lengths of its sides.

EXAMPLE 7

Find the perimeter of the polygon shown.

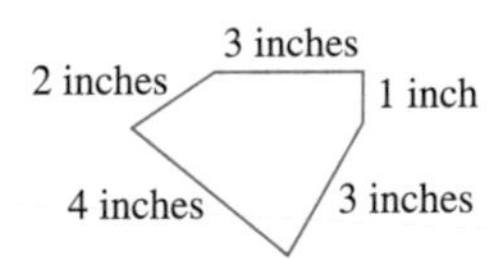

Solution:

To find the perimeter (distance around), add the lengths of the sides.

$$2 \text{ in.} + 3 \text{ in.} + 1 \text{ in.} + 3 \text{ in.} + 4 \text{ in.} = 13 \text{ in.}$$

The perimeter is 13 inches.

PRACTICE PROBLEM 7

Find the perimeter of the geometric figure shown. (A centimeter is a unit of length in the metric system.)

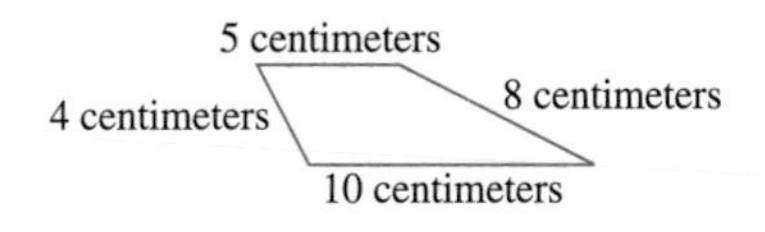

EXAMPLE 8

The largest commercial building in the world under one roof is the flower auction building of the cooperative VBA in Aalsmeer, Netherlands. The floor plan is a rectangle that measures 776 meters by 639 meters. Find the perimeter of this building. (The meter is a unit of length in the metric system.) (*Source: The Handy Science Answer Book,* Visible Ink Press, 1994)

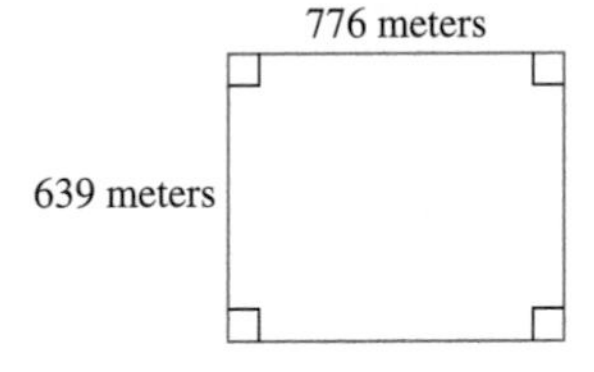

Solution:

Recall that opposite sides of a rectangle have the same lengths.

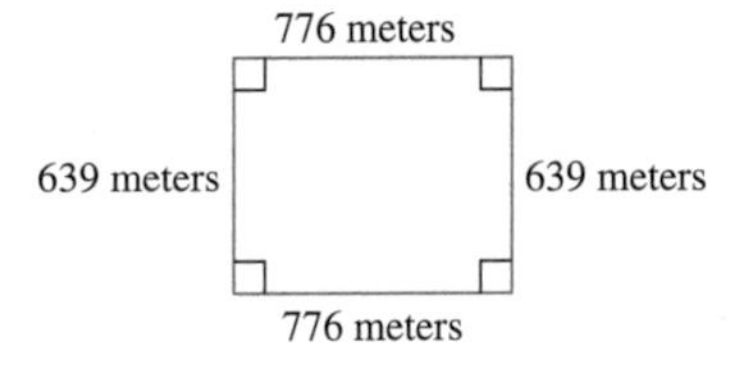

The sum of the lengths of its sides is

$$\begin{array}{r} 639 \\ 639 \\ 776 \\ +\ 776 \\ \hline 2830 \end{array}$$

The perimeter of the building is 2830 meters.

PRACTICE PROBLEM 8

The Pentagon Building in Washington, DC, is in the shape of a pentagon and is the world's largest office building in terms of ground space. Each of the Pentagon's five sides is 921 feet. Find the distance around the building.

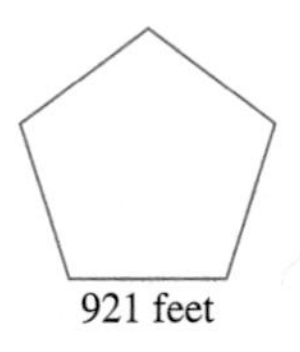

Answers:
7. 27 centimeters
8. 4605 feet

4 Often, real-life problems occur that can be solved by writing an addition statement. We can call this statement a **mathematical model** of the problem. The first step to solving any written problem is to *understand* the problem by reading it carefully. Descriptions of problems solved through addition **may** include any of these key words or phrases:

KEY WORDS OR PHRASES	EXAMPLE	SYMBOLS
added to	5 added to 7	$7 + 5$
plus	0 plus 78	$0 + 78$
increased by	12 increased by 6	$12 + 6$
more than	11 more than 25	$25 + 11$
total	the total of 8 and 1	$8 + 1$
sum	the sum of 4 and 133	$4 + 133$

Using the facts described, write an addition statement. Then write the corresponding solution of the real-life problem. It is sometimes helpful to write the statement in words and then translate to numbers.

PRACTICE PROBLEM 9

A new Janda Spirit-L motorbike costs $2431. The Spirit-X costs $486 more than the Spirit-L. How much does a Spirit-X motorbike cost?

EXAMPLE 9

Ivy is station manager of Smoky Hills Public Television. Her monthly salary of $2460 is increased by a raise of $427. What is her new salary?

Solution:

The key phrase here is "increased by" and suggests an addition model. To find Ivy's new salary, add her old salary, $2460, to the increase, $427.

IN WORDS		TRANSLATE TO NUMBERS
old salary	→	$2460
+ increase	→	+ 427
new salary	→	$2887

Her new salary is $2887.

PRACTICE PROBLEM 10

Elham Abo-Zahrah collects thimbles. She has 42 glass thimbles, 17 steel thimbles, 37 porcelain thimbles, 9 silver thimbles, and 15 plastic thimbles. How many thimbles are in her collection?

EXAMPLE 10

Alan collects baseball cards. He has 109 cards for New York Yankees, 96 for Chicago White Sox, 79 for Kansas City Royals, 42 for Seattle Mariners, 67 for Oakland Athletics, and 52 for California Angels. How many cards does he have in total?

Solution:

The key word here is "total." To find the total number of Alan's baseball cards, find the sum of the quantities from each team.

IN WORDS		TRANSLATE TO NUMBERS
		33
New York Yankee cards	→	109
Chicago White Sox cards	→	96
Kansas City Royal cards	→	79
Seattle Mariner cards	→	42
Oakland Athletic cards	→	67
+ California Angel cards	→	+ 52
Total cards	→	445

Alan has a total of 445 baseball cards.

Answers:
9. $2917
10. 120 thimbles

CALCULATOR EXPLORATIONS

ADDING NUMBERS

To add numbers on a calculator, find the keys marked [+] and [=].

For example, to add 5 and 7 on a calculator, press the keys [5] [+] [7] [=].

The display will read [12].

Thus, $5 + 7 = 12$.

To add 687 and 981 on a calculator, press the keys [687] [+] [981] [=].

The display will read [1668].

Thus, $687 + 981 = 1668$. (Although entering 687, for example, requires pressing more than one key, numbers are grouped together for easier reading.)

Use a calculator to add the following.

1. $89 + 45$

2. $76 + 97$

3. $285 + 55$

4. $8773 + 652$

5. $\begin{array}{r} 985 \\ 1210 \\ 562 \\ +\ \ 77 \\ \hline \end{array}$

6. $\begin{array}{r} 465 \\ 9888 \\ 620 \\ +\ 1550 \\ \hline \end{array}$

MENTAL MATH

Find each sum mentally.

1. $5 + 7$

2. $20 + 30$

3. $5000 + 4000$

4. $4300 + 26$

5. $1620 + 0$

6. $6 + 126 + 4$

EXERCISE SET 1.2

Add. See Examples 1 and 2.

1. $14 + 22$ **2.** $27 + 31$ **3.** $62 + 30$ **4.** $37 + 42$

5. $12 + 13 + 24$ **6.** $23 + 45 + 30$ **7.** $5267 + 132$ **8.** $236 + 6243$

9. $53 + 64$ **10.** $41 + 74$ **11.** $22 + 49$ **12.** $35 + 47$

13. $38 + 79$ **14.** $92 + 37$

Rewrite each sum using the commutative property of addition. See Example 3.

15. $9 + 4$ **16.** $11 + 6$ **17.** $2 + 13$ **18.** $7 + 8$

Rewrite each sum using the associative property of addition. See Example 4.

19. $3 + (2 + 9)$ **20.** $(5 + 3) + 8$ **21.** $(4 + 1) + 10$ **22.** $9 + (2 + 6)$

23. $8 + (4 + 6)$ **24.** $(7 + 3) + 11$

Add. Look for two or three numbers whose sum is 10, 20, and so on. See Example 5.

25. $8 + 9 + 2 + 5 + 1$ **26.** $3 + 5 + 8 + 5 + 7$ **27.** $6 + 21 + 14 + 9 + 12$ **28.** $12 + 4 + 8 + 26 + 10$

29. $81 + 17 + 23 + 79 + 12$ **30.** $64 + 28 + 56 + 25 + 32$

Add. See Examples 5 and 6.

31. $62 + 18 + 14$ **32.** $23 + 49 + 18$ **33.** $40 + 800 + 70$

ANSWERS

1. ______
2 ______
3. ______
4. ______
5. ______
6. ______
7. ______
8. ______
9. ______
10. ______
11. ______
12. ______
13. ______
14. ______
15. ______
16. ______
17. ______
18. ______
19. ______
20. ______
21. ______
22. ______
23. ______
24. ______
25. ______
26. ______
27. ______
28. ______
29. ______
30. ______
31. ______
32. ______
33. ______

ANSWERS

34. ______
35. ______
36. ______
37. ______
38. ______
39. ______
40. ______
41. ______
42. ______
43. ______
44. ______
45. ______
46. ______
47. ______
48. ______
49. ______
50. ______
51. ______
52. ______
53. ______
54. ______
55. ______
56. ______
57. ______
58. ______

34. 30 + 900 + 20

35. 7542 + 49 + 682

36. 1624 + 1832 + 1976

37. 24 + 9006 + 489 + 2407

38. 16 + 748 + 1056 + 770

39. $\begin{array}{r} 627 \\ 628 \\ +\ 629 \\ \hline \end{array}$

40. $\begin{array}{r} 427 \\ 383 \\ +\ 229 \\ \hline \end{array}$

41. $\begin{array}{r} 6820 \\ 4271 \\ +\ 5626 \\ \hline \end{array}$

42. $\begin{array}{r} 6789 \\ 4321 \\ +\ 5555 \\ \hline \end{array}$

43. $\begin{array}{r} 507 \\ 593 \\ +\ \ 10 \\ \hline \end{array}$

44. $\begin{array}{r} 864 \\ 733 \\ +\ 356 \\ \hline \end{array}$

45. $\begin{array}{r} 4200 \\ 2107 \\ +\ 2692 \\ \hline \end{array}$

46. $\begin{array}{r} 5000 \\ 400 \\ +\ 3021 \\ \hline \end{array}$

47. $\begin{array}{r} 49 \\ 628 \\ 5{,}762 \\ +\ 29{,}462 \\ \hline \end{array}$

48. $\begin{array}{r} 26 \\ 582 \\ 4{,}763 \\ +\ 62{,}511 \\ \hline \end{array}$

49. $\begin{array}{r} 121{,}742 \\ 57{,}279 \\ 6{,}586 \\ +\ 426{,}782 \\ \hline \end{array}$

50. $\begin{array}{r} 504{,}218 \\ 321{,}920 \\ 38{,}507 \\ +\ 594{,}687 \\ \hline \end{array}$

Find the perimeter of each figure. See Examples 7 and 8.

51.

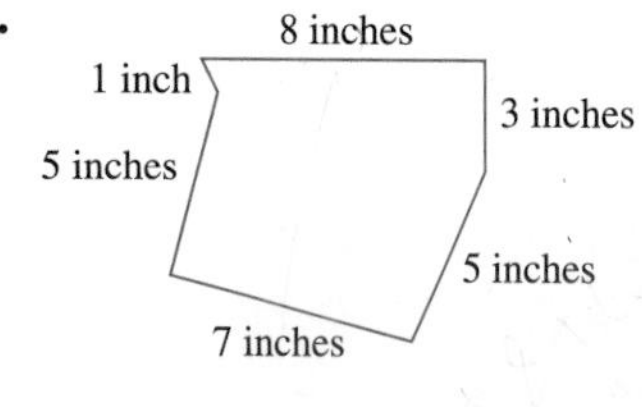

52.

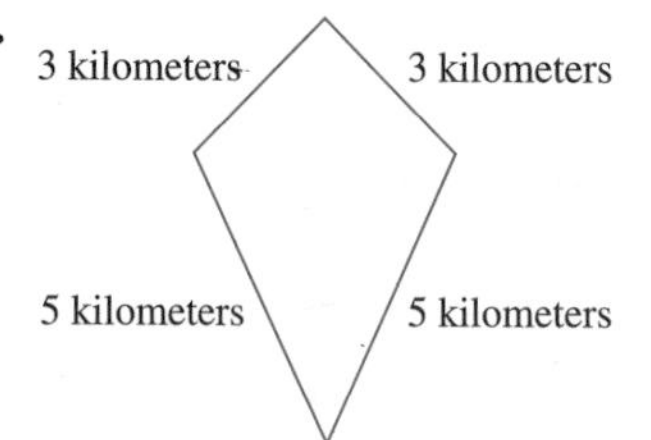

53.

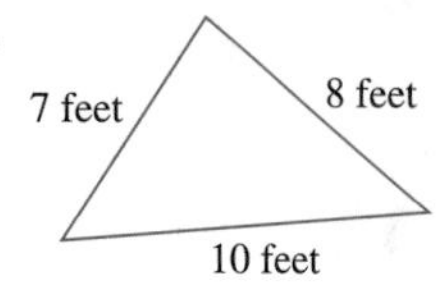

54.

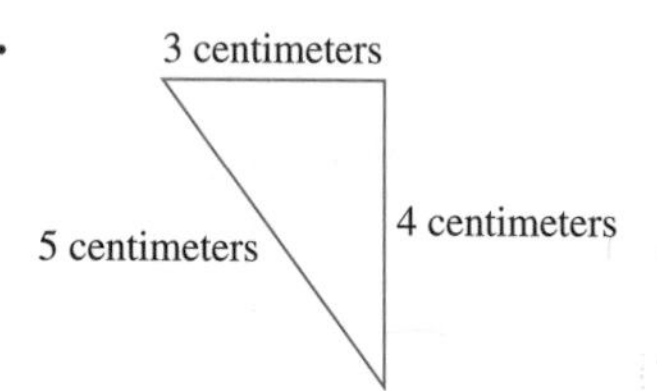

55.

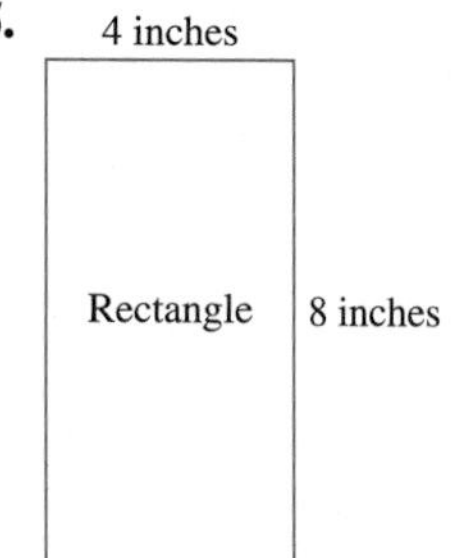

56.

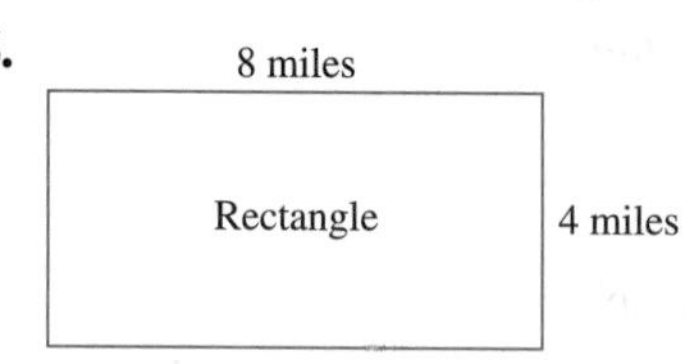

57.

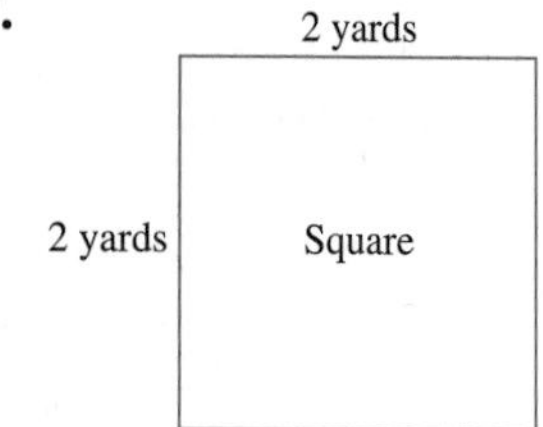

58.

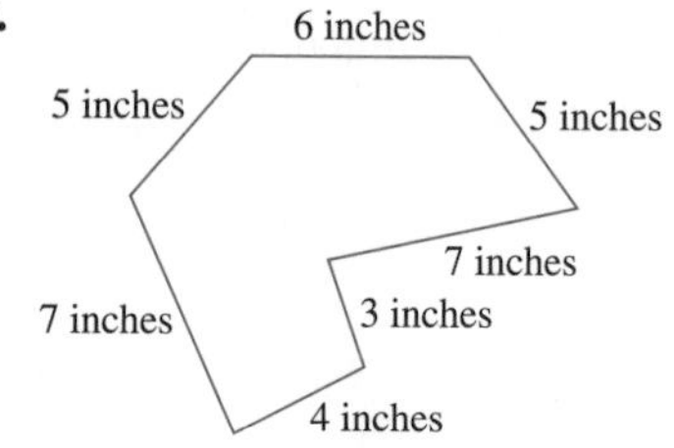

Solve. See Examples 8, 9, and 10.

59. Terry Zimmerman is executive assistant for the Sedgwick County Board of Commissioners. His salary of $24,762 was increased by $989. Find his new salary.

60. A beginning kindergarten teacher's salary is $19,265. Cecile Gould's salary is $2496 more than this. Find Cecile's salary.

61. Staffas Broussard is a retired postal employee. His yearly retirement pension is $26,826. He also receives $3622 per year from his investments. Find his total yearly income.

62. A dress sells for $89 plus a profit of $15. Find the total price of the dress.

63. The distance from Kansas City, Kansas, to Hays, Kansas, is 285 miles. Colby, Kansas, is 98 miles farther from Kansas City than Hays. Find how far it is from Kansas City to Colby.

64. The highest point in Kansas is Mt. Sunflower at 4039 feet above sea level. The highest mountain in the world is Mt. Everest in Asia. Its peak is 24,989 feet higher than Mt. Sunflower. Find how high Mt. Everest is. (*Sources:* U.S. Geological Survey and National Geographic Society)

65. Leo Callier is installing an invisible fence in his back yard. How many feet of wiring is needed to enclose the yard below?

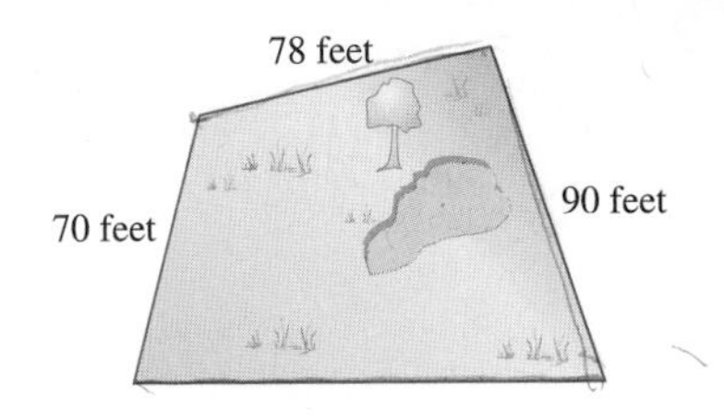

66. A homeowner is considering adding gutters around her home. Find the perimeter of her rectangular home.

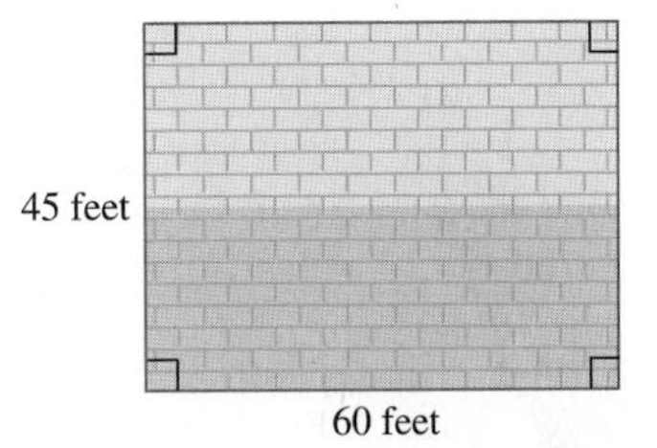

67. Elaine, Kay, and their mother are going to pool their money to buy a used car costing $2699. Elaine has $726, Kay has $1025, and their mother has $897. Determine whether they have enough money to buy the car.

68. It costs $75 per couple to attend the opening of the production of *Peter Pan* at the Audubon Playhouse in New Orleans. If Mark Martin has $37 and his friend Karen Callac has $43, determine whether they have enough money to purchase tickets.

69. The lean hamburger patty production for the Acme Hamburger Patty Manufacturing Co. during the years 1991 to 1996 is shown in the table. Find the total production for the **years 1993, 1994,** and **1995.**

YEAR	PATTIES PRODUCED
1991	34,243,879
1992	35,234,123
1993	29,345,209
1994	38,999,453
1995	40,232,345
1996	42,666,547

70. There are five automobiles on the big Value Used Car lot that another dealer is interested in.

CAR	VALUE
1965 Mustang	$6899
1957 Chevrolet	$8299
1983 Corvette	$12,989
1977 Dodge	$729
1969 Ford	$639

Find the total value of the cars manufactured **before 1970.**

ANSWERS

59. ______
60. ______
61. ______
62. ______
63. ______
64. ______
65. ______
66. ______
67. ______
68. ______
69. ______
70. ______

ANSWERS

71. ______

72. ______

73. ______

74. ______

75. ______

76. ______

71. The highest waterfall in the United States is Yosemite Falls in Yosemite National Park in California. Yosemite Falls is made up of three sections: Upper Yosemite Falls (1430 feet), Cascades in the middle section (675 feet), and Lower Yosemite Falls (320 feet). What is the total height of Yosemite Falls? (*Source:* U.S. Department of the Interior)

72. The average SAT verbal score in 1994 was 423. The average verbal score in 1995 was 5 points higher. Find the average verbal score in 1995. (*Source:* College Entrance Examination Board)

73. In your own words, explain the commutative law of addition.

74. In your own words, explain the associative law of addition.

75. Add: 78,962 + 129,968,350 + 36,462,880

76. Add: 56,468,980 + 1,236,785 + 986,768,000

1.3

SUBTRACTING WHOLE NUMBERS

OBJECTIVES

1. Subtract whole numbers.
2. Identify properties of subtraction.
3. Subtract whole numbers when borrowing is necessary.
4. Solve problems by subtracting whole numbers.

TAPE PA 1.3

1 If you have \$5 and someone gives you \$3, you have a total of \$8, since $5 + 3 = 8$.

Similarly, if you have \$8, and then someone borrows \$3, you have \$5 left. **Subtraction** is finding the **difference** of two numbers.

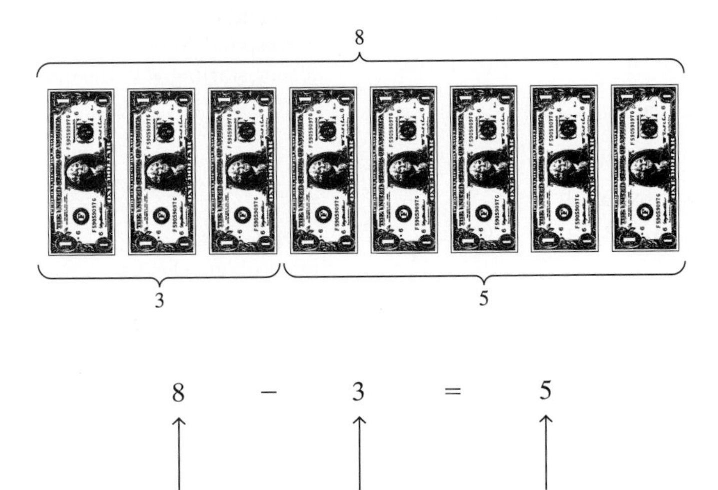

Notice that addition and subtraction are very closely related.

$$8 - 3 = 5 \text{ because } 5 + 3 = 8$$

This means that subtraction can be *checked* by addition, and we say that addition and subtraction are inverse operations.

EXAMPLE 1

Subtract, and then check each answer by adding.

a. $12 - 9$ **b.** $11 - 6$ **c.** $5 - 5$ **d.** $7 - 0$

Solution:

a. $12 - 9 = 3$ because $3 + 9 = 12$ **b.** $11 - 6 = 5$ because $5 + 6 = 11$

c. $5 - 5 = 0$ because $0 + 5 = 5$ **d.** $7 - 0 = 7$ because $7 + 0 = 7$

PRACTICE PROBLEM 1

Subtract:

a. $14 - 9$ **b.** $9 - 9$

c. $4 - 0$

Answers:
1a. 5 **b.** 0 **c.** 4

2 In Example 1(c), 5 is subtracted from 5, and the difference is 0. In Example 1(d), 0 is subtracted from 7, and the difference is 7. These two examples illustrate the subtraction properties of 0.

SUBTRACTION PROPERTIES OF 0

The difference of any number and that same number is 0. For example,

$$11 - 11 = 0$$

The difference of any number and 0 is that same number. For example,

$$45 - 0 = 45$$

When subtraction involves numbers of two or more digits, it is more convenient to subtract vertically. For example, subtract:

$893 - 52.$

$$\begin{array}{r} 8\ 9\ 3 \\ -\quad 5\ 2 \\ \hline 8\ 4\ 1 \end{array} \begin{array}{l} \leftarrow \text{minuend} \\ \leftarrow \text{subtrahend} \\ \leftarrow \text{difference} \end{array}$$

Line up numbers vertically so that the minuend is on top and the place values correspond. Subtract in corresponding places, starting with the ones place.

To check, add.

$$\begin{array}{r} \text{difference} \\ +\ \text{subtrahend} \\ \hline \text{minuend} \end{array} \quad \text{or} \quad \begin{array}{r} 841 \\ +\ 52 \\ \hline 893 \end{array} \longleftarrow$$

Since this is the original minuend, the problem checks.

PRACTICE PROBLEM 2

Find 4689 − 3253.

EXAMPLE 2

Find 7826 −505, and then check by adding.

Solution:

$$\begin{array}{r} 7826 \\ -\ 505 \\ \hline 7321 \end{array}$$

Check:

$$\begin{array}{r} 7321 \\ +\ 505 \\ \hline 7826 \end{array}$$

3 When a digit in the second number (subtrahend) is larger than the corresponding digit in the first number (minuend), *borrowing* is necessary. For example, find

$$\begin{array}{r} 81 \\ -\ 63 \\ \hline \end{array}$$

Since 3 in the ones place of 63 is larger than 1 in the ones place of 81, borrowing is necessary.

Answer:
2. 1436

BORROWING

8 tens − **1 ten** = 7 tens → 7 11 ← **1 ten** + 1 one = 11 ones

```
 7 11
 8̸  1̸
− 6  3
```

Next, subtract the ones-place digits and, last, subtract the tens-place digits.

```
  7  11
  8̸   1̸
− 6   3
  1   8  ← 11 − 3 = 8
  ↑
  └───── 7 − 6 = 1
```

Check:

```
  18
+ 63
  81
```

EXAMPLE 3

Subtract 43 − 29. Then check by adding.

Solution:

```
  3 13
  4̸  3̸
− 2  9
  1  4
```

Check:

```
  14
+ 29
  43
```

PRACTICE PROBLEM 3

Subtract 227 − 175. Then check by adding.

Sometimes we may have to borrow from more than one place. To subtract 7631 − 152, first borrow from the tens place.

```
      2 11
  7 6 3̸ 1̸
−   1 5 2
        9 ← 11 − 2 = 9
```

In the tens place, 5 is greater than 2, so borrow again. Borrow from the hundreds place.

6 hundreds − **1 hundred** = 5 hundreds

1 hundred + 2 tens or 10 tens + 2 tens = 12 tens

```
     12
    5 2̸ 11
  7 6̸ 3̸ 1̸
−   1 5  2
  7 4 7  9
```

Check:

```
  7479
+  152
  7631
```

Answer:
3. 52

PRACTICE PROBLEM 4
Subtract: 400 − 164.

EXAMPLE 4

Subtract: 900 −174. Then check by adding.

Solution:

In the ones place, 4 is larger than 0, so borrow from the tens place. But the tens place of 900 is 0, so to borrow from the tens place we must first borrow from the hundreds place.

$$\begin{array}{rrr} \scriptstyle 8 & \scriptstyle 10 & \\ \not 9 & \not 0 & 0 \\ -1 & 7 & 4 \\ \hline \end{array}$$

Now borrow from the tens place.

$$\begin{array}{rrr} & \scriptstyle 9 & \\ \scriptstyle 8 & \scriptstyle \not{10} & \scriptstyle 10 \\ \not 9 & \not 0 & \not 0 \\ -1 & 7 & 4 \\ \hline 7 & 2 & 6 \end{array}$$

Check:

$$\begin{array}{r} \scriptstyle 11 \\ 726 \\ +174 \\ \hline 900 \end{array}$$

4 Descriptions of real-life problems that suggest solving with a subtraction model include these key words or phrases:

KEY WORDS OR PHRASES	EXAMPLES	SYMBOLS
subtract	subtract 5 from 8	8 − 5
difference	the difference of 10 and 2	10 − 2
less	17 less 3	17 − 3
take away	14 take away 9	14 − 9
decreased by	7 decreased by 5	7 − 5

PRACTICE PROBLEM 5
DeWitt has 29 rings decreased by the 13 rings she gave to her granddaughters. How many rings does she have left?

EXAMPLE 5

Merlyn Schaferkotter has 95 shares of stock, less the 22 shares he sold. How many shares does Merlyn have left?

Solution:

The key words here are "less" and "left." Subtract the shares sold from the original number of shares.

WORDS		TRANSLATE TO NUMBERS
Shares started with	⟶	95
− Shares sold	⟶	−22
Shares left		73

Merlyn has 73 shares left.

Answers:
4. 236
5. 16 rings

EXAMPLE 6

A subcompact car gets 42 miles per gallon of gas. A full-size car gets 17 miles per gallon of gas. How many more miles per gallon does the subcompact car get than the full-size car?

Solution:

WORDS		TRANSLATE TO NUMBERS
		3 12
subcompact miles per gallon	⟶	~~4~~ ~~2~~
− full-size miles per gallon	⟶	− 1 7
more miles per gallon		2 5

The subcompact car gets 25 more miles per gallon than the full-size car.

PRACTICE PROBLEM 6

A new suit originally priced at $92 is now on sale for $47. How much money was taken off the original price?

Graphs can be used to visualize data. The graph shown next is called a **bar graph.**

EXAMPLE 7

A telephone survey was taken to identify favorite sport activities, and the results from the five most popular activities are shown below in the form of a bar graph. In this particular graph, each bar represents a different sport activity, and the height of each bar represents the number of people who responded that the particular sport was their favorite activity. Use this graph to answer the questions below.

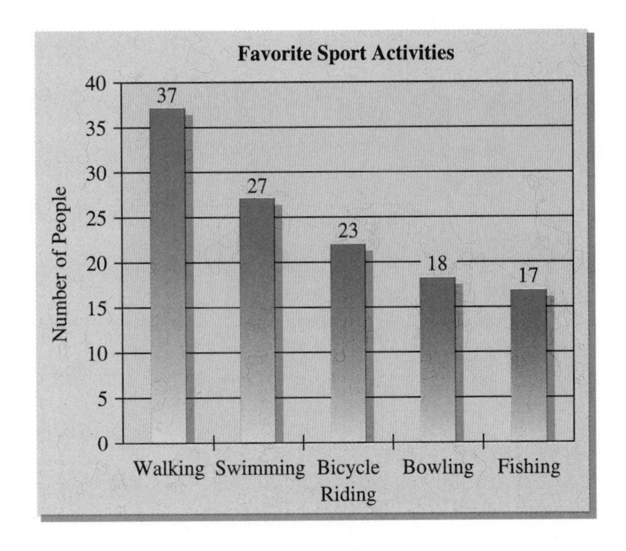

a. What activity was preferred by the most people?

b. How many more people responded that swimming is their favorite activity than fishing?

Solution:

a. The activity preferred by the most people is the one with the highest bar, which is walking.

b. The number of people who responded with swimming is 27. The number of

PRACTICE PROBLEM 7

Use the graph of Example 7 to answer the following.

a. Find the total number of people who responded to the five favorite sport activities. (Assume that each person could respond to only one activity.)

b. How many more people responded that walking rather than bowling is their favorite activity?

Answers:
6. $45
7a. 122 people
b. 19 people

people who responded with fishing is 17. To find how many more responded with swimming than with fishing, we find the difference:

$$27 - 17 = 10.$$

Ten more people responded with swimming than with fishing.

CALCULATOR EXPLORATIONS

SUBTRACTING NUMBERS

To subtract numbers on a calculator, find the keys marked [−] and [=].
For example, to find 83 − 49 on a calculator, press the keys [83] [−] [49] [=].
The display will read [34]. Thus, 83 − 49 = 34.

Use a calculator to subtract the following.

1. 865 − 95
2. 76 − 27
3. 147 − 38
4. 366 − 87
5. 9625 − 647
6. 10,711 − 8925

MENTAL MATH

Find the difference.

1. 9 − 2
2. 6 − 6
3. 5 − 0
4. 44 − 22
5. 93 − 93
6. 700 − 400
7. 700 − 300
8. 700 − 700
9. 600 − 100
10. 600 − 0

EXERCISE SET 1.3

Subtract and check by adding. See Examples 1 through 4.

1. $\begin{array}{r} 67 \\ -\ 23 \\ \hline \end{array}$ **2.** $\begin{array}{r} 72 \\ -\ 41 \\ \hline \end{array}$ **3.** $\begin{array}{r} 82 \\ -\ 22 \\ \hline \end{array}$ **4.** $\begin{array}{r} 27 \\ -\ 10 \\ \hline \end{array}$

5. $\begin{array}{r} 389 \\ -\ 124 \\ \hline \end{array}$ **6.** $\begin{array}{r} 572 \\ -\ 321 \\ \hline \end{array}$ **7.** $\begin{array}{r} 677 \\ -\ 423 \\ \hline \end{array}$ **8.** $\begin{array}{r} 766 \\ -\ 324 \\ \hline \end{array}$

9. $\begin{array}{r} 998 \\ -\ 453 \\ \hline \end{array}$ **10.** $\begin{array}{r} 912 \\ -\ 610 \\ \hline \end{array}$ **11.** $\begin{array}{r} 749 \\ -\ 149 \\ \hline \end{array}$ **12.** $\begin{array}{r} 257 \\ -\ 257 \\ \hline \end{array}$

13. $\begin{array}{r} 62 \\ -\ 37 \\ \hline \end{array}$ **14.** $\begin{array}{r} 55 \\ -\ 29 \\ \hline \end{array}$ **15.** $\begin{array}{r} 70 \\ -\ 25 \\ \hline \end{array}$ **16.** $\begin{array}{r} 80 \\ -\ 37 \\ \hline \end{array}$

17. $\begin{array}{r} 938 \\ -\ 792 \\ \hline \end{array}$ **18.** $\begin{array}{r} 436 \\ -\ 275 \\ \hline \end{array}$ **19.** $\begin{array}{r} 922 \\ -\ 634 \\ \hline \end{array}$ **20.** $\begin{array}{r} 674 \\ -\ 299 \\ \hline \end{array}$

21. $\begin{array}{r} 600 \\ -\ 432 \\ \hline \end{array}$ **22.** $\begin{array}{r} 300 \\ -\ 149 \\ \hline \end{array}$

23. $42 - 36$ **24.** $73 - 29$ **25.** $923 - 476$

26. $813 - 227$ **27.** $6283 - 560$ **28.** $5349 - 720$

ANSWERS

1. ________
2. ________
3. ________
4. ________
5. ________
6. ________
7. ________
8. ________
9. ________
10. ________
11. ________
12. ________
13. ________
14. ________
15. ________
16. ________
17. ________
18. ________
19. ________
20. ________
21. ________
22. ________
23. ________
24. ________
25. ________
26. ________
27. ________
28. ________

ANSWERS

29. ____
30. ____
31. ____
32. ____
33. ____
34. ____
35. ____
36. ____
37. ____
38. ____
39. ____
40. ____
41. ____
42. ____
43. ____
44. ____
45. ____
46. ____
47. ____
48. ____
49. ____
50. ____
51. ____
52. ____
53. ____
54. ____

29. 533 − 29

30. 724 − 16

31. 200 − 111

32. 300 − 211

33. 1983 − 1904

34. 1983 − 1914

35. 56,422 − 16,508

36. 76,652 − 29,498

37. 50,000 − 17,289

38. 40,000 − 23,582

39. 7020 − 1979

40. 6050 − 1878

41. 51,111 − 19,898

42. 62,222 − 39,898

43. Subtract 5 from 9.

44. Subtract 9 from 21.

45. Find the difference of 41 and 21.

46. Find the difference of 16 and 5.

47. Subtract 56 from 63.

48. Subtract 41 from 59.

Solve. See Examples 5 and 6.

49. Michelle and Ronnie Brower entered a crawfish eating contest. Michelle ate 63 crawfish, which was 7 more than Ronnie ate. Find how many crawfish Ronnie ate.

50. A stock worth $135 per share on October 10 dropped to $78 per share on October 30 of the same year. Find how much it lost in value from October 10th to the 30th.

51. Dyllis King is reading a 503-page book. If she has just finished reading page 239, how many more pages must she read to finish the book?

52. When Lou and Judy Zawislak began a trip, the odometer read 55,492. When the trip was over, the odometer read 59,320. How many miles did they drive on their trip?

53. The peak of Mt. McKinley in Alaska is 20,320 feet above sea level. The peak of Long's Peak in Colorado is 14,255 feet above sea level. How much higher is the peak of Mt. McKinley than Long's Peak? (*Source:* U.S. Geological Survey)

54. On one day in May the temperature in Paddin, Indiana, dropped 27 degrees from 2 P.M. to 4 P.M. If the temperature at 2 P.M. was 73 degrees, what was the temperature at 4 P.M.?

55. Buhler Gomez has a total of $539 in his checking account. If he writes a check for each of the items below, how much money will be left in his account?

South Central Bell	$27
Central LA Electric Co.	$101
Mellon Finance	$236

56. Pat Salanki's blood cholesterol level is 243. The doctor tells him it should be decreased to 185. How much of a decrease is this?

57. The distance from Kansas City to Denver is 645 miles. Hays, Kansas, lies on the road between the two and is 287 miles from Kansas City. What is the distance between Hays and Denver?

58. Alan Little is trading his car in on a new car. The new car costs $15,425. His car is worth $7998. How much more money does he need to buy the new car?

59. A new VCR with remote control costs $525. Prunella Pasch has $914 in her savings account. How much will she have left in her savings account after she buys the VCR?

60. Christopher Columbus landed in America in 1492. The Revolutionary War started in 1776. How many years after Columbus landed did the war begin?

61. A stereo that sells regularly for $547 is discounted by $99 in a sale. What is the sale price?

62. A meeting of the board of governors of the Mathematical Association of America was attended by 72 governors. Forty-six of the governors were men. How many were women?

63. The numbers of women enrolled in mathematics classes at FHSU during the fall semester were 78 in Basic Mathematics, 185 in College Algebra, and 23 in Calculus. The total number of students enrolled in these mathematics classes was 459. How many men were enrolled in these classes?

64. The number of cable TV systems operating in the United States in 1995 was 11,215. In 1990, there were 9575 cable TV systems in operation. How many more cable systems were operating in 1995 than in 1990? (*Source: Television and Cable Factbook,* Warren Publishing, Inc.)

65. Marilyn Gonzales and Sharon Waterbury were candidates for student government president. Who won the election if the votes were cast as follows?

	Candidate	
Class	Marilyn	Sharon
Freshman	276	295
Sophomore	362	122
Junior	201	312
Senior	179	182

66. Two students submitted advertising budgets for a student government fundraiser.

	Student A	Student B
Radio Ads	$600	$300
Newspaper Ads	$200	$400
Posters	$150	$240
Hand bills	$120	$170

If $1200 is available for advertising, how much excess would each budget have?

ANSWERS

55. ______

56. ______

57. ______

58. ______

59. ______

60. ______

61. ______

62. ______

63. ______

64. ______

65. ______

66. ______

ANSWERS

67. ____

68. ____

69. ____

70. ____

71. ____

72. ____

73. ____

74. ____

75. ____

76. ____

77. ____

78. ____

The following bar graph shows the number of aircraft departures in 1993 for the top five airports in the United States. Use this graph to answer the questions below. See Example 7

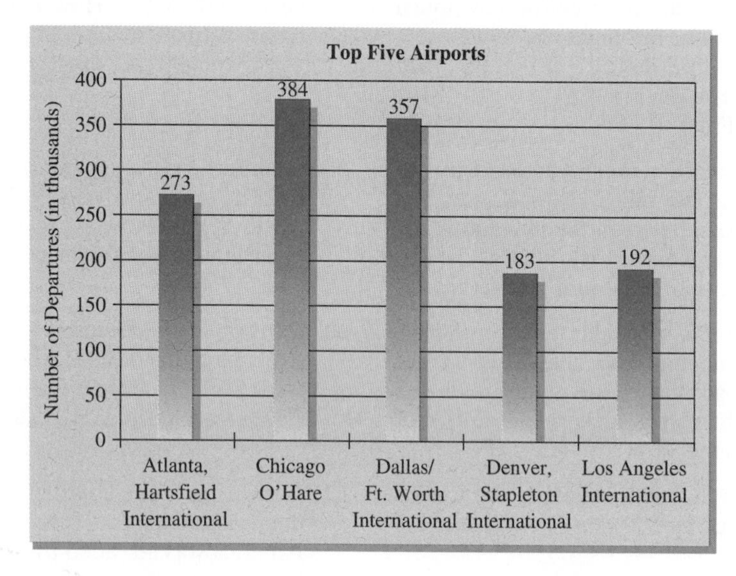

67. Which airport is busiest?

68. Which airports have more than 200 thousand departures per year?

69. How many more departures per year does Dallas/Ft. Worth International Airport have than Los Angeles International Airport?

70. How many more departures per year does Chicago O'Hare Airport have than Atlanta Hartsfield International Airport?

The following table shows the top 10 leading United States advertisers in 1994 and the amount of money spent in that year on ads. Use this table to answer Exercises 71 through 75. (Source: Competitive Media Reporting and Publishers Information Bureau)

Procter & Gamble	$1,464,994,000
General Motors	$1,398,806,000
Philip Morris	$1,307,661,000
Ford Motor	$920,264,000
Chrysler	$758,684,000
AT&T	$700,429,000
Pepsico	$669,686,000
Toyota Motor	$536,123,000
Sears Roebuck	$507,662,000
Walt Disney	$503,621,000

71. Which companies spent more than $1 billion on ads?

72. Which companies spent less than $700 million on ads?

73. How much more money did Philip Morris spend on ads than Walt Disney?

74. How much more money did AT&T spend on ads than Toyota Motor?

75. Find the total amount of money spent by these ten companies on ads.

76. The local college library is having a Million Pages of Reading promotion. The freshmen have read a total of 289,462 pages, the sophomores have read a total of 369,477 pages, the juniors have read a total of 218,287 pages, and the seniors have read a total of 121,685 pages. Have they reached the goal of a million pages? If not, how many more pages need to be read?

Fill in the missing digits.

77.
```
  526_
- 2_85
  28_4
```

78.
```
  10,_4_
-  85_4
   _710
```

1.4

ROUNDING AND ESTIMATING

OBJECTIVES

1. Round whole numbers.
2. Use rounding to estimate sums and differences.
3. Solve problems by estimating.

TAPE PA 1.4

1 **Rounding** a whole number means approximating it. A rounded whole number is often easier to use, understand, or remember than the precise whole number. For example, instead of trying to remember the Iowa state population as 2,829,252, it is much easier to remember it rounded to the nearest million: 3 million people.

To understand rounding, let's look at examples on a number line. The whole number 36 is closer to 40 than 30, so 36 rounded to the nearest ten is 40.

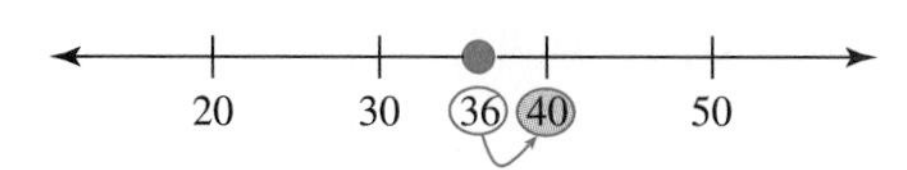

The whole number 52 rounded to the nearest ten is 50, because 52 is closer to 50 than to 60.

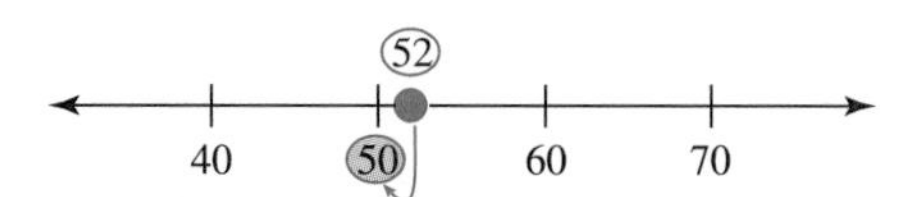

Trying to round 25 to the nearest ten, we see that 25 is halfway between 20 and 30. It is not closer to either number. In such a case, we round to the larger ten, that is, to 30.

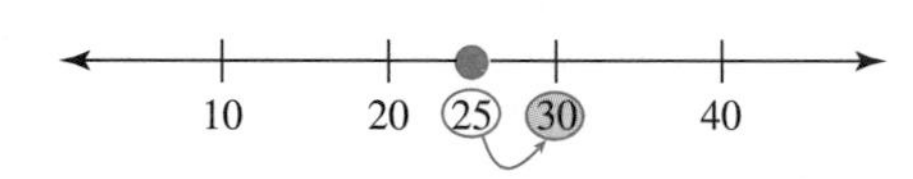

To round a whole number without using a number line, follow these steps.

ROUNDING WHOLE NUMBERS TO A GIVEN PLACE VALUE

Step 1. Locate the digit to the right of the given place value.

Step 2. If this digit is 5 or greater, add 1 to the digit in the given place value and replace each digit to its right by 0.

Step 3. If this digit is less than 5, replace it and each digit to its right by 0.

PRACTICE PROBLEM 1
Round 64,298 to the nearest thousand.

EXAMPLE 1
Round 73,568 to the nearest thousand.

Solution:

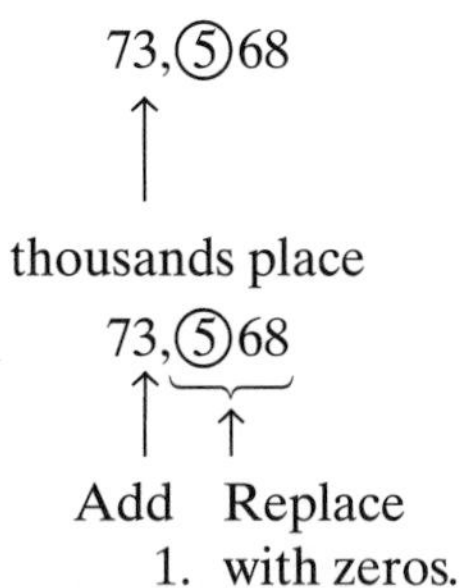

The digit to the right of the thousands place is the hundreds place, which is circled.

Since the circled digit is 5 or greater, add 1 to the 3 in the thousands place and replace each digit to the right by 0.

We find that 73,568 rounded to the nearest thousand is 74,000.

PRACTICE PROBLEM 2
Round 56,702 to the nearest ten thousand.

EXAMPLE 2
Round 278,362 to the nearest thousand.

Solution:

Thousands place

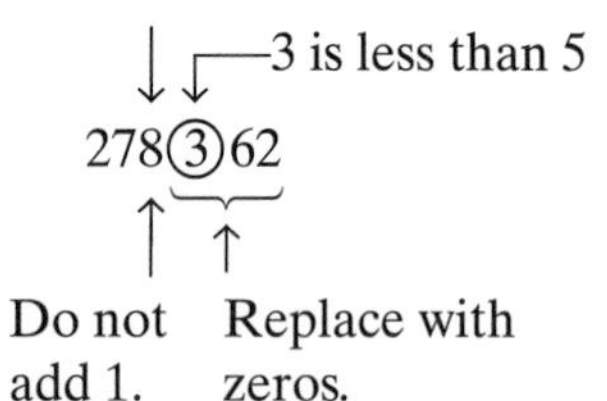

The number 278,362 rounded to the nearest thousand is 278,000.

PRACTICE PROBLEM 3
Round 265,489,263 to the nearest million.

EXAMPLE 3
Round 248,982 to the nearest hundred.

Solution:

Hundreds place

↓ ↓ 8 is greater than or equal to 5

248,9⑧2

↑

Add 1. $9 + 1 = 10$ so replace the digit 9 by 0 and carry 1 to the place value to the left.

2 4 8̸ (8 + 1), 9̸ (0) 8 2

Add 1. Replace with zeros.

The number 248,982 rounded to the nearest hundred is 249,000.

Answers:
1. 64,000
2. 60,000
3. 265,000,000

2 By rounding addends, we can estimate sums. An estimated sum is appropriate when an exact sum is not necessary. To estimate the sum shown, round each addend to the nearest hundred and then add.

768	rounds to	800
1952	rounds to	2000
225	rounds to	200
+ 149	rounds to	+ 100
		3100

The estimated sum is 3100, which is close to the exact sum of 3094.

EXAMPLE 4

Round each addend to the nearest hundred to find an estimated sum.

$$\begin{array}{r} 294 \\ 625 \\ 1071 \\ +\ 349 \\ \hline \end{array}$$

Solution:

294	rounds to	300
625	rounds to	600
1071	rounds to	1100
+ 349	rounds to	+ 300
		2300

The estimated sum is 2300. (The exact sum is 2339.)

PRACTICE PROBLEM 4

Round each addend to the nearest ten to find an estimated sum.

$$\begin{array}{r} 79 \\ 35 \\ 42 \\ 21 \\ +\ 98 \\ \hline \end{array}$$

EXAMPLE 5

Round each number to the nearest hundred to find an estimated difference.

$$\begin{array}{r} 4725 \\ -\ 2879 \\ \hline \end{array}$$

Solution:

4725	rounds to	4700
− 2879	rounds to	− 2900
		1800

The estimated difference is 1800. (The exact difference is 1846.)

PRACTICE PROBLEM 5

Round each number to the nearest thousand to find an estimated difference.

$$\begin{array}{r} 4725 \\ -2879 \\ \hline \end{array}$$

3 Making estimates is often the quickest way to solve real-life problems when their solutions do not need to be exact.

Answers:
4. 280
5. 2000

PRACTICE PROBLEM 6

Vivian is trying to estimate how far it is from Gove, Kansas, to Hays, Kansas. Round each given distance on the map to the nearest ten to estimate the total distance.

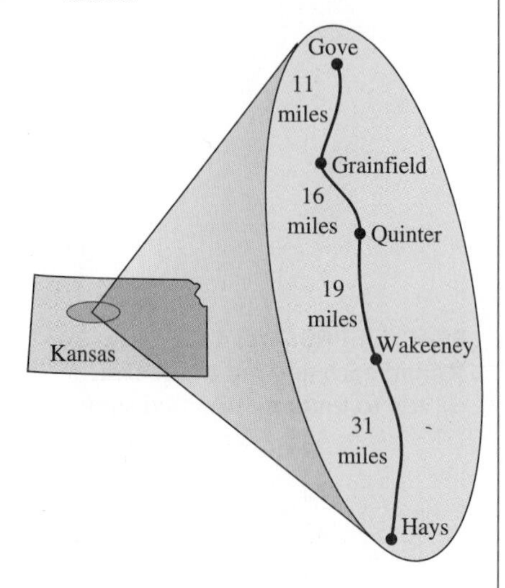

EXAMPLE 6

Jose Guillermo is trying to estimate quickly the distance from Temple, Texas, to Brenham, Texas. Round each distance to the nearest ten and then estimate the total distance.

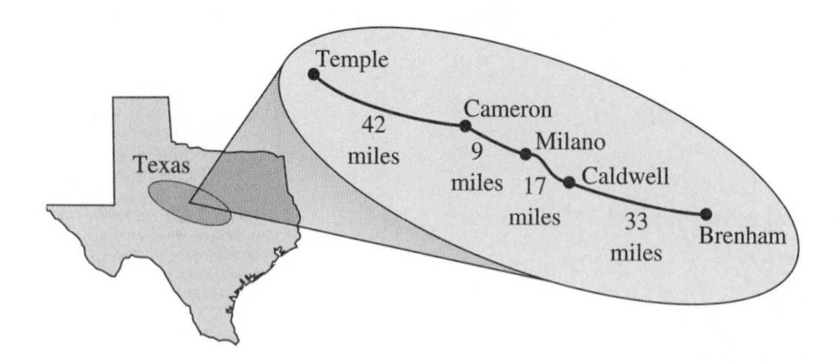

Solution:

DISTANCE		ESTIMATION
42	rounds to	40
9	rounds to	10
17	rounds to	20
+ 33	rounds to	+ 30
		100

It is approximately 100 miles from Temple to Brenham. (The exact distance is 101 miles.)

PRACTICE PROBLEM 7

The Johnson family traveled 15,329 miles last year and the Liu family traveled 19,572 miles. Round each distance to the nearest thousand and estimate how much farther the Lius traveled than the Johnsons.

EXAMPLE 7

A new TransX automobile sells for $27,895 and a new Petrol-Z automobile sells for $12,329. Round each amount to the nearest thousand and estimate how much more the TransX costs than the Petrol-Z.

Solution:

TransX cost	—	27,895	rounds to	28,000
− Petrol-Z cost	—	− 12,329	rounds to	− 12,000
Difference in cost	—			16,000

The TransX costs approximately $16,000 more than the Petrol-Z.

Answers:
6. 80 miles
7. 5,000 miles

EXERCISE SET 1.4

Round each whole number to the given place. See Examples 1 through 3.

1. 632 to the nearest ten

2. 273 to the nearest ten

3. 635 to the nearest ten

4. 275 to the nearest ten

5. 792 to the nearest ten

6. 394 to the nearest ten

7. 395 to the nearest ten

8. 582 to the nearest ten

9. 1096 to the nearest ten

10. 2198 to the nearest ten

11. 42,682 to the nearest thousand

12. 42,682 to the nearest ten-thousand

13. 248,695 to the nearest hundred

14. 179,406 to the nearest hundred

15. 36,499 to the nearest thousand

16. 96,501 to the nearest thousand

17. 99,995 to the nearest ten

18. 39,994 to the nearest ten

19. 59,725,642 to the nearest ten-million

20. 39,523,698 to the nearest million

ANSWERS

1. ______
2. ______
3. ______
4. ______
5. ______
6. ______
7. ______
8. ______
9. ______
10. ______
11. ______
12. ______
13. ______
14. ______
15. ______
16. ______
17. ______
18. ______
19. ______
20. ______

ANSWERS

27. ______

28. ______

29. ______

30. ______

31. ______

32. ______

33. ______

34. ______

35. ______

36. ______

37. ______

38. ______

39. ______

40. ______

41. ______

Fill in the chart by rounding each number to the given place value.

		TEN	HUNDRED	THOUSAND
21.	5281			
22.	7619			
23.	9444			
24.	7777			
25.	14,876			
26.	85,049			

27. Estimate to the nearest thousand the 1994–95 enrollment of East Tennessee State University: 11,512. (*Source:* East Tennessee State University)

28. Estimate to the nearest hundred the hourly cost of operating a B747-400 aircraft: $6592. (*Source:* Air Transport Association of America)

Estimate the sum or difference by rounding each number to the nearest ten. See Examples 4 and 5.

29. 29 + 35 + 42 + 16

30. 62 + 72 + 15 + 19

31. 649 − 272

32. 555 − 235

33. 729 + 634 + 825

34. 123 + 138 + 147

35. 805 − 797

36. 995 − 492

37. 164 + 295 + 149

38. 809 + 206 + 491

Estimate the sum or difference by rounding each number to the nearest hundred. See Examples 4 and 5.

39. 1812 + 1776 + 1945

40. 2010 + 2001 + 1984

41. 1774 − 1492

42.
$$\begin{array}{r} 1989 \\ -1870 \\ \hline \end{array}$$

43.
$$\begin{array}{r} 2995 \\ 1649 \\ +3940 \\ \hline \end{array}$$

44.
$$\begin{array}{r} 799 \\ 1655 \\ +271 \\ \hline \end{array}$$

Two of the given calculator answers below are incorrect. Find them by estimating each sum.

45. 362 + 419 781

46. 522 + 785 1307

47. 432 + 679 + 198 1139

48. 229 + 443 + 606 1278

49. 7806 + 5150 12,956

50. 5233 + 4988 9,011

51. 31,439 + 18,781 50,220

52. 68,721 + 52,335 121,056

Round each given number to the indicated place and solve. See Examples 6 and 7.

53. Campo appliance store advertises three refrigerators on sale at $799, $1299, and $999. Estimate the total cost of these three refrigerators to the nearest hundred dollars.

54. Jared Nuss scored 89, 92, 100, 67, 75, and 79 on his calculus tests. Estimate his total point score to the nearest ten.

55. Joe Puri has five rare dolls in his collection. The five dolls are worth $895, $675, $425, $750, and $350. Estimate, to the nearest hundred dollars, the value of his collection.

56. Carmelita Watkins is pricing new stereo systems. One system sells for $1895 and another system sells for $1524. Estimate, to the nearest hundred dollars, the difference in price of these systems.

57. Arlene Neville wants to estimate quickly the distance from Stockton to LaCrosse. Estimate this distance to the nearest 10 miles.

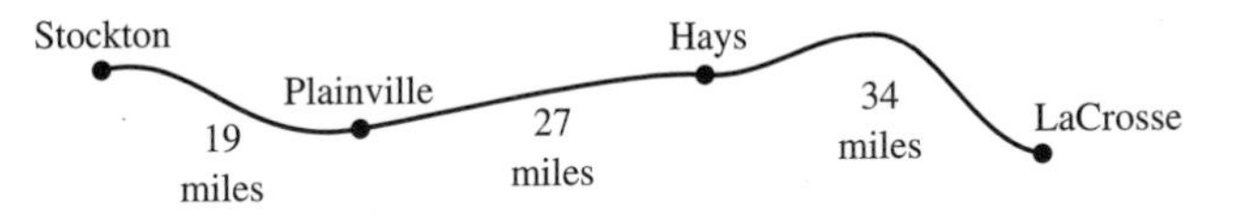

58. Barbara, Leon, and Rheem have $826, $1025, and $897, respectively. Estimate to the nearest $100 the total amount of money they have.

ANSWERS

42. ______

43. ______

44. ______

45. ______

46. ______

47. ______

48. ______

49. ______

50. ______

51. ______

52. ______

53. ______

54. ______

55. ______

56. ______

57. ______

58. ______

ANSWERS

59. ______

60. ______

61. ______

62. ______

63. ______

64. ______

65. ______

59. The peak of Mt. Everest, in Asia, is 29,028 feet above sea level. The top of Mt. Sunflower, in Kansas, is 4039 feet above sea level. Estimate, to the nearest thousand feet, the difference in elevation of these two peaks. (*Sources:* U.S. Geological Survey and National Geographic Society)

60. The Gonzales family took a trip and traveled 458, 489, 377, 243, 69, and 702 miles on six consecutive days. Estimate, to the nearest hundred miles, the distance they traveled.

61. In 1990 the population of New York City was 7,322,564 and the population of Houston, Texas, was 1,629,902. Estimate, to the nearest one hundred thousand, how much larger New York was than Houston. (*Source:* Bureau of the Census, 1990 census)

62. The distance from Kansas City to Boston is 1429 miles and from Kansas City to Chicago is 530 miles. Estimate, to the nearest hundred miles, how much farther Boston is from Kansas City than Chicago is.

63. In the 1964 Presidential election, Lyndon Johnson received 41,126,233 votes and Barry Goldwater received 27,174,898 votes. Estimate, to the nearest million, the number of votes by which Johnson won the election.

64. Enrollment figures at Normal State University showed an increase from 49,713 credit hours in 1988 to 51,746 credit hours in 1989. Estimate the increase to the nearest thousand credit hours.

65. The U.S. Commerce Department reported that the five states with highest per capita income in 1994 were Connecticut with $29,402, New Jersey with $28,038, Massachusetts with $25,616, Maryland with $24,933, and New York with $25,999. Estimate the total of these incomes to the nearest ten thousand dollars.

The following table from the previous section shows the top 10 leading United States advertisers in 1994 and the amount of money spent in that year on ads. Use this table to answer Exercises 66 through 69. (Source: Competitive Media Reporting and Publishers Information Bureau)

Procter & Gamble	$1,464,994,000
General Motors	$1,398,806,000
Philip Morris	$1,307,661,000
Ford Motor	$ 920,264,000
Chrysler	$ 758,684,000
AT&T	$ 700,429,000
Pepsico	$ 669,686,000
Toyota Motor	$ 536,123,000
Sears Roebuck	$ 507,662,000
Walt Disney	$ 503,621,000

66. Approximate the amount of money spent by Pepsico to the nearest hundred million.

67. Approximate the amount of money spent by General Motors to the nearest hundred million.

68. Approximate the amount of money spent by Sears Roebuck to the nearest million.

69. Approximate the amount of money spent by Ford Motor to the nearest million.

70. On August 23, 1989, it was estimated that one million five hundred thousand people joined hands in a human chain stretching 370 miles to protest the fiftieth anniversary of the pact that allowed the then Soviet Union to annex the Baltic nations in 1939. If the estimate of the number of people is to the nearest one hundred thousand, determine the largest possible number of people in the chain.

71. Explain how to round a number to the nearest thousand.

ANSWERS

66. ________

67. ________

68. ________

69. ________

70. ________

71. ________

1.5

MULTIPLYING WHOLE NUMBERS AND AREA

TAPE PA 1.5

OBJECTIVES

1. Identify factor and product.
2. Identify properties of multiplication.
3. Multiply whole numbers.
4. Find the area of a rectangle.
5. Solve problems by multiplying whole numbers.

1 Suppose that we wish to count the number of desks in a classroom. The desks are arranged in 5 rows and each row has 6 desks.

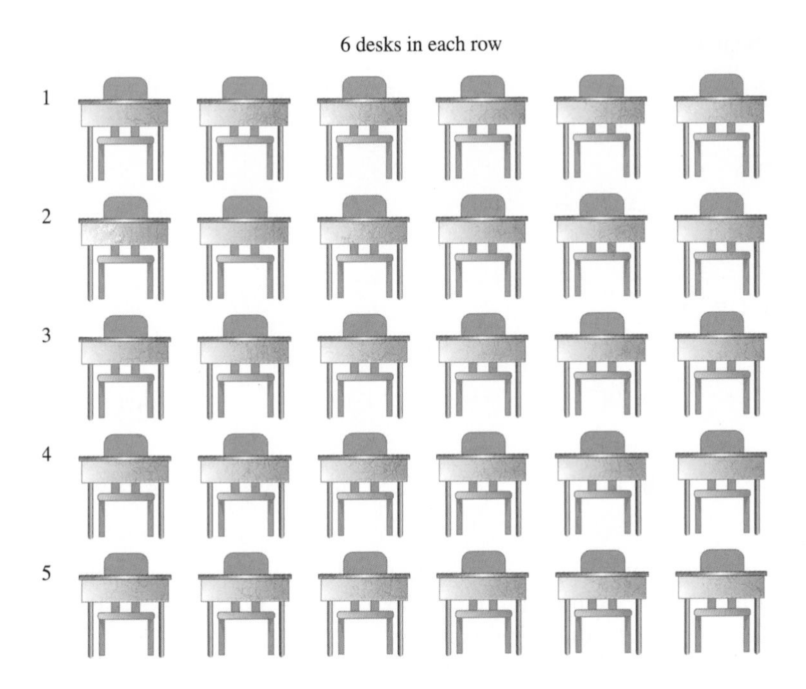

Adding 5 sixes gives the total number of desks: $6 + 6 + 6 + 6 + 6 = 30$ desks. When each addend is the same, we refer to this as **repeated addition.**

Multiplication is repeated addition but with different notation.

$$\underbrace{6 + 6 + 6 + 6 + 6}_{\text{5 sixes}} = \underset{\text{factor}}{\underset{\uparrow}{5}} \times \underset{\text{factor}}{\underset{\uparrow}{6}} = \underset{\text{product}}{\underset{\uparrow}{30}}$$

The $\times$ is called a multiplication sign. The numbers 5 and 6 are called **factors.** The number 30 is called the **product.** The notation 5×6 is read as "five times six." The symbols $\cdot$ and ( ) can also be used to indicate multiplication.

$$5 \times 6 = 30, \quad 5 \cdot 6 = 30, \quad (5)(6) = 30, \quad \text{and } 5(6) = 30$$

2 As for addition, we memorize products of one-digit whole numbers and then use certain properties of multiplication to multiply larger numbers. (If necessary, review the multiplication of one-digit numbers in Appendix B.) Notice in

the appendix that, when a number is multiplied by zero, the product is always zero. We can describe this result as a property, the **multiplication property of 0.**

MULTIPLICATION PROPERTY OF 0

The product of 0 and any number is 0. For example,

$$5 \cdot 0 = 0$$
$$0 \cdot 8 = 0$$

Also notice in the appendix that when a number is multiplied by one, the result is always the original number. We call this result the **multiplication property of 1.**

MULTIPLICATION PROPERTY OF 1

The product of 1 and any number is that same number. For example,

$$1 \cdot 9 = 9$$
$$6 \cdot 1 = 6$$

PRACTICE PROBLEM 1

Multiply:

a. 3×0 **b.** $4(1)$
c. $(0)(34)$ **d.** $1 \cdot 76$

EXAMPLE 1

Multiply.

a. 6×1 **b.** $0(8)$ **c.** $1 \cdot 45$ **d.** $(75)(0)$

Solution:

a. $6 \times 1 = 6$ **b.** $0(8) = 0$ **c.** $1 \cdot 45 = 45$ **d.** $(75)(0) = 0$

Like addition, multiplication is commutative and associative. Notice that, when multiplying two numbers, the order of these numbers can be changed without changing the product. For example,

$$3 \cdot 5 = 5 \cdot 3$$

or

$$15 = 15$$

This property is the **commutative property of multiplication.**

COMMUTATIVE PROPERTY OF MULTIPLICATION

Changing the **order** of two factors does not change their product. For example,

$$9 \cdot 2 = 2 \cdot 9$$

Another property that can help us when multiplying is the **associative property of multiplication.** This property states that, when multiplying numbers, the grouping of the numbers can be changed without changing the product. For example,

Answers:
1a. 0 **b.** 4 **c.** 0 **d.** 76

$$2 \cdot (3 \cdot 4) = (2 \cdot 3) \cdot 4$$

or

$$2 \cdot 12 = 6 \cdot 4$$

or

$$24 = 24$$

ASSOCIATIVE PROPERTY OF MULTIPLICATION

Changing the **grouping** of factors does not change their product. For example,

$$5 \cdot (3 \cdot 2) = (5 \cdot 3) \cdot 2$$

EXAMPLE 2

Rewrite each using the commutative property of multiplication or the associative property of multiplication.

a. $3 \cdot 6$ **b.** $2 \cdot (6 \cdot 4)$ **c.** $11 \cdot 5$ **d.** $(8 \cdot 10) \cdot 7$

Solution:

a. By the commutative property of multiplication, $3 \cdot 6 = 6 \cdot 3$.

b. By the associative property of multiplication, $2 \cdot (6 \cdot 4) = (2 \cdot 6) \cdot 4$.

c. In multiplication, the **order** of factors can be changed without changing their product: $11 \cdot 5 = 5 \cdot 11$.

d. In multiplication, the **grouping** of factors can be changed without changing their product: $(8 \cdot 10) \cdot 7 = 8 \cdot (10 \cdot 7)$.

PRACTICE PROBLEM 2

Rewrite each using the commutative property of multiplication or the associative property of multiplication.

a. $3 \cdot (4 \cdot 7)$ **b.** $8 \cdot 6$

c. $(2 \cdot 9) \cdot 5$ **d.** $7 \cdot 9$

With these properties, along with the **distributive property,** we can find the product of any whole numbers. The distributive property says that multiplication **distributes** over addition. For example, notice that $3(2 + 5)$ is the same as $3 \cdot 2 + 3 \cdot 5$.

$$3(2 + 5) = 3 \cdot 2 + 3 \cdot 5$$

or

$$3(7) = 6 + 15$$

or

$$21 = 21$$

Notice in $3(2 + 5) = 3 \cdot 2 + 3 \cdot 5$ that each number inside the parentheses is multiplied by 3.

DISTRIBUTIVE PROPERTY

Multiplication distributes over addition.
For example,

$$2(3 + 4) = 2 \cdot 3 + 2 \cdot 4$$

Answers:
2a. $3 \cdot (4 \cdot 7) = (3 \cdot 4) \cdot 7$
b. $8 \cdot 6 = 6 \cdot 8$
c. $(2 \cdot 9) \cdot 5 = 2 \cdot (9 \cdot 5)$
d. $7 \cdot 9 = 9 \cdot 7$

PRACTICE PROBLEM 3

Rewrite each using the distributive property.

a. $5(2 + 3)$ **b.** $9(8 + 7)$

c. $3(6 + 1)$

EXAMPLE 3

Rewrite each using the distributive property.

a. $3(4 + 5)$ **b.** $10(6 + 8)$ **c.** $2(7 + 3)$

Solution:

By the distributive property, we have

a. $3(4 + 5) = 3 \cdot 4 + 3 \cdot 5$

b. $10(6 + 8) = 10 \cdot 6 + 10 \cdot 8$

c. $2(7 + 3) = 2 \cdot 7 + 2 \cdot 3$

3 Let's use the distributive property to multiply 7(48). To do so, we begin by writing the expanded form of 48 and then applying the distributive property.

$$\begin{aligned} 7(48) &= 7(40 + 8) & \\ &= 7 \cdot 40 + 7 \cdot 8 & \text{Apply the distributive property.} \\ &= 280 + 56 & \text{Multiply.} \\ &= 336 & \text{Add.} \end{aligned}$$

This is how we multiply whole numbers. When multiplying whole numbers, we will use the following notation.

$$\begin{array}{r} {}^{5} \\ 48 \\ \times \ \ 7 \\ \hline 336 \end{array}$$

$7 \cdot 8 = 56$ — Write 6 in the ones place and carry 5 to the tens place.

$7 \cdot 4 = 28$ and $28 + 5 = 33$

PRACTICE PROBLEM 4

Multiply: 36×4.

EXAMPLE 4

Multiply: $\begin{array}{r} 25 \\ \times \ 8 \\ \hline \end{array}$

Solution:

$$\begin{array}{r} {}^{4} \\ 25 \\ \times \ 8 \\ \hline 200 \end{array}$$

To multiply larger whole numbers, use the following similar notation. Multiply 89×52.

STEP 1	STEP 2	STEP 3
$\begin{array}{r} {}^{1} \\ 89 \\ \times\ 52 \\ \hline 178 \end{array}$ ← Multiply 89×2.	$\begin{array}{r} {}^{4} \\ 89 \\ \times\ 52 \\ \hline 178 \\ 4450 \\ \hline \end{array}$ ← Multiply 89×50.	$\begin{array}{r} 89 \\ \times\ 52 \\ \hline 178 \\ 4450 \\ \hline 4628 \end{array}$ Add.

The numbers 178 and 4450 are called **partial products.** The sum of the partial products, 4628, is the product of 89 and 52.

Answers:

3a. $5(2 + 3) = 5 \cdot 2 + 5 \cdot 3$
b. $9(8 + 7) = 9 \cdot 8 + 9 \cdot 7$
c. $3(6 + 1) = 3 \cdot 6 + 3 \cdot 1$
4. 144

EXAMPLE 5

Multiply: 236×86.

Solution:

$$\begin{array}{rl} 236 & \\ \times\ \ 86 & \\ \hline 1{,}416 & \longleftarrow 6(236) \\ 18{,}880 & \longleftarrow 80(236) \\ \hline 20{,}296 & \text{Add.} \end{array}$$

PRACTICE PROBLEM 5

Multiply: 594×72.

4 A special application of multiplication is finding the area of a region. Area measures the amount of surface of a region. For example, we measure a plot of land or the living area of a home by area. The figures show two examples of units of area measure. (A centimeter is a unit of length in the metric system.)

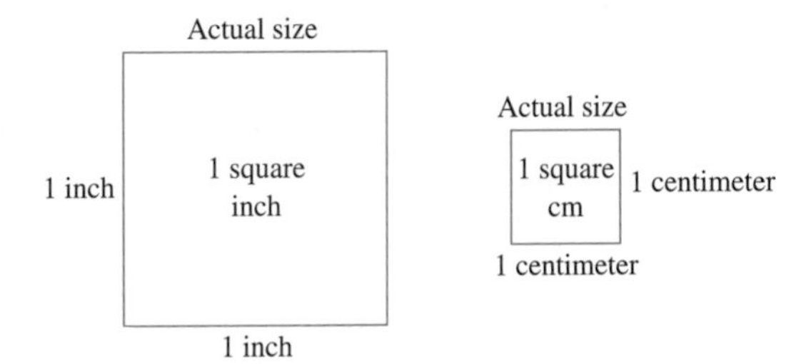

To measure the area of a geometric figure such as the rectangle shown, count the number of square units that cover the region.

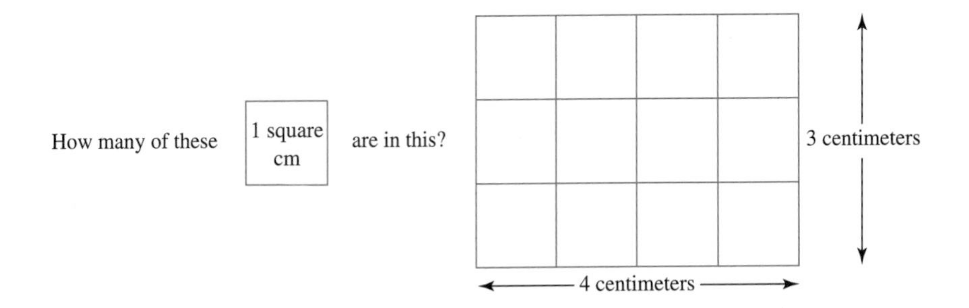

This rectangular region contains 12 square units, each 1 square centimeter. Thus, the area is 12 square centimeters. This total number of squares can be found by counting or by multiplying **4 · 3** (length · width) as shown.

$$\begin{aligned} \text{Area of a rectangle} &= \text{length} \cdot \text{width} \\ &= (4 \text{ centimeters})(3 \text{ centimeters}) \\ &= 12 \text{ square centimeters} \end{aligned}$$

In this section, we find the areas of rectangles only. In later sections, we find the areas of other geometric regions.

Answer:
5. 42,768

PRACTICE PROBLEM 6

The state of Wyoming is in the shape of a rectangle whose length is 360 miles and whose width is 280 miles. Find its area.

EXAMPLE 6

The state of Colorado is in the shape of a rectangle whose length is 380 miles and whose width is 280 miles. Find its area.

Solution:

The area of a rectangle is the product of its length and its width.

$$\begin{aligned}\text{Area} &= \text{ength} \cdot \text{width}\\ &= (\ 80 \text{ miles})(280 \text{ miles})\\ &= 106{,}400 \text{ square miles}\end{aligned}$$

The area of Colorado is 106,400 square miles.

5 There are several words or phrases that indicate the operation of multiplication. Some of these are as follows:

KEY WORDS OR PHRASES	EXAMPLE	SYMBOLS
multiply	multiply 5 by 7	$5 \cdot 7$
product	the product of 3 and 2	$3 \cdot 2$
times	10 times 13	$10 \cdot 13$

Many key words or phrases describing real-life problems that might suggest an addition model might be better solved by a multiplication model instead. For example, to find the **total** cost of 8 shirts, each selling for \$27, we can either add $27 + 27 + 27 + 27 + 27 + 27 + 27 + 27$, or we can multiply $8(27)$.

PRACTICE PROBLEM 7

A computer printer can print 240 characters per second in draft mode. How many total characters can it print in 15 seconds?

EXAMPLE 7

A single-density computer disk can hold 370 thousand bytes of information. How many total bytes can 42 such disks hold?

Solution:

Forty-two disks will hold 42×370 thousand bytes.

WORDS		TRANSLATE TO NUMBERS
bytes per disk	—	370
× disks	—	× 42
		740
		14,800
total bytes		15,540

Forty-two disks will hold 15,540 thousand bytes.

PRACTICE PROBLEM 8

Softball T-shirts come in two styles: plain at \$6 each and striped at \$7 each. The team orders 4 plain shirts and 5 striped shirts. Find the total cost of the order.

EXAMPLE 8

Earline Martin agrees to take her children and their cousins to the San Antonio Zoo. The ticket price for each child is \$4 and for each adult, \$6. If 8 children and 1 adult plan to go, how much money is needed for admission?

Solution:

If the price of one child's ticket is \$4, the price for 8 children is $8 \cdot 4 = \$32$. The price of one adult ticket is \$6, so the total cost is

Answers:

6. 100,800 square miles
7. 3600 characters
8. \$59

WORDS		TRANSLATE TO NUMBERS
price of 8 children	⟶	\$32
+ price of adult	⟶	+ 6
total cost		\$38

The total cost is \$38.

EXAMPLE 9

The average page of a book contains 259 words. Estimate, rounding to the nearest ten, the total number of words contained on 22 pages.

Solution:

The exact number of words is 259 $\times$ 22. Estimate this product by rounding each factor to the nearest ten.

259	rounds to	⟶	$\overset{1}{260}$
$\times$ 22	rounds to	⟶	$\times$ 20
			5200

There are approximately 5200 words contained on 22 pages.

PRACTICE PROBLEM 9

If an average page in a book contains 259 words, estimate, rounding to the nearest hundred, the total number of words contained on 195 pages.

Answer:

9. 60,000 words

CALCULATOR EXPLORATIONS

MULTIPLYING NUMBERS

To multiply numbers on a calculator, find the keys marked [×] and [=]. For example, to find $31 \cdot 66$ on a calculator, press the keys [31] [×] [66] [=]. The display will read [2046]. Thus, $31 \cdot 66 = 2046$.

Use a calculator to multiply the following.

1. 72×48

2. 81×92

3. $163 \cdot 94$

4. $285 \cdot 144$

5. 983(277)

6. 1562(843)

MENTAL MATH

Multiply. See Example 1.

1. $1 \cdot 24$

2. $55 \cdot 1$

3. $0 \cdot 19$

4. $27 \cdot 0$

5. $8 \cdot 0 \cdot 9$

6. $7 \cdot 6 \cdot 0$

7. $87 \cdot 1$

8. $1 \cdot 41$

EXERCISE SET 1.5

Use the commutative or associative property to rewrite each product. See Example 2.

1. $6 \cdot 9$ **2.** $3 \cdot (2 \cdot 8)$ **3.** $(4 \cdot 8) \cdot 10$ **4.** $21 \cdot 13$

5. $5 \cdot (7 \cdot 12)$ **6.** $(4 \cdot 1) \cdot 15$ **7.** $32 \cdot 89$ **8.** $2 \cdot 32$

Use the distributive property to rewrite each expression. See Example 3.

9. $4(3 + 9)$ **10.** $5(8 + 2)$ **11.** $2(4 + 6)$

12. $6(1 + 4)$ **13.** $10(11 + 7)$ **14.** $12(12 + 3)$

Multiply. See Example 4.

15. 42×6 **16.** 79×3 **17.** 624×3 **18.** 638×5

19. 277×6 **20.** 882×2 **21.** 1062×5 **22.** 9021×3

Multiply. See Example 5.

23. 298×14 **24.** 591×72 **25.** 231×47 **26.** 526×23

27. 809×14 **28.** 307×16 **29.** $(620)(40)$ **30.** $(720)(80)$

31. $(998)(12)(0)$ **32.** $(593)(47)(0)$ **33.** $(590)(1)(10)$ **34.** $(240)(1)(20)$

35. 1234×48 **36.** 1357×79 **37.** 609×234 **38.** 505×127

ANSWERS

1. ______
2. ______
3. ______
4. ______
5. ______
6. ______
7. ______
8. ______
9. ______
10. ______
11. ______
12. ______
13. ______
14. ______
15. ______
16. ______
17. ______
18. ______
19. ______
20. ______
21. ______
22. ______
23. ______
24. ______
25. ______
26. ______
27. ______
28. ______
29. ______
30. ______
31. ______
32. ______
33. ______
34. ______
35. ______
36. ______
37. ______
38. ______

ANSWERS

39. ______
40. ______
41. ______
42. ______
43. ______
44. ______
45. ______
46. ______
47. ______
48. ______
49. ______
50. ______
51. ______
52. ______
53. ______
54. ______
55. ______
56. ______
57. ______
58. ______
59. ______
60. ______
61. ______
62. ______

39. 5621×324

40. 1234×567

41. 1941×235

42. 1876×437

43. 589×110

44. 426×110

Estimate the products by rounding each factor to the nearest hundred. See Example 9.

45. 576×354

46. 982×650

47. 604×451

48. 111×999

Find the area of each rectangle. See Example 6.

49.

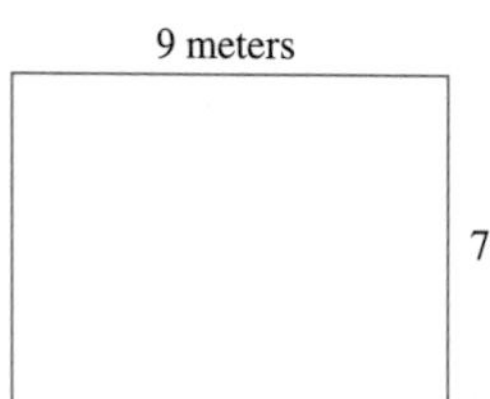

50.

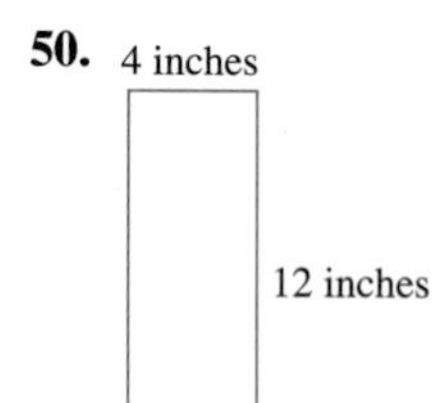

51.

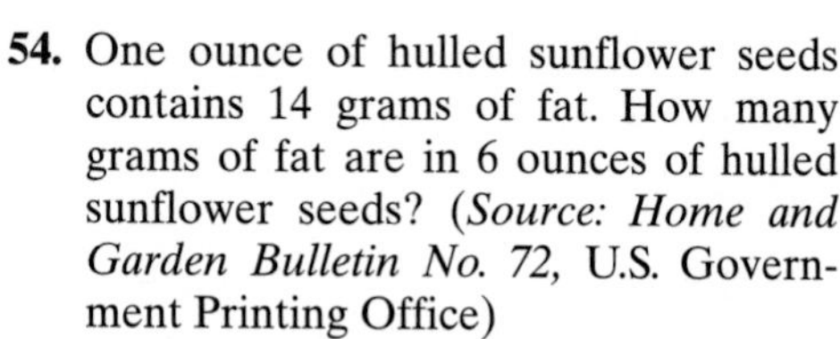

52.

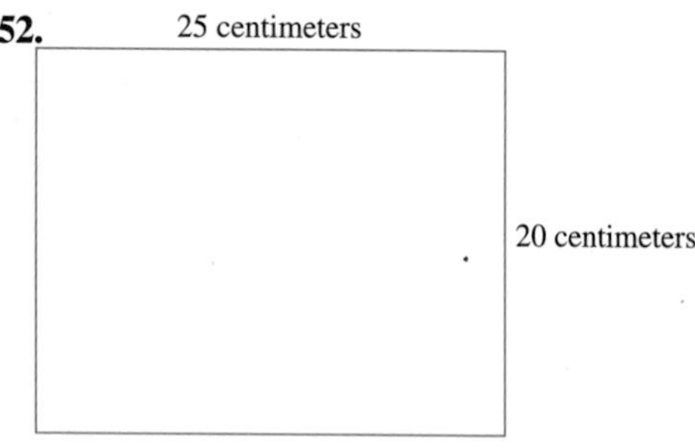

Solve. See Examples 6 through 9.

53. One tablespoon of olive oil contains 125 calories. How many calories are in 3 tablespoons of olive oil? (*Source: Home and Garden Bulletin No. 72,* U.S. Government Printing Office)

54. One ounce of hulled sunflower seeds contains 14 grams of fat. How many grams of fat are in 6 ounces of hulled sunflower seeds? (*Source: Home and Garden Bulletin No. 72,* U.S. Government Printing Office)

55. The textbook for a course in Civil War history costs $37. There are 35 students in the class. Find the total cost of the history books for the class.

56. The seats in the mathematics lecture hall are arranged in 12 rows with 6 seats in each row. Find how many seats are in this room.

57. A case of canned peas has **two layers** of cans. In each layer are 8 rows with 12 cans in each row. Find how many cans are in a case.

58. An apartment building has **three floors.** Each floor has five rows of apartments with four apartments in each row. Find how many apartments there are.

59. A plot of land measures 90 feet by 110 feet. Find its area.

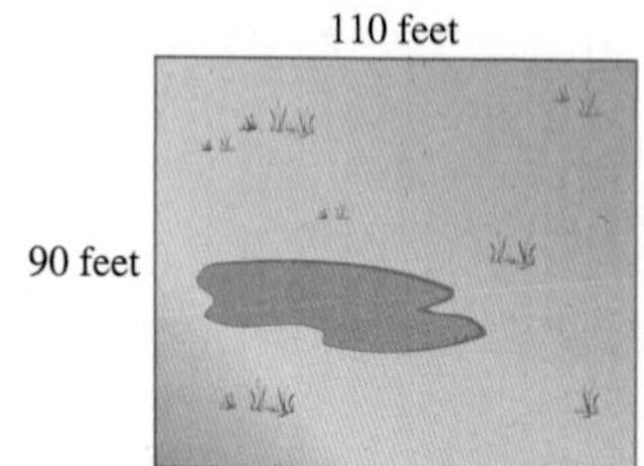

60. A house measures 45 feet by 60 feet. Find the floor area of the house.

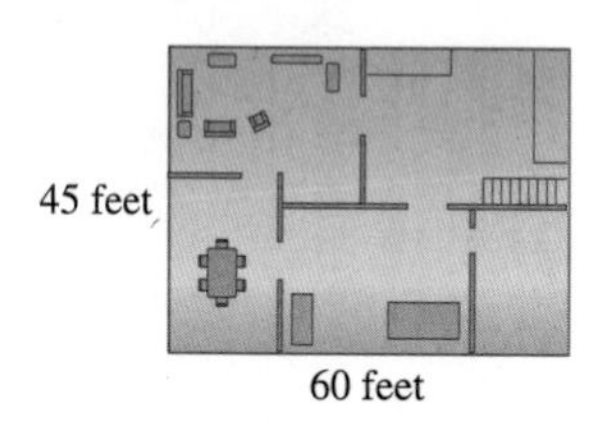

61. Gerald Potter's calculator has six rows with seven keys in each row. Find how many keys there are on his calculator.

62. An advanced scientific calculator has **two sides.** Each side has seven rows with five keys in each row. Find how many keys there are on the calculator.

EXERCISE SET 1.6

Divide and then check by multiplying. See Examples 4 and 5.

1. $108 \div 9$

2. $85 \div 5$

3. $6\overline{)222}$

4. $8\overline{)640}$

5. $1014 \div 3$

6. $504 \div 4$

Divide and then check by multiplying. See Examples 6 and 7.

7. $98 \div 6$

8. $422 \div 7$

9. $2\overline{)1127}$

10. $3\overline{)1240}$

11. $186 \div 5$

12. $167 \div 3$

13. $2121 \div 8$

14. $333 \div 4$

Divide and then check by multiplying. See Examples 8 and 9.

15. $1127 \div 23$

16. $2016 \div 42$

17. $715 \div 55$

18. $1856 \div 32$

19. $9449 \div 97$

20. $1938 \div 44$

21. $3708 \div 18$

22. $7224 \div 12$

23. $6578 \div 13$

24. $5670 \div 14$

25. $9299 \div 46$

26. $2539 \div 64$

27. $\frac{10{,}620}{236}$

28. $\frac{5781}{123}$

29. $\frac{10{,}194}{103}$

30. $\frac{23{,}048}{240}$

31. $20{,}619 \div 102$

32. $40{,}803 \div 203$

33. $45{,}046 \div 223$

34. $164{,}592 \div 543$

ANSWERS

1. ______
2. ______
3. ______
4. ______
5. ______
6. ______
7. ______
8. ______
9. ______
10. ______
11. ______
12. ______
13. ______
14. ______
15. ______
16. ______
17. ______
18. ______
19. ______
20. ______
21. ______
22. ______
23. ______
24. ______
25. ______
26. ______
27. ______
28. ______
29. ______
30. ______
31. ______
32. ______
33. ______
34. ______

ANSWERS

35. ______
36. ______
37. ______
38. ______
39. ______
40. ______
41. ______
42. ______
43. ______
44. ______
45. ______
46. ______
47. ______
48. ______
49. ______
50. ______
51. ______

Find the average of each list of numbers. See Example 13.

35. 14, 22, 45, 18, 30, 27

36. 38, 26, 15, 29, 29, 51, 22

37. 204, 968, 552, 268

38. 121, 200, 185, 176, 163

39. 86, 79, 81, 69, 80

40. 92, 96, 90, 85, 92, 79

Solve. See Examples 10 through 13.

41. Kathy Gomez teaches Spanish lessons for $85 per student for a 5-week session. From one group of students, she collects $4930. Find how many students are in the group.

42. Martin Thieme teaches American Sign Language classes for $55 per student for a 7-week session. He collects $1430 from the group of students. Find how many students are in the group.

43. Twenty-one people pooled their money and bought lottery tickets. One ticket won a prize of $5,292,000. Find how many dollars each person receives.

44. The gravity of Jupiter is 318 times as strong as the gravity of Earth, so objects on Jupiter weigh 318 times as much as they weigh on Earth. If a person would weigh 52,470 pounds on Jupiter, find how much he weighs on Earth.

45. A truck hauls wheat to a storage granary. It carries a total of 5810 bushels of wheat in 14 trips. How much does the truck haul each trip, if each trip it hauls the same amount?

46. The white stripes dividing the lanes on a highway are 25 feet long and the spaces between them are 25 feet long. Find how many whole stripes there are in 1 mile of highway. (A mile is 5280 feet.)

47. There is a bridge over highway I-35 every three miles. Find how many bridges there are over 265 miles of I-35.

48. An 18-hole golf course is 5580 yards long. Find the average distance between holes.

49. Wendy Holladay has a piece of rope 185 feet long that she cuts into pieces for an experiment in her second-grade class. Each piece of rope is to be 8 feet long. Determine whether she has enough rope for her 22-student class. Determine the amount extra or the amount short.

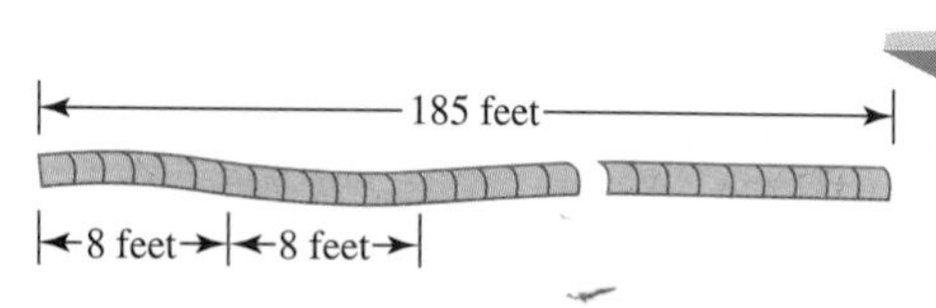

50. Jesse White is in the requisitions department of Central Electric Lighting Company. Light poles along a highway are placed 492 feet apart. Find how many poles he should order for a 1-mile strip of highway. (A mile is 5280 feet.)

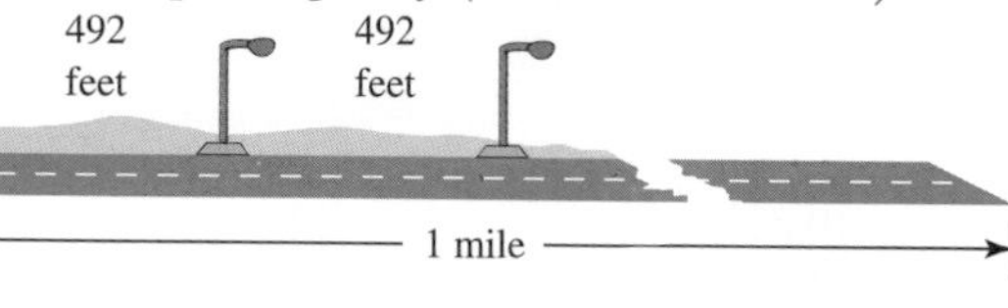

51. Mehdi receives a paycheck every four weeks, on Friday. Find how many paychecks he receives in a year. (A year has 52 weeks.)

52. Find how many yards there are in 1 mile. (A mile is 5280 feet; a yard is 3 feet.)

1 foot	1 foot	1 foot	1 foot	1 foot	1 foot	

1 yard 1 yard

5280 feet
? yards

53. Find how many whole feet there are in 1 rod. (A mile is 5280 feet; 1 mile is 320 rods.)

The monthly normal temperature in degrees Fahrenheit for Minneapolis, Minnesota, is given in the table. Use this table to answer Exercises 54 and 55. (Source: National Climatic Data Center)

January	12°	July	74°
February	18°	August	71°
March	31°	Sept.	61°
April	46°	Oct.	49°
May	59°	Nov.	33°
June	68°	Dec.	18°

54. Find the average temperature for December, January, and February.

55. Find the average temperature for the entire year.

The following table from previous sections shows the top 10 leading United States advertisers in 1994 and the amount of money spent in that year on ads. Use this table to answer Exercises 56 and 57.

Procter & Gamble	$1,464,994,000
General Motors	$1,398,806,000
Philip Morris	$1,307,661,000
Ford Motor	$920,264,000
Chrysler	$758,684,000
AT&T	$700,429,000
Pepsico	$669,686,000
Toyota Motor	$536,123,000
Sears Roebuck	$507,662,000
Walt Disney	$503,621,000

56. Find the average amount of money spent on ads for the year by the top three companies.

57. Find the average amount of money spent on ads for the year by Pepsico, Toyota Motor, Sears Roebuck, and Walt Disney.

ANSWERS

52. ____

53. ____

54. ____

55. ____

56. ____

57. ____

ANSWERS

58. ______________

59. ______________

60. ______________

61. ______________

In Example 13 in this section, we found that the average of 93, 86, 71, *and* 82 *is* 83. *Use this information to answer Exercises 58 and 59.*

58. If the number 71 is removed from the list of numbers, does the average increase or decrease?

59. If the number 93 is removed from the list of numbers, does the average increase or decrease?

60. Without computing, tell whether the average of 126, 135, 198, 113 is 86. Explain why it is or why it is not.

61. If the area of a rectangle is 30 square feet and its width is 3 feet, what is its length?

30 square feet

3 feet

?

EXERCISE SET 1.7

Write using exponential notation. See Example 1.

1. $3 \cdot 3 \cdot 3 \cdot 3$

2. $5 \cdot 5 \cdot 5$

3. $7 \cdot 7 \cdot 7 \cdot 7 \cdot 7 \cdot 7 \cdot 7 \cdot 7$

4. $6 \cdot 6 \cdot 6 \cdot 6 \cdot 6$

5. $12 \cdot 12 \cdot 12$

6. $10 \cdot 10$

7. $6 \cdot 6 \cdot 5 \cdot 5 \cdot 5$

8. $4 \cdot 4 \cdot 4 \cdot 3 \cdot 3$

9. $9 \cdot 9 \cdot 9 \cdot 8$

10. $7 \cdot 7 \cdot 7 \cdot 4$

11. $3 \cdot 2 \cdot 2 \cdot 2 \cdot 2 \cdot 2$

12. $4 \cdot 6 \cdot 6 \cdot 6 \cdot 6$

13. $3 \cdot 2 \cdot 2 \cdot 5 \cdot 5 \cdot 5$

14. $6 \cdot 6 \cdot 2 \cdot 9 \cdot 9 \cdot 9 \cdot 9$

Evaluate. See Example 2.

15. 5^2

16. 6^2

17. 5^3

18. 6^3

19. 2^6

20. 2^7

21. 2^{10}

22. 1^{12}

23. 7^1

24. 8^1

25. 3^5

26. 5^4

ANSWERS

1. ______
2. ______
3. ______
4. ______
5. ______
6. ______
7. ______
8. ______
9. ______
10. ______
11. ______
12. ______
13. ______
14. ______
15. ______
16. ______
17. ______
18. ______
19. ______
20. ______
21. ______
22. ______
23. ______
24. ______
25. ______
26. ______

ANSWERS

27. ______
28. ______
29. ______
30. ______
31. ______
32. ______
33. ______
34. ______
35. ______
36. ______
37. ______
38. ______
39. ______
40. ______
41. ______
42. ______
43. ______
44. ______
45. ______
46. ______
47. ______
48. ______
49. ______
50. ______
51. ______
52. ______

27. 2^8 **28.** 3^3 **29.** 4^3 **30.** 4^4

31. 9^2 **32.** 8^2 **33.** 9^3 **34.** 8^3

35. 10^2 **36.** 10^3 **37.** 10^4 **38.** 10^5

39. 10^1 **40.** 14^1 **41.** 1920^1 **42.** 6849^1

43. 3^6 **44.** 4^5

Simplify. See Examples 3, 4, and 6.

45. $15 + 3 \cdot 2$ **46.** $24 + 6 \cdot 3$ **47.** $20 - 4 \cdot 3$ **48.** $17 - 2 \cdot 6$

49. $5 \cdot 9 - 16$ **50.** $8 \cdot 4 - 10$ **51.** $28 \div 4 - 3$ **52.** $42 \div 7 - 6$

53. $14 + \frac{24}{8}$

54. $32 + \frac{8}{2}$

55. $6 \cdot 5 + 8 \cdot 2$

56. $3 \cdot 4 + 9 \cdot 1$

57. $0 \div 6 + 4 \cdot 7$

58. $0 \div 8 + 7 \cdot 6$

59. $6 + 8 \div 2$

60. $6 + 9 \div 3$

61. $(6 + 8) \div 2$

62. $(6 + 9) \div 3$

63. $(6^2 - 4) \div 8$

64. $(7^2 - 7) \div 7$

65. $(3 + 5^2) \div 2$

66. $(13 + 6^2) \div 7$

67. $6^2 \cdot (10 - 8)$

68. $5^3 \div (10 + 15)$

69. $\frac{18 + 6}{2^4 - 4}$

70. $\frac{15 + 17}{5^2 - 3^2}$

71. $(2 + 5) \cdot (8 - 3)$

72. $(9 - 7) \cdot (12 + 18)$

73. $\frac{7(9 - 6) + 3}{3^2 - 3}$

74. $\frac{5(12 - 7) - 4}{5^2 - 2^3 - 10}$

75. $5 \div 0 + 24$

76. $18 - 7 \div 0$

Simplify. See Example 5.

77. $3^4 - [35 - (12 - 6)]$

78. $[40 - (8 - 2)] - 2^5$

79. $(7 \cdot 5) + [9 \div (3 \div 3)]$

80. $(18 \div 6) + [(3 + 5) \cdot 2]$

ANSWERS

53. ______

54. ______

55. ______

56. ______

57. ______

58. ______

59. ______

60. ______

61. ______

62. ______

63. ______

64. ______

65. ______

66. ______

67. ______

68. ______

69. ______

70. ______

71. ______

72. ______

73. ______

74. ______

75. ______

76. ______

77. ______

78. ______

79. ______

80. ______

ANSWERS

81. ______
82. ______
83. ______
84. ______
85. ______
86. ______
87. ______
88. ______
89. ______
90. ______
91. ______
92. ______
93. ______
94. ______

81. $8 \cdot [4 + (6 - 1) \cdot 2] - 50 \cdot 2$

82. $35 \div [3^2 + (9 - 7) - 2^2] + 10 \cdot 3$

83. $7^2 - \{18 - [40 \div (4 \cdot 2) + 2] + 5^2\}$

84. $29 - \{5 + 3[8 \cdot (10 - 8)] - 50\}$

Find the area of each square. See Example 7.

85.

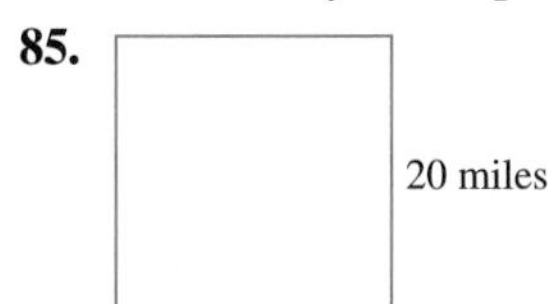

86.

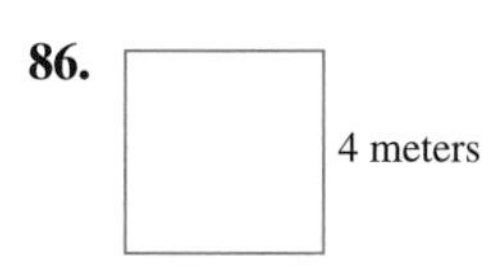

87.

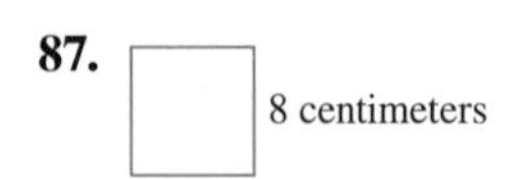

88.

31 feet

89. The Eiffel Tower stands on a square base measuring 100 meters on each side. Find the area of the base.

90. A square lawn that measures 72 feet on each side is to be fertilized. If 5 bags of fertilizer are available and each bag can fertilize 1000 square feet, is there enough fertilizer to cover the lawn?

Insert grouping symbols so that the given expression evaluates to the given number.

91. $2 + 3 \cdot 6 - 2; 28$

92. $2 + 3 \cdot 6 - 2; 20$

93. $24 \div 3 \cdot 2 + 2 \cdot 5; 14$

94. $24 \div 3 \cdot 2 + 2 \cdot 5; 15$

95. A building contractor is bidding on a contract to install gutters on seven homes in a retirement community all in the following shape. To estimate his cost of materials, he needs to know the total perimeter of all seven homes. Find the total perimeter.

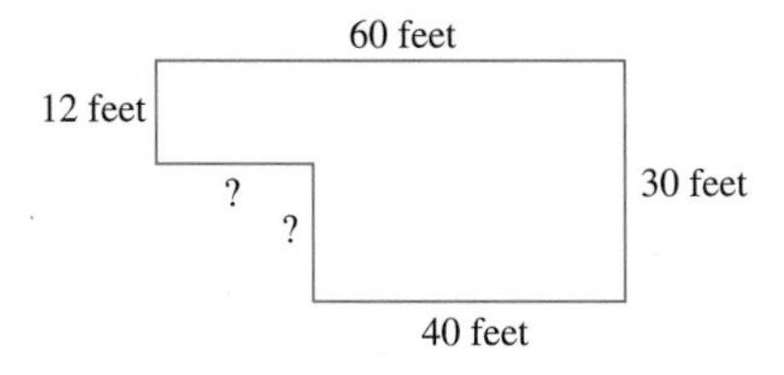

Simplify.

96. $25^3 \cdot (45 - 7 \cdot 5) \cdot 5$

97. $(7 + 2^4)^5 - (3^5 - 2^4)^2$

98. Explain why $2 \cdot 3^2$ is not the same as $(2 \cdot 3)^2$.

ANSWERS

95. ______

96. ______

97. ______

98. ______

1.8

TAPE PA 1.8

INTRODUCTION TO VARIABLES AND ALGEBRAIC EXPRESSIONS

OBJECTIVES

1 Evaluate algebraic expressions given replacement values.

2 Translate phrases into variable expressions.

Perhaps the most important quality of mathematics is that it is a science of pattern. Communicating about patterns is made possible by using a letter to represent all the numbers fitting a pattern. We call such a letter a **variable.** For example, in Section 1.2 we presented the addition property of 0, which states that the sum of 0 and any number is that number. We might write

$$0 + 1 = 1$$
$$0 + 2 = 2$$
$$0 + 3 = 3$$
$$0 + 4 = 4$$
$$0 + 5 = 5$$
$$0 + 6 = 6$$
$$\vdots$$

continuing indefinitely. This is a pattern, and all whole numbers fit the pattern. We can communicate this pattern for all whole numbers by letting a letter, such as a, represent all whole numbers. We can then write

$$0 + a = a$$

Learning to use variable notation is a primary goal of learning **algebra.** We now take some important beginning steps in learning to use variable notation.

A combination of operations on letters (variables) and numbers is called an **algebraic expression** or simply an **expression.**

ALGEBRAIC EXPRESSIONS

$$3 + x \qquad 5 \cdot y \qquad 2 \cdot z - 1 + x$$

If two variables or a number and a variable are next to each other, with no operation sign between them, the operation is multiplication. For example,

$$2x \quad \text{means} \quad 2 \cdot x$$

and

$$xy \text{ or } x(y) \quad \text{means} \quad x \cdot y$$

Also, the meaning of an exponent remains the same when the base is a variable. For example,

$$x^2 = \underbrace{x \cdot x}_{\text{2 factors of } x} \quad \text{and} \quad y^5 = \underbrace{y \cdot y \cdot y \cdot y \cdot y}_{\text{5 factors of } y}$$

1 Algebraic expressions such as $3x$ have different values depending on replacement values for x. For example, if x is 2, then $3x$ becomes

$$3x = 3 \cdot 2$$
$$= 6$$

If x is 7, then $3x$ becomes

$$3x = 3 \cdot 7$$
$$= 21$$

Replacing a variable in an expression by a number and then finding the value of the expression is called **evaluating the expression** for the variable. When finding the value of an expression, remember to follow the order of operations given in Section 1.7.

PRACTICE PROBLEM 1
Evaluate $x - 2$ if x is 5.

EXAMPLE 1

Evaluate $x + 7$ if x is 8.

Solution:

Replace x with 8 in the expression $x + 7$.

$x + 7 = 8 + 7$ Replace x with 8.
$= 15$ Add.

When we write a statement such as "x is 5," we can use an equals symbol to represent "is" so that

x is 5 can be written as $x = 5$.

PRACTICE PROBLEM 2
Evaluate $y(x - 3)$ for $x = 3$ and $y = 7$.

EXAMPLE 2

Evaluate $2(x - y)$ for $x = 8$ and $y = 4$.

Solution:

$2(x - y) = 2(8 - 4)$ Replace x with 8 and y with 4.
$= 2(4)$ Subtract.
$= 8$ Multiply.

PRACTICE PROBLEM 3
Evaluate $\frac{y + 6}{x}$ for $x = 2$ and $y = 8$.

EXAMPLE 3

Evaluate $\frac{x - 5y}{y}$ for $x = 21$ and $y = 3$.

Solution:

$\frac{x - 5y}{y} = \frac{21 - 5(3)}{3}$ Replace x with 21 and y with 3.
$= \frac{21 - 15}{3}$ Multiply.
$= \frac{6}{3}$ Subtract.
$= 2$ Divide.

Answers:
1. 3
2. 0
3. 7

EXAMPLE 4

Evaluate $x^2 + z - 3$ for $x = 5$ and $z = 4$.

Solution:

$$x^2 + z - 3 = 5^2 + 4 - 3 \quad \text{Replace } x \text{ with 5 and } z \text{ with 4.}$$
$$= 25 + 4 - 3 \quad \text{Evaluate } 5^2.$$
$$= 26 \quad \text{Add and subtract from left to right.}$$

PRACTICE PROBLEM 4

Evaluate $25 - z^3 + x$ for $z = 2$ and $x = 1$.

EXAMPLE 5

The expression $\frac{5(F - 32)}{9}$ can be used to write degrees Fahrenheit F as degrees Celsius. Find the value of this expression for $F = 86$.

Solution:

$$\frac{5(F - 32)}{9} = \frac{5(86 - 32)}{9}$$
$$= \frac{5(54)}{9}$$
$$= \frac{270}{9}$$
$$= 30$$

Thus 86°F is the same temperature as 30°Celsius.

PRACTICE PROBLEM 5

Evaluate $\frac{5(F - 32)}{9}$ for $F = 41$.

2 To aid us in solving problems later, we practice translating verbal phrases into algebraic expressions. Certain key words and phrases suggesting addition, subtraction, multiplication, or division are reviewed next.

ADDITION	SUBTRACTION	MULTIPLICATION	DIVISION
sum	difference	product	quotient
plus	minus	times	divided by
added to	subtracted from	multiply	into
more than	less than	twice	per
increased by	decreased by	of	
total	less	double	

Answers:
4. 18
5. 5

PRACTICE PROBLEM 6

Write as an algebraic expression. Use x to represent "a number."

a. twice a number

b. 8 increased by a number

c. 10 minus a number

d. 10 subtracted from a number

e. the quotient of 6 and a number

EXAMPLE 6

Write as an algebraic expression. Use x to represent "a number."

a. 7 more than a number

b. 15 decreased by a number

c. the product of 2 and a number

d. the quotient of a number and 5

e. 2 subtracted from a number

Solution:

a. In words: 7 | more than | a number

Translate: $7 \quad + \quad x$

b. In words: 15 | decreased by | a number

Translate: $15 \quad - \quad x$

c. In words: The product of — 2 | and | a number

Translate: $2 \quad \cdot \quad x$ or $2x$

d. In words: The quotient of — a number | and | 5

Translate: $x \quad \div \quad 5$ or $\frac{x}{5}$

e. In words: 2 | subtracted from | a number

Translate: $x \quad - \quad 2$

Answers:

6a. $2x$ **b.** $x + 8$ **c.** $10 - x$ **d.** $x - 10$ **e.** $6 \div x$ or $\frac{6}{x}$

EXERCISE SET 2.1

Represent each quantity by an integer. See Example 1.

1. A worker in a silver mine in Nevada works 1445 feet underground.

2. A scuba diver is swimming 35 feet below the surface of the water in the Gulf of Mexico.

3. The peak of Mount Whitney in California is 14,494 feet above sea level. (*Source:* U.S. Geological Survey)

4. The average depth of the Atlantic Ocean is 11,730 feet below the surface of the ocean.

5. The Virginia Cavaliers football team lost 15 yards on a play.

6. The record high temperature in Arizona is 128 degrees Fahrenheit above zero. (*Source:* National Climatic Data Center)

Graph each integer in the list on the same number line. See Example 2.

7. 1, 2, 4, 6

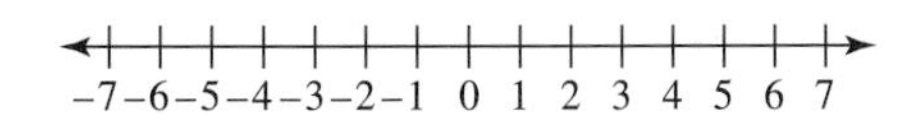

8. 3, 5, 2, 0

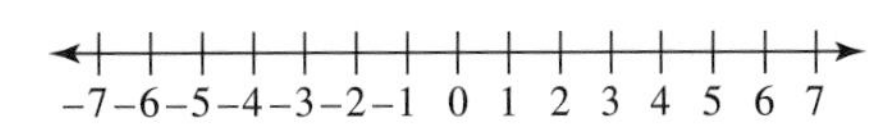

9. 1, −1, −2, −4, −7

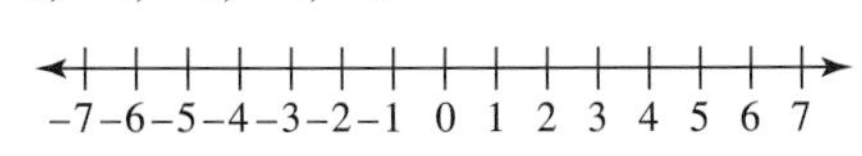

10. 2, −2, −4, 6

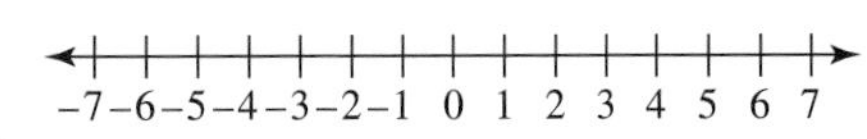

11. 0, 2, 5, 7

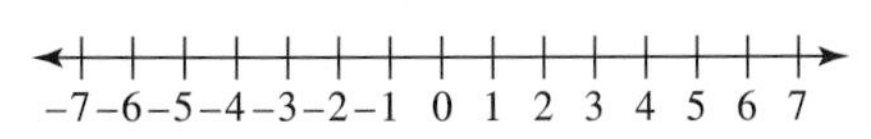

12. 0, 3, 6, 10

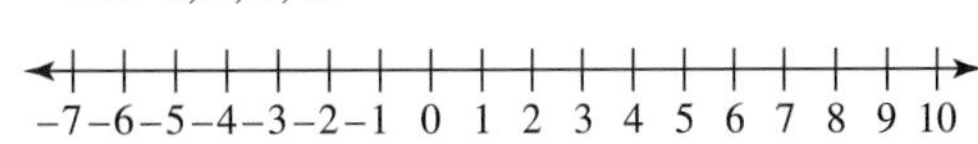

13. 0, −2, −7, −5

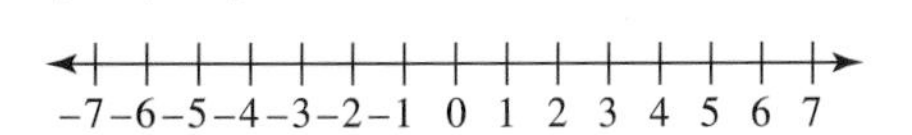

14. 0, −7, 3, −6

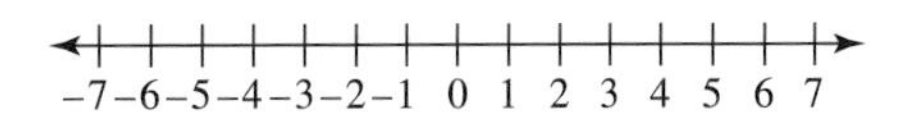

15. 4, −4, 1, −1

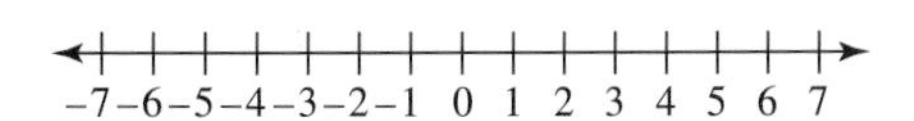

16. −1, −2, −5, 5

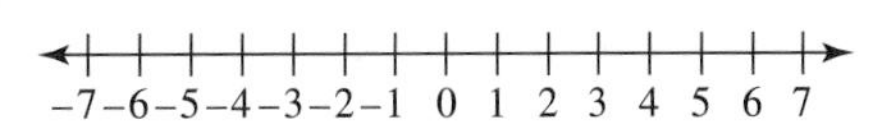

Insert $<$ or $>$ between each pair of integers to make a true statement. See Example 3.

17. 5 7 **18.** 16 10 **19.** 4 0 **20.** 8 0

21. −5 −7 **22.** −12 −10 **23.** 0 −3 **24.** 0 −7

25. −26 26 **26.** 13 −13

ANSWERS

1. ______
2. ______
3. ______
4. ______
5. ______
6. ______
17. ______
18. ______
19. ______
20. ______
21. ______
22. ______
23. ______
24. ______
25. ______
26. ______

ANSWERS

27. ______
28. ______
29. ______
30. ______
31. ______
32. ______
33. ______
34. ______
35. ______
36. ______
37. ______
38. ______
39. ______
40. ______
41. ______
42. ______
43. ______
44. ______
45. ______
46. ______
47. ______
48. ______
49. ______
50. ______
51. ______
52. ______
53. ______
54. ______
55. ______
56. ______
57. ______
58. ______
59. ______
60. ______
61. ______
62. ______

Simplify. See Example 4.

27. $|5|$ **28.** $|7|$ **29.** $|-8|$ **30.** $|-19|$

31. $|0|$ **32.** $|100|$ **33.** $|-5|$ **34.** $|-10|$

Find the opposite of each integer. See Example 5.

35. 5 **36.** 8 **37.** -4 **38.** -6

39. 23 **40.** 123 **41.** -10 **42.** -23

Simplify. See Example 6.

43. $|-7|$ **44.** $|-11|$ **45.** $-|20|$ **46.** $-|43|$

47. $-|-3|$ **48.** $-|-18|$ **49.** $-(-8)$ **50.** $-(-7)$

51. $|-14|$ **52.** $-(-14)$ **53.** $-(-29)$ **54.** $-|-29|$

Insert $<$, $>$, or $=$ between each pair of numbers to make a true statement.

55. 2 7 **56.** 1 11 **57.** -3 -5 **58.** -17 -6

59. $|-9|$ $|-14|$ **60.** $|-8|$ $|-4|$ **61.** $|-33|$ $-(-33)$ **62.** $-|17|$ $-(-17)$

EXAMPLE 3

Add.

a. $-2 + 5$ **b.** $3 + (-7)$

Solution:

a. $-2 + 5$

Step 1. $|-2| = 2, |5| = 5$, and $5 - 2 = 3$.

Step 2. 5 has the larger absolute value and its sign is an understood + :

$-2 + 5 = +3$ or 3

b. $3 + (-7)$

Step 1. $|3| = 3, |-7| = 7$, and $7 - 3 = 4$.

Step 2. -7 has the larger absolute value and its sign is $-$:

$3 + (-7) = -4$

When we add three or more numbers, follow the order of operations and add from left to right. That is, start at the left and add the first two numbers. Then add their sum to the next number. Continue this process until the addition is completed.

EXAMPLE 4

Add: $(-3) + 4 + (-11)$.

Solution:

$$\begin{aligned}(-3) + 4 + (-11) &= 1 + (-11)\\ &= -10\end{aligned}$$

EXAMPLE 5

Add: $1 + (-10) + (-8) + 9$.

Solution:

$$\begin{aligned}1 + (-10) + (-8) + 9 &= -9 + (-8) + 9\\ &= -17 + 9\\ &= -8\end{aligned}$$

2 We can continue our work with algebraic expressions by evaluating expressions given integer replacement values.

EXAMPLE 6

Evaluate $x + y$ if $x = 3$ and $y = -5$.

Solution:

Replace x with 3 and y with -5 in $x + y$.

$$\begin{aligned}x + y &= 3 + (-5)\\ &= -2\end{aligned}$$

PRACTICE PROBLEM 3

Add.

a. $-3 + 9$ **b.** $2 + (-8)$

PRACTICE PROBLEM 4

Add: $8 + (-3) + (-13)$.

PRACTICE PROBLEM 5

Add: $5 + (-3) + 12 + (-14)$.

PRACTICE PROBLEM 6

Evaluate $x + y$ if $x = -4$ and $y = 1$.

Answers:
3a. 6 **b.** -6
4. -8
5. 0
6. -3

PRACTICE PROBLEM 7

Evaluate $x + y$ if $x = -11$ and $y = -6$.

EXAMPLE 7

Evaluate $x + y$ if $x = -2$ and $y = -10$.

Solution:

$$x + y = (-2) + (-10)$$
$$= -12$$

3 Next, we practice solving problems that require adding integers.

PRACTICE PROBLEM 8

If the temperature was $-8°$ Fahrenheit at 6 A.M., and it rose 4 degrees by 7 A.M., and then rose another 7 degrees in the hour from 7 A.M. to 8 A.M., what was the temperature at 8 A.M.?

EXAMPLE 8

On January sixth the temperature in Caribou, Maine at 8 A.M. was $-12°$ Fahrenheit. By 9 A.M., the temperature had risen by 4 degrees, and by 10 A.M. it had risen 6 degrees from the 9 A.M. temperature. What was the temperature at 10 A.M.?

Solution:

In words:	8 A.M. temperature	+	rise of 4°	+	rise of 6°	=	temperature at 10 A.M.
	↓		↓		↓		↓
Translate:	-12	+	$(+4)$	+	$(+6)$	=	-2

The temperature was $-2°$F at 10 A.M.

MENTAL MATH

Add.

1. $5 + 0$ **2.** $(-2) + 0$ **3.** $0 + (-35)$ **4.** $0 + 3$

Answers:
7. -17
8. 3°F

EXERCISE SET 2.2

Add using a number line. See Example 1.

1. $8 + 2$

−7 −6 −5 −4 −3 −2 −1 0 1 2 3 4 5 6 7 8 9 10

2. $9 + (-4)$

−7 −6 −5 −4 −3 −2 −1 0 1 2 3 4 5 6 7 8 9 10

3. $-4 + 7$

−7 −6 −5 −4 −3 −2 −1 0 1 2 3 4 5 6 7 8 9 10

4. $10 + (-3)$

−7 −6 −5 −4 −3 −2 −1 0 1 2 3 4 5 6 7 8 9 10

5. $-13 + 7$

−13 −12 −11 −10 −9 −8 −7 −6 −5 −4 −3 −2 −1 0 1 2 3 4

6. $(-6) + (-5)$

−13 −12 −11 −10 −9 −8 −7 −6 −5 −4 −3 −2 −1 0 1 2 3 4

Add. See Examples 2 and 3.

7. $23 + 12$ **8.** $15 + 42$ **9.** $-6 + (-2)$ **10.** $-5 + (-4)$

11. $-43 + 43$ **12.** $-62 + 62$ **13.** $6 + (-2)$ **14.** $8 + (-3)$

15. $-6 + 8$ **16.** $-8 + 12$ **17.** $3 + (-5)$ **18.** $5 + (-9)$

Add. See Examples 4 and 5.

19. $-4 + 2 + (-5)$ **20.** $-1 + 5 + (-8)$ **21.** $-5 + (-7) + (-11)$

22. $-10 + (-3) + (-2)$ **23.** $12 + (-4) + (-4) + 12$ **24.** $18 + (-9) + 5 + (-2)$

Evaluate $x + y$ given the following replacement values. See Examples 6 and 7.

25. $x = -2$ and $y = 3$ **26.** $x = -7$ and $y = 11$ **27.** $x = -20$ and $y = -50$

28. $x = -1$ and $y = -29$ **29.** $x = 3$ and $y = -30$ **30.** $x = 13$ and $y = -17$

ANSWERS

7. ______
8. ______
9. ______
10. ______
11. ______
12. ______
13. ______
14. ______
15. ______
16. ______
17. ______
18. ______
19. ______
20. ______
21. ______
22. ______
23. ______
24. ______
25. ______
26. ______
27. ______
28. ______
29. ______
30. ______

ANSWERS

31. ______
32. ______
33. ______
34. ______
35. ______
36. ______
37. ______
38. ______
39. ______
40. ______
41. ______
42. ______
43. ______
44. ______
45. ______
46. ______
47. ______
48. ______
49. ______
50. ______
51. ______
52. ______
53. ______
54. ______
55. ______
56. ______
57. ______
58. ______
59. ______
60. ______
61. ______
62. ______
63. ______
64. ______

Add.

31. $-2 + (-9)$ **32.** $-6 + (-1)$ **33.** $-12 + (-12)$ **34.** $-23 + (-23)$

35. $-25 + (-32)$ **36.** $-45 + (-90)$ **37.** $-123 + (-100)$ **38.** $-500 + (-230)$

39. $-7 + 7$ **40.** $-10 + 10$ **41.** $12 + (-5)$ **42.** $24 + (-10)$

43. $-6 + 3$ **44.** $-8 + 2$ **45.** $-12 + 3$ **46.** $-15 + 5$

47. $56 + (-26)$ **48.** $89 + (-37)$ **49.** $-37 + 57$ **50.** $-25 + 65$

51. $-42 + 93$ **52.** $-64 + 164$

53. $-8 + (-14) + (-11)$ **54.** $-10 + (-6) + (-1)$

55. $5 + (-1) + 17$ **56.** $3 + (-23) + 6$

57. $(-10) + 14 + 25 + (-16)$ **58.** $34 + (-12) + (-11) + 213$

59. $34 + (-67)$ **60.** $42 + (-83)$

61. $124 + (-144)$ **62.** $325 + (-375)$

63. $-82 + (-43)$ **64.** $-56 + (-33)$

EXAMPLE 5

Evaluate $x - y$ if $x = -3$ and $y = 9$.

Solution:

Replace x with -3 and y with 9 in $x - y$.

$$\begin{aligned} &x \;-\; y \\ &\downarrow \;\downarrow\; \downarrow \\ &= (-3) - 9 \\ &= (-3) + (-9) \\ &= -12 \end{aligned}$$

PRACTICE PROBLEM 5

Evaluate $x - y$ if $x = -2$ and $y = 14$.

EXAMPLE 6

Evaluate $a - b$ if $a = 8$ and $b = -6$.

Solution:

Watch your signs carefully!

$$\begin{aligned} &a \;-\; b \\ &\downarrow \;\downarrow\; \downarrow \\ &= 8 - (-6) \quad \text{Replace } a \text{ with 8 and } b \text{ with } -6. \\ &= 8 + (+6) \\ &= 14 \end{aligned}$$

PRACTICE PROBLEM 6

Evaluate $y - z$ if $y = -3$ and $z = -4$.

Solving problems often requires subtracting integers.

EXAMPLE 7

The highest point in the United States is the top of Mount McKinley at a height of 20,320 feet above sea level. The lowest point is Death Valley, California, which is 282 feet below sea level. How much higher is Mount McKinley than Death Valley? (*Source:* U.S. Geological Survey)

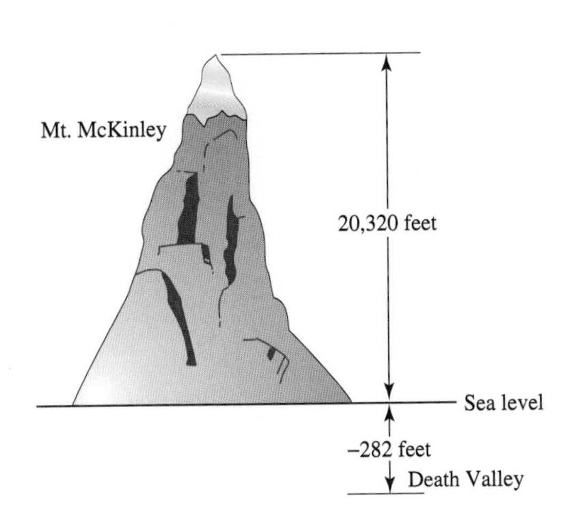

PRACTICE PROBLEM 7

The highest point in Asia is the top of Mount Everest at a height of 29,028 feet above sea level. The lowest point is the Dead Sea which is 1312 feet below sea level. How much higher is Mount Everest than the Dead Sea? (*Source:* National Geographic Society)

Answers:
5. −16
6. 1
7. 30,340 feet

Solution:

To find how much higher, we subtract. Don't forget that if Death Valley is 282 feet *below* sea level, we represent its height by -282.

In words:	Height of Mt. McKinley	minus	Height of Death Valley	
Translate:	20,320	$-$	(-282)	
	$= 20{,}320$	$+$	282	$= 20{,}602$ feet

Mount McKinley is 20,602 feet higher than Death Valley.

EXERCISE SET 2.3

Perform the indicated operations. See Examples 1 and 2.

1. $5 - 5$ **2.** $-6 - (-6)$ **3.** $8 - 3$ **4.** $5 - 2$

5. $3 - 8$ **6.** $2 - 5$ **7.** $7 - (-7)$ **8.** $12 - (-12)$

9. $-5 - (-8)$ **10.** $-25 - (-25)$ **11.** $-14 - 4$ **12.** $-2 - 42$

13. $2 - 16$ **14.** $8 - 9$

15. Subtract 18 from -20. **16.** Subtract 10 from -22.

17. Find the difference of -20 and -3. **18.** Find the difference of -8 and -13.

19. Subtract -11 from 2. **20.** Subtract -50 from -50.

Simplify. See Examples 3 and 4.

21. $7 - 3 - 2$ **22.** $8 - 4 - 1$ **23.** $12 - 5 - 7$

24. $30 - 7 - 12$ **25.** $-5 - 8 - (-12)$ **26.** $-10 - 6 - (-9)$

27. $-10 + (-5) - 12$ **28.** $-15 + (-8) - 4$ **29.** $12 - (-34) + (-6)$

30. $23 - (-17) + (-9)$

Evaluate $x - y$ given the following replacement values. See Examples 5 and 6.

31. $x = -3$ and $y = 5$ **32.** $x = -7$ and $y = 1$ **33.** $x = 6$ and $y = -30$

ANSWERS

1. ______
2. ______
3. ______
4. ______
5. ______
6. ______
7. ______
8. ______
9. ______
10. ______
11. ______
12. ______
13. ______
14. ______
15. ______
16. ______
17. ______
18. ______
19. ______
20. ______
21. ______
22. ______
23. ______
24. ______
25. ______
26. ______
27. ______
28. ______
29. ______
30. ______
31. ______
32. ______
33. ______

ANSWERS

34. _____
35. _____
36. _____
37. _____
38. _____
39. _____
40. _____
41. _____
42. _____
43. _____
44. _____
45. _____
46. _____
47. _____
48. _____
49. _____
50. _____
51. _____
52. _____
53. _____
54. _____
55. _____
56. _____
57. _____
58. _____
59. _____
60. _____
61. _____
62. _____
63. _____
64. _____
65. _____
66. _____
67. _____
68. _____
69. _____
70. _____

34. $x = 9$ and $y = -2$ **35.** $x = -4$ and $y = -4$ **36.** $x = -8$ and $y = -10$

37. $x = 1$ and $y = -18$ **38.** $x = 14$ and $y = -12$

Perform the indicated operations.

39. $-10 - (-10)$ **40.** $-5 - (-5)$ **41.** $-15 - (-15)$

42. $-24 - (-24)$ **43.** $3 - 7$ **44.** $4 - 12$

45. $30 - 45$ **46.** $29 - 56$ **47.** $-4 - 10$

48. $-5 - 8$ **49.** $-230 - 870$ **50.** $-15 - 26$

51. $4 - (-6)$ **52.** $6 - (-9)$ **53.** $-(-6) - 12 + (-16)$

54. $-(-9) - 7 + (-23)$ **55.** $-9 - (-12) + (-7) - 4$ **56.** $-6 - (-8) + (-12) - 7$

57. $-7 - (-3)$ **58.** $-12 - (-5)$ **59.** $-16 - (-23)$

60. $-45 - (-16)$ **61.** $-25 - (-13)$ **62.** $-230 - (-170)$

63. $|-3| - |-7|$
(Evaluate absolute values first.)

64. $|-12| - |-5|$ **65.** $|-6| - |6|$

66. $|-9| - |9|$ **67.** $|-17| - |-29|$ **68.** $|-23| - |-42|$

69. $-3 + 4 - (-23) - 10$ **70.** $5 + (-18) - (-21) - 2$

Evaluate each expression for the given replacement values.

71. $x - y - z$ if $x = -4, y = 3$, and $z = 15$

72. $x - y - z$ if $x = -14, y = 8$, and $z = -6$

73. $a + b - c$ if $a = -16, b = 14$, and $c = -22$

74. $a + b - c$ if $a = -1, b = -1$, and $c = 100$

Solve. See Example 7.

75. The coldest temperature ever recorded on earth was $-126°F$ in Antarctica. The warmest temperature ever recorded was 136°F in the Sahara Desert. How many degrees warmer is 136°F than $-126°F$? (*Source: Questions Kids Ask*, Grolier Limited, 1991)

76. The coldest temperature ever recorded in the United States was $-80°F$ in Alaska. The warmest temperature ever recorded was 134°F in California. How many degrees warmer is 134°F than $-80°F$? (*Source: The World Almanac and Book of Facts, 1996*)

77. Aaron has $125 in his checking account. He writes a check for $117, makes a deposit of $45, and then writes another check for $69. Find the amount left in his account. (Write the amount as an integer.)

78. In canasta, it is possible to have a negative score. If Juan Santanilla's score is 15, what is his new score if he loses 20 points?

79. The temperature on a February morning is $-6°$ Celsius at 6 A.M. If the temperature drops 3 degrees by 7 A.M., rises 4 degrees between 7 A.M. and 8 A.M., and then drops 7 degrees between 8 A.M. and 9 A.M., find the temperature at 9 A.M.

80. A mountain peak in Colorado has an elevation of 14,393 feet above sea level. A deep-sea trench in the Pacific Ocean has an elevation of 12,456 feet below sea level. Find the difference in elevation between these two points.

Let a and b be positive numbers. Answer the following true or false.

81. $a - b$ is always a positive number.

82. $-a - b$ is always a negative number.

83. $|-8 - 3| = 8 - 3$

84. $|-2 - (-6)| = |-2| - |-6|$

The following bar graph shows heights of selected lakes. For Exercises 85–88, find the difference in elevation for the lakes listed.

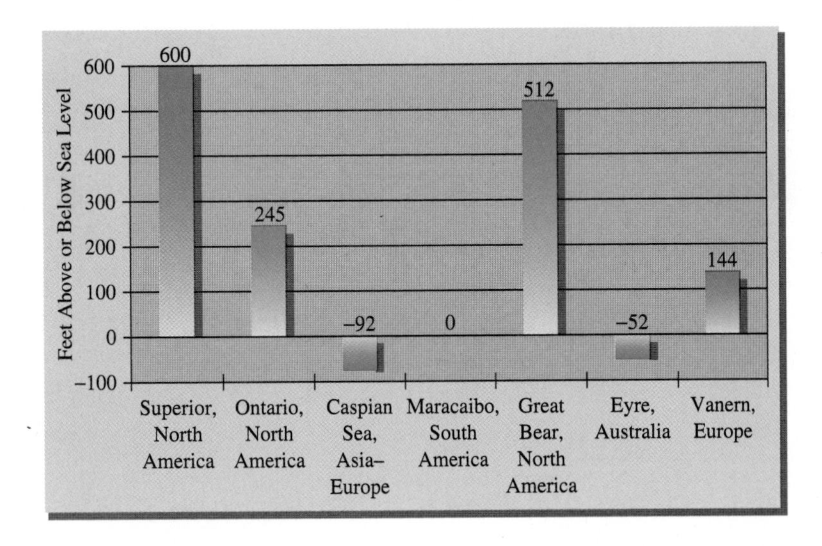

85. Lake Superior and Lake Eyre.

86. Great Bear Lake and Caspian Sea.

87. Lake Maracaibo and Lake Vanern.

88. Lake Eyre and Caspian Sea.

89. In your own words, explain how to subtract one integer from another.

ANSWERS

71. ______
72. ______
73. ______
74. ______
75. ______
76. ______
77. ______
78. ______
79. ______

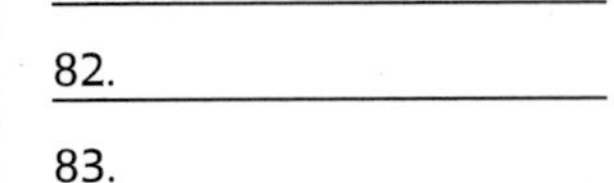

80. ______
81. ______
82. ______
83. ______
84. ______
85. ______
86. ______
87. ______
88. ______
89. ______

ANSWERS

90. ______

91. ______

92. ______

93. ______

94. ______

95. ______

Review Exercises

Multiply. See Section 1.5.

90. $8 \cdot 0$

91. $0 \cdot 8$

92. $1 \cdot 8$

93. $8 \cdot 1$

94. $\begin{array}{r} 23 \\ \times\ 46 \\ \hline \end{array}$

95. $\begin{array}{r} 51 \\ \times\ 89 \\ \hline \end{array}$

TAPE PA 2.4

2.4

MULTIPLYING AND DIVIDING INTEGERS

OBJECTIVES

1. Multiply integers.
2. Find the value of an algebraic expression by multiplying.
3. Divide integers.
4. Find the value of an algebraic expression by dividing.

Multiplying and dividing integers is similar to multiplying and dividing whole numbers. One difference is that we need to determine whether the result is a positive number or a negative number.

Consider the following pattern of products.

First factor decreases by 1 each time.

$$3 \cdot 2 = 6$$
$$2 \cdot 2 = 4$$
$$1 \cdot 2 = 2$$
$$0 \cdot 2 = 0$$

Product decreases by 2 each time.

This pattern can be continued, as follows.

$$-1 \cdot 2 = -2$$
$$-2 \cdot 2 = -4$$
$$-3 \cdot 2 = -6$$

This suggests that the product of a negative number and a positive number is a negative number.

What is the sign of the product of two negative numbers? To find out, we form another pattern of products. Again, we decrease the first factor by 1 each time, but this time the second factor is negative.

$$2 \cdot (-3) = -6$$
$$1 \cdot (-3) = -3$$
$$0 \cdot (-3) = 0$$

Product increases by 3 each time.

This pattern continues as:

$$-1 \cdot (-3) = 3$$
$$-2 \cdot (-3) = 6$$
$$-3 \cdot (-3) = 9$$

This suggests that the product of two negative integers is a positive integer. Thus we can determine the sign of a product when we know the signs of the factors.

MULTIPLYING INTEGERS

The product of two integers having the same sign is a positive number.
The product of two integers having unlike signs is a negative number.

PRACTICE PROBLEM 1

Multiply.

a. $-2 \cdot 6$ **b.** $-4(-3)$

c. $0 \cdot (-10)$

EXAMPLE 1

Multiply.

a. $-7 \cdot 3$ **b.** $-2(-5)$ **c.** $0 \cdot (-4)$

Solution:

a. $-7 \cdot 3 = -21$ **b.** $-2(-5) = 10$ **c.** $0 \cdot (-4) = 0$

To find the product of more than two numbers, we multiply from left to right.

PRACTICE PROBLEM 2

Multiply the following.

a. $(-2)(-1)(-2)(-1)$

b. $7(0)(-4)$

EXAMPLE 2

Multiply the following.

a. $7(-6)(-2)$ **b.** $(-2)(-3)(-4)$

Solution:

a. $7(-6)(-2) = -42(-2) = 84$ **b.** $(-2)(-3)(-4) = 6(-4) = -24$

Recall from our study of exponents that $2^3 = 2 \cdot 2 \cdot 2 = 8$. We can now work with bases that are negative numbers. For example,

$$(-2)^3 = (-2)(-2)(-2) = -8$$

PRACTICE PROBLEM 3

Evaluate $(-3)^4$.

EXAMPLE 3

Evaluate $(-5)^2$.

Solution:

$(-5)^2 = (-5)(-5) = 25$

2 Next, we practice evaluating expressions given integer replacement values.

PRACTICE PROBLEM 4

Evaluate $x \cdot y$ if $x = 5$ and $y = -9$.

EXAMPLE 4

Evaluate $x \cdot y$ if $x = -2$ and $y = 7$.

Solution:

Replace x with -2 and y with 7 in $x \cdot y$.

$$x \cdot y = (-2) \cdot 7$$
$$= -14$$

3 Division of integers is related to multiplication of integers. The sign rules for division can be discovered by writing a related multiplication problem. For example,

$$\frac{6}{2} = 3 \quad \text{because} \quad 3 \cdot 2 = 6$$

$$\frac{-6}{2} = -3 \quad \text{because} \quad -3 \cdot 2 = -6$$

Answers:
1a. -12 **b.** 12 **c.** 0
2a. 4 **b.** 0
3. 81
4. -45

$$\frac{6}{-2} = -3 \quad \text{because } -3 \cdot (-2) = 6$$

$$\frac{-6}{-2} = 3 \quad \text{because } 3 \cdot (-2) = -6$$

DIVIDING INTEGERS

The quotient of two integers having the same sign is a positive number.

The quotient of two integers having unlike signs is a negative number.

Thus the rules for division of integers are similar to the rules for multiplication of integers.

EXAMPLE 5

Find each quotient.

a. $\frac{-12}{6}$ **b.** $-20 \div (-4)$

Solution:

a. $\frac{-12}{6} = -2$ **b.** $-20 \div (-4) = 5$

PRACTICE PROBLEM 5

Find each quotient.

a. $\frac{28}{-7}$ **b.** $-18 \div (-2)$

EXAMPLE 6

Divide the following if possible.

a. $\frac{0}{-5}$ **b.** $\frac{-7}{0}$

Solution:

Recall properties of 0 for division.

a. $\frac{0}{-5} = 0$ because $0 \cdot -5 = 0$.

b. $\frac{-7}{0}$ is undefined because there is no number that gives a product of -7 when multiplied by 0.

PRACTICE PROBLEM 6

Divide the following if possible.

a. $\frac{-1}{0}$ **b.** $\frac{0}{-2}$

4 Next, we evaluate $\frac{x}{y}$ given replacement values. Remember that if the replacement value makes the denominator 0, the resulting expression is undefined.

EXAMPLE 7

Evaluate $\frac{x}{y}$ if $x = -24$ and $y = 6$.

Solution:

$$\frac{x}{y} = \frac{-24}{6}$$
$$= -4$$

PRACTICE PROBLEM 7

Evaluate $\frac{x}{y}$ if $x = -9$ and $y = -3$.

Answers:
5a. -4 **b.** 9
6a. undefined **b.** 0
7. 3

EXERCISE SET 2.4

Multiply. See Example 1.

1. $-2(-3)$ **2.** $5(-3)$ **3.** $-4(9)$ **4.** $-7(-2)$

5. $8(-8)$ **6.** $-9(9)$ **7.** $0(-14)$ **8.** $-6(0)$

Multiply. See Example 2.

9. $6(-4)(2)$ **10.** $-2(3)(-7)$ **11.** $-1(-2)(-4)$ **12.** $8(-3)(3)$

13. $-4(4)(-5)$ **14.** $-2(-5)(-4)$ **15.** $10(-5)(0)$ **16.** $2(-1)(3)(-2)$

17. $-5(3)(-1)(-1)$ **18.** $3(0)(-4)(-8)$

Evaluate. See Example 3.

19. $(-2)^2$ **20.** $(-2)^4$ **21.** $(-3)^3$ **22.** $(-1)^4$

23. $(-5)^2$ **24.** $(-4)^3$ **25.** $(-2)^3$ **26.** $(-3)^2$

Evaluate $a \cdot b$ for the given replacement values. See Example 4.

27. $a = -4$ and $b = 7$ **28.** $a = 5$ and $b = -1$ **29.** $a = 3$ and $b = -2$

30. $a = -9$ and $b = -6$ **31.** $a = -5$ and $b = -5$ **32.** $a = -8$ and $b = 8$

33. $a = 0$ and $b = -10$ **34.** $a = -7$ and $b = 0$

ANSWERS

1. ______
2. ______
3. ______
4. ______
5. ______
6. ______
7. ______
8. ______
9. ______
10. ______
11. ______
12. ______
13. ______
14. ______
15. ______
16. ______
17. ______
18. ______
19. ______
20. ______
21. ______
22. ______
23. ______
24. ______
25. ______
26. ______
27. ______
28. ______
29. ______
30. ______
31. ______
32. ______
33. ______
34. ______

ANSWERS

35. ______
36. ______
37. ______
38. ______
39. ______
40. ______
41. ______
42. ______
43. ______
44. ______
45. ______
46. ______
47. ______
48. ______
49. ______
50. ______
51. ______
52. ______
53. ______
54. ______
55. ______
56. ______
57. ______
58. ______
59. ______
60. ______
61. ______
62. ______
63. ______
64. ______
65. ______
66. ______
67. ______
68. ______

Find each quotient. See Examples 5 and 6.

35. $-24 \div 6$ **36.** $90 \div (-9)$ **37.** $\frac{-30}{6}$ **38.** $\frac{56}{-8}$

39. $\frac{-88}{-11}$ **40.** $\frac{-32}{4}$ **41.** $\frac{0}{14}$ **42.** $\frac{-13}{0}$

43. $\frac{39}{-3}$ **44.** $\frac{-24}{-12}$

Evaluate $\frac{x}{y}$ for the given replacement values. See Example 7.

45. $x = 5$ and $y = -5$ **46.** $x = 9$ and $y = -3$ **47.** $x = -12$ and $y = 0$

48. $x = -10$ and $y = -10$ **49.** $x = -36$ and $y = -6$ **50.** $x = 0$ and $y = -5$

51. $x = -100$ and $y = 10$ **52.** $x = -26$ and $y = 2$

Multiply or divide the following.

53. $-12(0)$ **54.** $0(-100)$ **55.** $-4(3)$ **56.** $-6 \cdot 2$

57. $-9 \cdot 6$ **58.** $-12(13)$ **59.** $-7(-6)$ **60.** $-9(-5)$

61. $-3(-4)(-2)$ **62.** $-7(-5)(-3)$ **63.** $(-4)^2$ **64.** $(-5)^2$

65. $-\frac{10}{5}$ **66.** $-\frac{25}{5}$ **67.** $-\frac{56}{8}$ **68.** $-\frac{49}{7}$

69. $-12 \div 3$ **70.** $-15 \div 3$ **71.** $4(-4)(-3)$ **72.** $6(-5)(-2)$

73. $-30(6)(-2)(-3)$ **74.** $-20 \cdot 5 \cdot (-5) \cdot (-3)$ **75.** $3 \cdot (-2) \cdot 0$

76. $-5(4)(0)$ **77.** $\frac{100}{-20}$ **78.** $\frac{45}{-9}$ **79.** $240 \div (-40)$

80. $\frac{480}{-8}$ **81.** $\frac{-12}{-4}$ **82.** $\frac{-36}{-3}$ **83.** $(-1)^4$

84. $(-2)^3$ **85.** $(-3)^5$ **86.** $(-9)^2$ **87.** $-2(3)(5)(-6)$

88. $-1(2)(7)(-3)$ **89.** $-1 \cdot (-1) \cdot (-1) \cdot (-1)$ **90.** $-1 \cdot (-1) \cdot (-1)$

91. $-2(-2)(-2)$ **92.** $-2(-2)(-2)(-2)$ **93.** $-42 \cdot 23$

94. $-56 \cdot 43$ **95.** $25 \cdot (-82)$ **96.** $70 \cdot (-23)$

Evaluate $x \cdot y$ and also $\frac{x}{y}$ for the given replacement values.

97. $x = -4$ and $y = -2$ **98.** $x = 20$ and $y = -5$

99. $x = 0$ and $y = -6$ **100.** $x = -3$ and $y = 0$

101. A football team lost 4 yards on each of three consecutive plays. Represent the total loss as a product of integers, and find the total loss.

102. Joe Norstrom lost $400 on each of seven consecutive days in the stock market. Represent his total loss as a product of integers and find his total loss.

103. A deep-sea diver must move up or down in the water in short steps in order to keep from getting a physical condition called the bends. Suppose the diver moves down from the surface in five steps of 20 feet each. Represent his total movement as a product of integers, and find the product.

104. A weather forecaster predicts that the temperature will drop 5 degrees each hour for the next six hours. Represent this drop as a product of integers and find the total drop in temperature.

ANSWERS

69. ____
70. ____
71. ____
72. ____
73. ____
74. ____
75. ____
76. ____
77. ____
78. ____
79. ____
80. ____
81. ____
82. ____
83. ____
84. ____
85. ____
86. ____
87. ____
88. ____
89. ____
90. ____
91. ____
92. ____
93. ____
94. ____
95. ____
96. ____
97. ____
98. ____
99. ____
100. ____
101. ____
102. ____
103. ____
104. ____

ANSWERS

105. ____
106. ____
107. ____
108. ____
109. ____
110. ____
111. ____
112. ____
113. ____
114. ____
115. ____
116. ____

Let a and b be positive numbers. Answer the following true or false.

❑ **105.** $a(-b)$ is a negative number.

❑ **106.** $(-a)(-b)$ is a negative number.

❑ **107.** $(-a)(-a)$ is a positive number.

❑ **108.** $(-a)(-a)(-a)$ is a positive number.

❑ **109.** In your own words, explain how to multiply two integers.

❑ **110.** In your own words, explain how to divide two integers.

Review Exercises

Perform the indicated operation. See Section 1.7.

111. $(3 \cdot 5)^2$

112. $(12 - 3)^2(18 - 10)$

113. $90 + 12^2 - 5^3$

114. $3 \cdot (7 - 4) + 2 \cdot 5^2$

115. $12 \div 4 - 2 + 7$

116. $12 \div (4 - 2) + 7$

2.5

ORDER OF OPERATIONS

TAPE PA 2.5

OBJECTIVES

1. Simplify expressions by using the order of operations.
2. Find the value of an algebraic expression.

1 We first discussed the order of operations in Chapter 1. In this section, you are given an opportunity to practice using the order of operations when expressions contain integers. The rules for order of operations are repeated here.

> **ORDER OF OPERATIONS**
>
> Simplify expressions using the following order. If grouping symbols such as parentheses ( ) or brackets [ ] are present, simplify expressions within those first, starting with the innermost set. If fraction bars are present, simplify above and below the fraction bar separately.
>
> **1.** Simplify any expressions with exponents.
>
> **2.** Perform multiplications or divisions in order from left to right.
>
> **3.** Perform additions or subtractions in order from left to right.

EXAMPLE 1

Find the value of each expression.

a. $(-3)^2$ **b.** -3^2

Solution:

a. $(-3)^2 = (-3)(-3) = 9$

b. Without parentheses, only the 3 is squared.

$$-3^2 = -(3)(3) = -9$$

PRACTICE PROBLEM 1

Find the value of each expression.

a. $(-2)^4$ **b.** -2^4

> REMINDER When simplifying expressions with exponents, notice that parentheses make an important difference.
>
> $(-3)^2$ and -3^2 do not mean the same thing.
>
> $(-3)^2$ means $(-3)(-3) = 9$.
>
> -3^2 means the opposite of $3 \cdot 3$, or -9.
>
> Only with parentheses is the -3 squared.

Answers:
1a. 16 **b.** −16

PRACTICE PROBLEM 2

Simplify $\frac{25}{5(-1)}$.

EXAMPLE 2

Simplify: $\frac{-6(2)}{-3}$.

Solution:

First, multiply -6 and 2. Then divide.

$$\frac{-6(2)}{-3} = \frac{-12}{-3}$$
$$= 4$$

PRACTICE PROBLEM 3

Simplify $\frac{-18 + 6}{-3 - 1}$.

EXAMPLE 3

Simplify $\frac{12 - 16}{-1 + 3}$.

Solution:

Simplify above and below the fraction bar separately. Then divide.

$$\frac{12 - 16}{-1 + 3} = \frac{-4}{2}$$
$$= -2$$

PRACTICE PROBLEM 4

Simplify $20 + 50 + (-4)^3$.

EXAMPLE 4

Simplify $60 + 30 + (-2)^3$.

Solution:

$60 + 30 + (-2)^3 = 60 + 30 + (-8)$ Write $(-2)^3$ as -8.

$= 82$ Add from left to right.

PRACTICE PROBLEM 5

Simplify $-2^3 + (-4)^2 + 1^5$.

EXAMPLE 5

Simplify $-4^2 + (-3)^2 - 1^3$.

Solution:

$-4^2 + (-3)^2 - 1^3 = -16 + 9 - 1$ Simplify expressions with exponents.

$= -8$ Add or subtract from left to right.

PRACTICE PROBLEM 6

Simplify $2(2 - 8) + (-12) - 3$.

EXAMPLE 6

Simplify $3(4 - 7) + (-2) - 5$.

Solution:

$3(4 - 7) + (-2) - 5 = \underbrace{3(-3)} + (-2) - 5$ Simplify inside parentheses.

$= -9 + (-2) - 5$ Multiply.

$= -11 + (-5)$ Add or subtract from left to right.

$= -16$

Answers:
2. -5
3. 3
4. 6
5. 9
6. -27

EXAMPLE 7

Simplify $(-3) \cdot |-5| - (-2) + 4^2$.

Solution:

$$\begin{aligned}(-3) \cdot |-5| - (-2) + 4^2 &= (-3) \cdot 5 - (-2) + 4^2 && \text{Write } |-5| \text{ as } 5.\\ &= (-3) \cdot 5 - (-2) + 16 && \text{Write } 4^2 \text{ as } 16.\\ &= -15 - (-2) + 16 && \text{Multiply.}\\ &= -13 + 16 && \text{Add or subtract from left to right.}\\ &= 3\end{aligned}$$

PRACTICE PROBLEM 7

Simplify $(-5) \cdot |-4| + (-3) + 2^3$.

EXAMPLE 8

Simplify $-2[-3 + 2(-1 + 6)] - 5$.

Solution:

Here we begin with the innermost set of parentheses.

$$\begin{aligned}-2[-3 + 2(-1 + 6)] - 5 &= -2[-3 + 2(5)] - 5 && \text{Write } -1 + 6 \text{ as } 5.\\ &= -2[-3 + 10] - 5 && \text{Multiply.}\\ &= -2(7) - 5 && \text{Add.}\\ &= -14 - 5 && \text{Multiply.}\\ &= -19 && \text{Subtract.}\end{aligned}$$

PRACTICE PROBLEM 8

Simplify $4(-6) \div [3(5 - 7)^2]$.

2

EXAMPLE 9

Evaluate $x^2 - y$ if $x = -3$ and $y = 10$.

Solution:

Replace x with -3 and y with 10 and simplify.

$$\begin{aligned}&x^2 - y\\ &= (-3)^2 - 10\\ &= 9 - 10\\ &= -1\end{aligned}$$

PRACTICE PROBLEM 9

Evaluate $x^2 + y$ if $x = -5$ and $y = -2$.

Answers:

7. -15
8. -2
9. 23

CALCULATOR EXPLORATIONS

ENTERING NEGATIVE NUMBERS

To enter a negative number on a calculator, find the key marked [+/−]. (On some calculators, this key may be marked [CHS].) To enter the number −2, for example, press the keys [2] [+/−]. The display will read [−2].

To find $-32 + (-131)$, press the keys

[32] [+/−] [+] [131] [+/−] [=]

The display will read [−163]. Thus, $-32 + (-131) = -163$.

Use a calculator to perform the indicated operations.

1. $-256 + 97$

2. $811 + (-1058)$

3. $6(-15) + (-46)$

4. $-129 + 10(48)$

5. $\dfrac{-813}{3} - 288$

6. $-2(-35) - 14$

SIMPLIFYING AN EXPRESSION CONTAINING A FRACTION BAR

Even though a calculator follows the order of operations, parentheses must sometimes be inserted. For example, to simplify $\dfrac{-8+6}{-2}$ on a calculator, enter parentheses about the expression above the fraction bar so that it is simplified separately.

To simplify $\dfrac{-8+6}{-2}$, press the keys

[(] [8] [+/−] [+] [6] [)] [÷] [2] [+/−] [=]

The display will read [1].

Thus $\dfrac{-8+6}{-2} = 1$.

Use a calculator to simplify.

7. $\dfrac{-12 - 36}{-6}$

8. $\dfrac{475}{-2 + (-17)}$

9. $\dfrac{-316 + (-458)}{28 + (-25)}$

10. $\dfrac{-234 + 86}{-18 + 16}$

EXERCISE SET 2.5

Simplify. See Examples 1–8.

1. $-1(-2) + 1$
2. $3 + (-8) \div 2$
3. $3 - 6 + 2$
4. $5 - 9 + 2$
5. $9 - 12 - 4$
6. $10 - 23 - 12$
7. $4 + 3(-6)$
8. $8 + 4(3)$
9. $5(-9) + 2$
10. $7(-6) + 3$
11. $(-10) + 4 \div 2$
12. $(-12) + 6 \div 3$
13. $25 \div (-5) + 12$
14. $28 \div (-7) + 10$
15. $\dfrac{16 - 13}{-3}$
16. $\dfrac{20 - 15}{-1}$
17. $\dfrac{24}{10 + (-4)}$
18. $\dfrac{88}{-8 - 3}$
19. $5(-3) - (-12)$
20. $7(-4) - (-6)$
21. $(-19) - 12(3)$
22. $(-24) - 14(2)$
23. $8 + 4^2$
24. $12 + 3^3$
25. $[8 + (-4)]^2$
26. $[9 + (-2)]^3$
27. $3^3 - 12$
28. $5^2 - 100$

ANSWERS

1. ______
2. ______
3. ______
4. ______
5. ______
6. ______
7. ______
8. ______
9. ______
10. ______
11. ______
12. ______
13. ______
14. ______
15. ______
16. ______
17. ______
18. ______
19. ______
20. ______
21. ______
22. ______
23. ______
24. ______
25. ______
26. ______
27. ______
28. ______

ANSWERS

29. ______
30. ______
31. ______
32. ______
33. ______
34. ______
35. ______
36. ______
37. ______
38. ______
39. ______
40. ______
41. ______
42. ______
43. ______
44. ______
45. ______
46. ______
47. ______
48. ______
49. ______
50. ______
51. ______
52. ______
53. ______
54. ______
55. ______

29. $16 - (-3)^4$

30. $20 - (-5)^2$

31. $|5 + 3| \cdot 2^3$

32. $|-3 + 7| \cdot 7^2$

33. $7 \cdot 8^2 + 4$

34. $10 \cdot 5^4 + 7$

35. $5^3 - (4 - 2^3)$

36. $8^2 - (5 - 2)^4$

37. $(3 - 12) \div 3$

38. $(12 - 19) \div 7$

39. $5 + 2^3 - 4^2$

40. $12 + 5^2 - 2^4$

41. $(5 - 9)^2 \div (4 - 2)^2$

42. $(2 - 7)^2 \div (4 - 3)^4$

43. $|8 - 24| \cdot (-2) \div (-2)$

44. $|3 - 15| \cdot (-4) \div (-16)$

45. $(-12 - 20) \div 16 - 25$

46. $(-20 - 5) \div 5 - 15$

47. $5(5 - 2) + (-5)^2 - 6$

48. $3 \cdot (8 - 3) + (-4) - 10$

49. $(2 - 7) \cdot (6 - 19)$

50. $(4 - 12) \cdot (8 - 17)$

51. $2 - 7 \cdot 6 - 19$

52. $4 - 12 \cdot 8 - 17$

53. $(-36 \div 6) - (4 \div 4)$

54. $(-4 \div 4) - (8 \div 8)$

55. $-5^2 - 6^2$

EXAMPLE 11

Find the area of the rectangular deck.

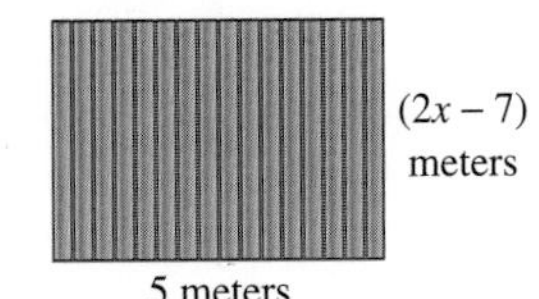

Solution:

Recall how to find the area of a rectangle.

$$A = \text{length} \cdot \text{width}$$
$$= 5(2x - 7) \quad \text{Let length} = 5 \text{ and width} = (2x - 7).$$
$$= 10x - 35 \quad \text{Multiply.}$$

The area is $(10x - 35)$ *square* meters.

PRACTICE PROBLEM 11

Find the area of the garden in the shape of a rectangle.

Answer:
11. $(36y + 27)$ square yards

EXERCISE SET 3.1

Find the numerical coefficient of each variable term. See Example 1.

1. $5y$ **2.** $-2z$ **3.** z

4. $3xy^2$ **5.** $11a$ **6.** $-x$

Simplify the following by combining like terms. See Examples 2 and 3.

7. $3x + 5x$ **8.** $8y + 3y$ **9.** $5n - 9n$

10. $7z - 10z$ **11.** $4c + c - 7c$ **12.** $5b - 8b - b$

13. $5x - 7x + x - 3x$ **14.** $8y + y - 2y - y$ **15.** $4a + 3a + 6a - 8$

16. $5b - 4b + b - 15$

Multiply. See Examples 4 and 5.

17. $6(5x)$ **18.** $4(4x)$ **19.** $-2(11y)$

20. $-3(21z)$ **21.** $12(6a)$ **22.** $9(7b)$

Multiply. See Examples 6 and 7.

23. $2(y + 2)$ **24.** $3(x + 1)$ **25.** $5(a - 8)$

26. $4(y - 6)$ **27.** $4(3x + 7)$ **28.** $8(8y + 10)$

ANSWERS

1. ________
2. ________
3. ________
4. ________
5. ________
6. ________
7. ________
8. ________
9. ________
10. ________
11. ________
12. ________
13. ________
14. ________
15. ________
16. ________
17. ________
18. ________
19. ________
20. ________
21. ________
22. ________
23. ________
24. ________
25. ________
26. ________
27. ________
28. ________

ANSWERS

29. ____
30. ____
31. ____
32. ____
33. ____
34. ____
35. ____
36. ____
37. ____
38. ____
39. ____
40. ____
41. ____
42. ____
43. ____
44. ____
45. ____
46. ____
47. ____
48. ____
49. ____
50. ____
51. ____
52. ____
53. ____
54. ____
55. ____
56. ____

Simplify the following. Use the distributive property to remove parentheses first. See Examples 8 and 9.

29. $2(x + 4) + 7$

30. $5(6 - y) - 2$

31. $4(6n - 5) + 3n$

32. $3(5 - 2b) - 4b$

33. $5(3c - 1) + 8$

34. $4(6d - 2) + 10$

35. $3 + 6(w + 2) + w$

36. $8z + 5(6 + z) + 20$

37. $2(3x + 1) + 5(x - 2)$

38. $3(5x - 2) + 2(3x + 1)$

Simplify the following.

39. $18y - 20y$

40. $x + 12x$

41. $z - 8z$

42. $12x - 8x$

43. $9d - 3c - d$

44. $8r + s - 7s$

45. $4x + 5y - y - 9x$

46. $a + 4b - 7a - 5b$

47. $5q + p - 6q - p$

48. $m - 8n + m + 8n$

49. $2(x + 1) + 20$

50. $5(x - 1) + 18$

51. $5(x - 7) - 8x$

52. $3(x + 2) - 11x$

53. $-5(z + 3) + 2z$

54. $-8(1 + v) + 6v$

55. $8 - x + 4x - 2 - 9x$

56. $5y - 4 + 9y - y + 15$

53. $x - 76{,}862 = 86{,}102$

54. $-36{,}109 + x = -14{,}995$

55. $-968 + 432 = 86y - 508 - 85y$

56. $-432 + (-449) = 102y - 529 - 101y$

Review Exercises

Round each whole number to the given place value. See Section 1.4.

57. 586 to the nearest ten

58. 82 to the nearest ten

59. 1026 to the nearest hundred

60. 52,333 to the nearest thousand

61. 2986 to the nearest thousand

62. 101,552 to the nearest hundred

Use the following bar graph to answer Exercises 63–66.

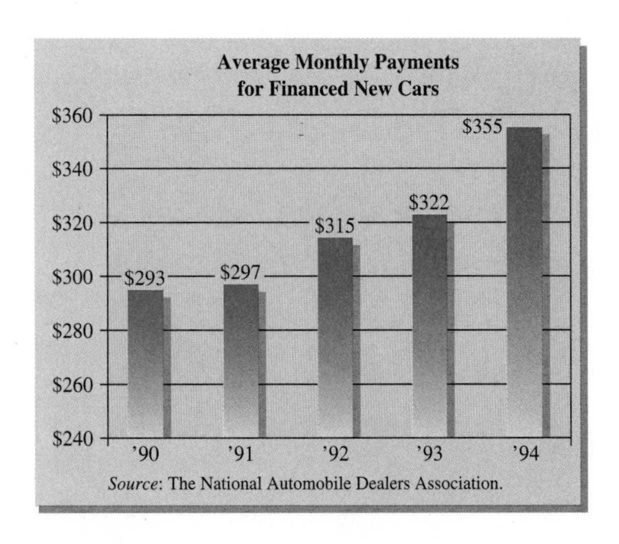

Source: The National Automobile Dealers Association.

63. Find the average monthly payment for new cars in 1993.

64. Find the average monthly payment for new cars in 1994.

65. How much more is the average monthly payment for financed new cars in 1994 than in 1990?

66. Describe any trends shown in this graph.

ANSWERS

53. ______

54. ______

55. ______

56. ______

57. ______

58. ______

59. ______

60. ______

61. ______

62. ______

63. ______

64. ______

65. ______

66. ______

TAPE PA 3.3

3.3

SOLVING EQUATIONS: THE DIVISION PROPERTY OF EQUALITY

OBJECTIVES

1. Use the division property of equality to solve equations.
2. Translate word phrases to algebraic expressions.

1 Although the addition property of equality is a powerful tool for helping us solve equations, it cannot help us solve all types of equations. For example, it cannot help us solve an equation such as $2x = 6$. To solve this equation, we use a second property of equality called the **division property of equality.**

DIVISION PROPERTY OF EQUALITY

If $a = b$, then $\frac{a}{c} = \frac{b}{c}$ as long as c is not 0.

In other words, both sides of an equation may be divided by the same nonzero number without changing the solution of the equation.

Picturing our balanced scale, if we divide the weight on each side by the same nonzero number, the scale or equation remains balanced.

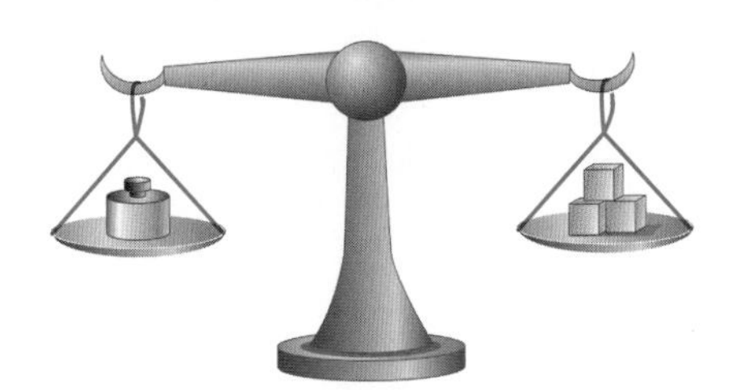

To solve $2x = 6$ for x, use the division property of equality, divide both sides of the equation by 2, and simplify as follows:

$$2x = 6$$

$$\frac{2 \cdot x}{2} = \frac{6}{2} \quad \text{Divide both sides by 2.}$$

Then it can be shown that an expression such as $\frac{2 \cdot x}{2}$ is equivalent to $\frac{2}{2} \cdot x$, so

$$\frac{2 \cdot x}{2} = \frac{6}{2}$$

can be written as

$$\frac{2}{2} \cdot x = \frac{6}{2}$$

$$1 \cdot x = 3 \quad \text{or} \quad x = 3$$

PRACTICE PROBLEM 1
Solve for y: $3y = -18$.

EXAMPLE 1

Solve for x: $-5x = 15$.

Solution:

To isolate x, divide both sides by -5.

$$-5x = 15 \quad \text{Original equation.}$$

$$\frac{-5x}{-5} = \frac{15}{-5} \quad \text{Divide both sides by } -5.$$

$$\frac{-5}{-5} \cdot x = \frac{15}{-5}$$

$$1x = -3 \quad \text{or} \quad x = -3 \quad \text{Simplify.}$$

To check, replace x with -3 in the original equation.

$$-5x = 15 \quad \text{Original equation.}$$

$$-5(-3) = 15 \quad \text{Let } x = -3.$$

$$15 = 15 \quad \text{True.}$$

The solution is -3.

PRACTICE PROBLEM 2
Solve for x: $-16 = 8x$.

EXAMPLE 2

Solve for y: $-8 = 2y$.

Solution:

To isolate y, divide both sides of the equation by 2.

$$-8 = 2y$$

$$\frac{-8}{2} = \frac{2y}{2} \quad \text{Divide both sides by 2.}$$

$$\frac{-8}{2} = \frac{2}{2} \cdot y$$

$$-4 = 1y \quad \text{or} \quad y = -4$$

Check to see that -4 is the solution.

PRACTICE PROBLEM 3
Solve for y: $-3y = -27$.

EXAMPLE 3

Solve for x: $-12x = -36$.

Solution:

Divide both sides of the equation by the coefficient of x, which is -12.

$$-12x = -36$$

$$\frac{-12x}{-12} = \frac{-36}{-12}$$

$$\frac{-12}{-12} \cdot x = \frac{-36}{-12}$$

$$x = 3$$

Answers:
1. -6
2. -2
3. 9

4. SOLVE the equation. To solve the equation, first add $-x$ to both sides.

$$2x + 3 = x - 6$$
$$2x + 3 + (-x) = x - 6 + (-x)$$
$$x + 3 = -6 \quad \text{Simplify.}$$
$$x + 3 + (-3) = -6 + (-3) \quad \text{Add } -3 \text{ to both sides.}$$
$$x = -9 \quad \text{Simplify.}$$

5. CHECK the proposed solution in the stated problem. Twice "-9" is -18 and $-18 + 3$ is -15. This is equal to the number minus 6 or "-9" $- 6$ or -15.
6. STATE your conclusion. The unknown number is -9.

EXAMPLE 3

In an election, the incumbent city council member, Laura Hartley, receives 275 **more** votes than her challenger. If a total of 7857 votes is cast in the election, find how many votes the incumbent, Laura Hartley, received.

Solution:

1. UNDERSTAND the problem. To do so, read and reread the problem.
2. ASSIGN a variable to an unknown. Use this variable to represent any other unknown quantities. Let $x =$ the number of challenger votes. Then $x + 275 =$ the number of incumbent votes since she received 275 more votes.
3. TRANSLATE the problem into an equation.

In words:	challenger votes	+	incumbent votes	=	total votes
Translate:	x	+	$x + 275$	=	7857

4. SOLVE the equation.

$$x + x + 275 = 7857$$
$$2x + 275 = 7857 \quad \text{Combine like terms.}$$
$$2x + 275 + (-275) = 7857 + (-275) \quad \text{Add } -275 \text{ to both sides.}$$
$$2x = 7582 \quad \text{Simplify.}$$
$$\frac{2x}{2} = \frac{7582}{2} \quad \text{Divide both sides by 2.}$$
$$x = 3791 \quad \text{Simplify.}$$

5. CHECK the proposed solution in the stated problem. Since x represents the number of votes the challenger received, the challenger received 3791 votes. The incumbent received $x + 275 = 3791 + 275 = 4066$ votes. To check, notice that the total number of challenger votes and incumbent votes is $3791 + 4066 = 7857$ votes, the given total of votes cast. Also, 4066 is 275 more votes than 3791, so the solution checks.
6. STATE your conclusion. The incumbent, Laura Hartley, received 4066 votes.

PRACTICE PROBLEM 3

At a recent United States/Japan summit meeting, 121 delegates attended. If the United States sent 19 more delegates than Japan, find how many the United States sent.

Answer:
3. 70 delegates

PRACTICE PROBLEM 4

A woman's $21,000 estate is to be divided so that her husband receives twice as much as her son. How much did each receive?

EXAMPLE 4

Leo Leal sold a used computer system and software for $2100, receiving four times as much money for the computer system as for the software. Find the price of each.

Solution:

1. UNDERSTAND the problem. To do so, read and reread the problem.
2. ASSIGN a variable to an unknown. Use this variable to represent any other unknown quantities. Let x = the software price. Then $4x$ = the computer system price.
3. TRANSLATE the problem into an equation.

In words:	software price	+	computer price	is	2100
Translate:	x	+	$4x$	=	2100

4. SOLVE the equation.

$$x + 4x = 2100$$

$$5x = 2100 \quad \text{Combine like terms.}$$

$$\frac{5x}{5} = \frac{2100}{5} \quad \text{Divide both sides by 5.}$$

$$x = 420 \quad \text{Simplify.}$$

5. CHECK the proposed solution in the stated problem. The software sold for $420. The computer system sold for $4x = 4(\$420) = \$1680$. Check: Since $\$420 + \$1680 = \$2100$, the total price, and $1680 is four times $420, the solution checks.
6. STATE your conclusion. The software sold for $420 and the computer system sold for $1680.

Answer:

4. husband, $14,000; son, $7,000

EXERCISE SET 3.5

Write the sentences as equations. Use x to represent "a number." See Example 1.

1. A number added to -5 is -7.

2. Five subtracted from a number equals 10.

3. Three times a number yields 27.

4. The quotient of 8 and a number is -2.

5. A number subtracted from -20 amounts to 104.

6. Two added to twice a number gives -14.

7. Twice the sum of a number and -1 is equal to 50.

8. Three times the difference of 7 and a number amounts to -40.

Solve. See Example 2.

9. Three times a number added to 9 is 33. Find the number.

10. Twice a number subtracted from 60 is 20. Find the number.

11. The product of 9 and a number gives 54. Find the number.

12. The product of 7 and a number is 14. Find the number.

13. The sum of 3, 4, and a number amounts to 16. Find the number.

14. A number less 5 is 11. Find the number.

15. Seventy-two is 8 times a number added to 24. Find the number.

16. The product of 11 and a number is 121. Find the number.

17. The difference of a number and 3 is the same as 45 less the number. Find the number.

18. Sixty-four less half of 64 is equal to the sum of some number and five times 4. Find the number.

19. Three times the difference of some number and 5 amounts to the quotient of 108 and 12. Find the number.

20. The product of some number plus 2 and 5 is 11 less than the number times 8. Find the number.

21. Eight decreased by some number equals the quotient of 15 and 5. Find the number.

22. Ten less some number is twice the sum of that number and 5. Find the number.

23. Thirteen added to the product of 3 and some number amounts to the product of 5 and the same number added to 3.

24. The product of 4 and a number is the same as 30 less twice that same number.

25. Five times a number less 40 is 8 more than the number.

26. Thirty less a number is equal to the product of 3 and the sum of the number and 6.

ANSWERS

1. ________
2. ________
3. ________
4. ________
5. ________
6. ________
7. ________
8. ________
9. ________
10. ________
11. ________
12. ________
13. ________
14. ________
15. ________
16. ________
17. ________
18. ________
19. ________
20. ________
21. ________
22. ________
23. ________
24. ________
25. ________
26. ________

ANSWERS

27. ______
28. ______
29. ______
30. ______
31. ______
32. ______
33. ______
34. ______
35. ______
36. ______
37. ______
38. ______
39. ______

27. The difference of a number and 3 is equal to the quotient of 10 and 5.

28. The product of a number and 3 is twice the sum of that number and 5.

Solve. See Examples 3 and 4.

29. In the 1992 presidential election, Clinton received 202 more electoral votes than Bush. If a total of 538 votes were cast for the two candidates, find how many votes each received. (*Source: The World Almanac, 1996*)

30. California has 22 more electoral votes for president than Texas. If the total number of electoral votes for these two states is 86, find the number for each state. (*Source: The World Almanac, 1996*)

31. Mark and Stuart Martin collect comic books. Mark has twice the number of books Stuart has. Together they have 120 comic books. Find how many books Mark has.

32. Heather and Mary Gamber collect baseball cards. Heather's collection is three times as large as Mary's 100-card collection. Find the total number of baseball cards in both collections.

33. A Toyota Camry is traveling twice as fast as a Dodge truck. If their combined speed is 105 miles per hour, find the speed of the car and find the speed of the truck.

34. A crow will eat five more ounces of food a day than a finch. If together they eat 13 ounces of food, find how many ounces of food the crow consumes and how many ounces of food the finch consumes.

35. Anthony Tedesco sold his used mountain bike and accessories for \$270. If he received five times as much money for the bike as he did for the accessories, find how much money he received for the bike.

36. A tractor and a plow attachment are worth \$1200. The tractor is worth seven times as much money as the plow. Find the value of the tractor and the value of the plow.

37. During the 1996 Rose Bowl college football game between the University of Southern California Trojans and the Northwestern University Wildcats, the total number of points scored by both teams was 73. The Trojans scored 9 more points than the Wildcats. How many points did the Trojans score? (Data from *USA Today*)

38. During the six games of the 1996 NBA Finals, the Chicago Bulls scored 23 more points than the Seattle SuperSonics. If the Bulls scored 558 points during these six basketball games, find the total number of points scored by both teams during the playoffs. (Data from the National Basketball Association)

39. Solve Example 3 again, but this time let x be the number of incumbent votes. Did you get the same results? Explain why or why not.

EXAMPLE 7

The length of the rectangular playground shown is twice its width. If the distance around the playground is 186 feet, find the width and length of the playground.

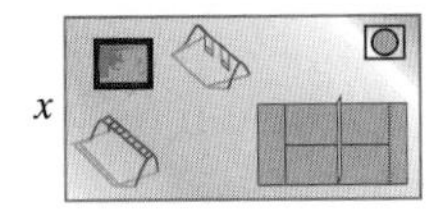

PRACTICE PROBLEM 7

The length of a rectangle is three times its width. If the perimeter is 64 centimeters, find the width and the length.

Solution:

1. UNDERSTAND. Read and reread the problem. Recall that the phrase "distance around" in the problem means perimeter and that the formula for perimeter of a rectangle is $P = 2l + 2w$.
2. ASSIGN a variable to an unknown. From the figure above, the variable x is given to be the width of the playground. Thus,

x = width of the playground and
$2x$ = length of the playground

since the length is twice the width.

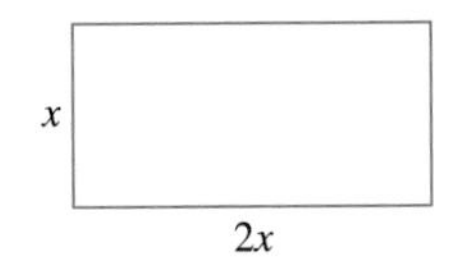

3. TRANSLATE.

Formula: $P = 2l + 2w$

Translate: $186 = 2(2x) + 2x$ Let $P = 186$, $l = 2x$, and $w = x$.

4. SOLVE the equation.

$186 = 2(2x) + 2x$
$186 = 4x + 2x$ Multiply $2(2x)$.
$186 = 6x$ Combine like terms.
$\frac{186}{6} = \frac{6x}{6}$ Divide both sides by 6.
$31 = x$

5. CHECK the proposed solution. If width $x = 31$ feet, then length $2x = 2(31 \text{ feet}) = 62$ feet.

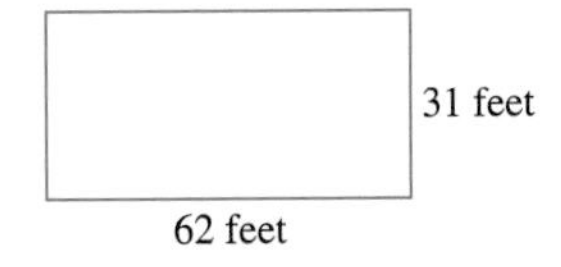

The perimeter of a rectangle with these dimensions is

$P = 2l + 2w$
$= 2(62 \text{ feet}) + 2(31 \text{ feet})$
$= 124 \text{ feet} + 62 \text{ feet}$
$= 186$ feet, the required perimeter.

6. STATE. The width of the playground is 31 feet and the length is 62 feet.

Answer:
7. Width: 8 centimeters, length: 24 centimeters

EXERCISE SET 3.6

Find the perimeter of each figure. See Examples 1–5.

1.

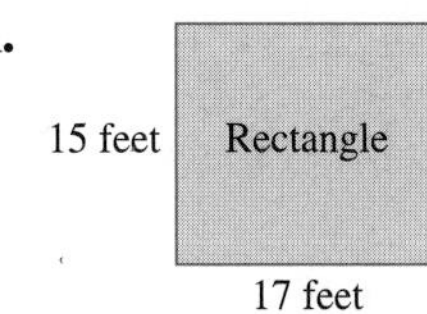

2.

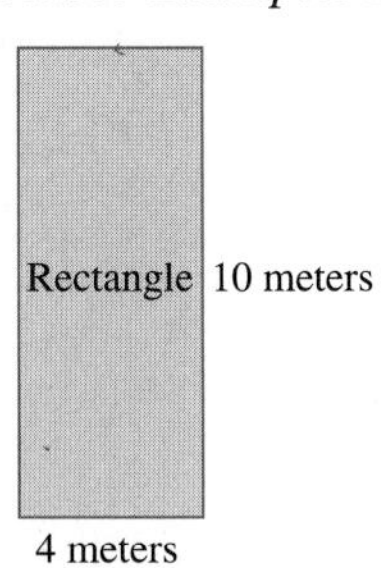

3.

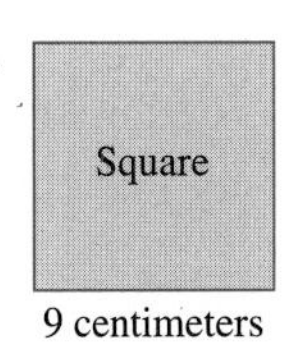

4.

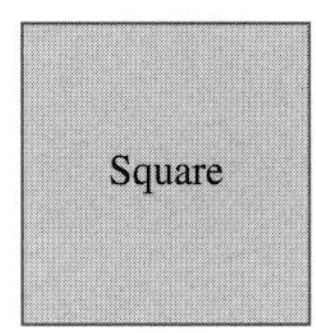

5.

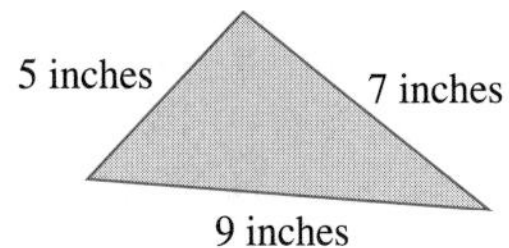

6.

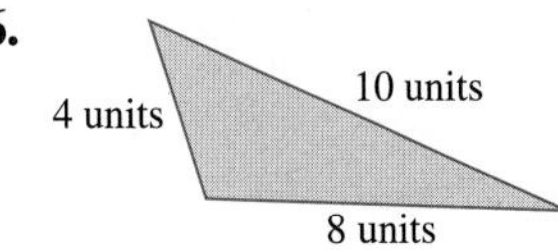

7.

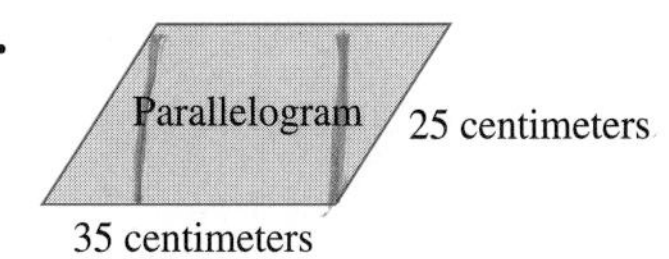

8.

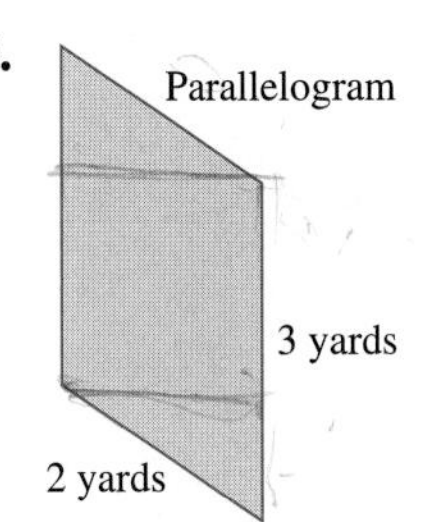

9.

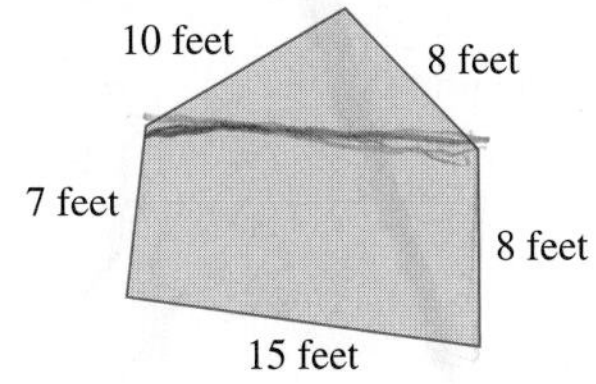

10.

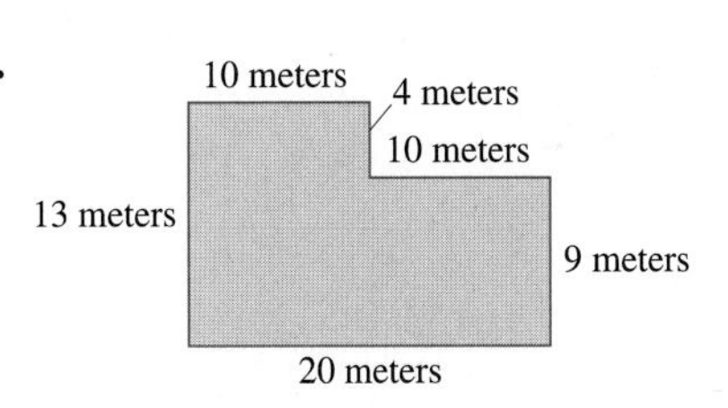

11.

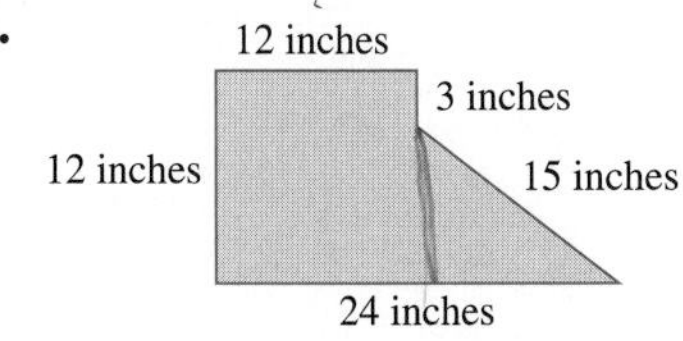

12.

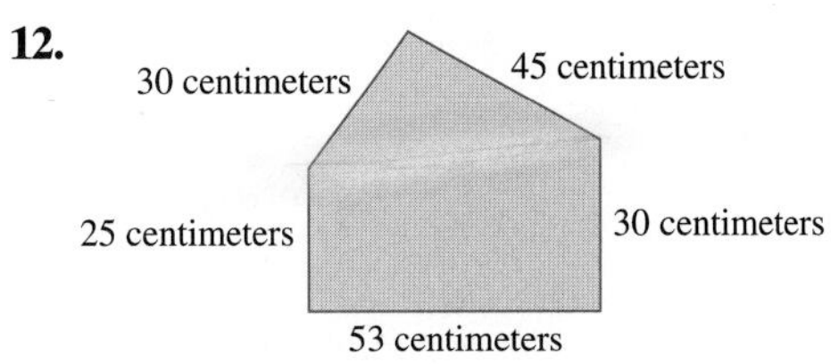

ANSWERS

1. ______
2. ______
3. ______
4. ______
5. ______
6. ______
7. ______
8. ______
9. ______
10. ______
11. ______
12. ______

ANSWERS

13. ______
14. ______
15. ______
16. ______
17. ______
18. ______
19. ______
20. ______
21. ______
22. ______
23. ______
24. ______
25. ______
26. ______
27. ______
28. ______

Solve. See Examples 1–5.

13. A polygon has sides of length 5 feet, 3 feet, 2 feet, 7 feet, and 4 feet. Find its perimeter.

14. A triangle has sides of length 8 inches, 12 inches, and 10 inches. Find its perimeter.

15. Baseboard is to be installed in a square room that measures 15 feet on one side. Find how much baseboard is needed.

16. Find how much fencing is needed to enclose a rectangular rose garden 85 feet by 15 feet.

17. If a football field is 53 yards wide and 120 yards long, what is the perimeter?

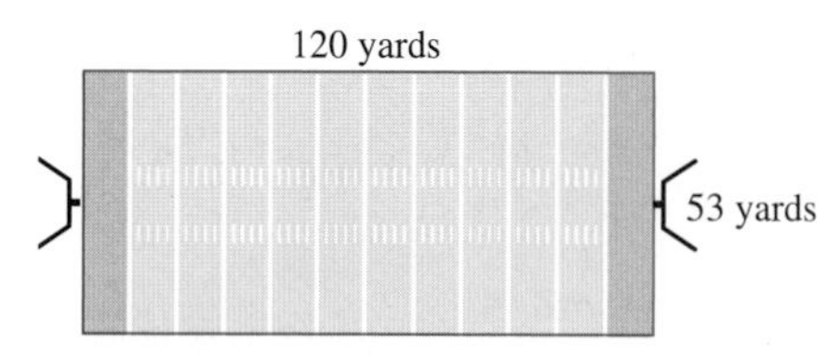

18. A stop sign has eight equal sides of length 12 inches. Find its perimeter.

19. A metal strip is being installed around a workbench that is 8 feet long and 3 feet wide. Find how much stripping is needed.

20. Find how much fencing is needed to enclose a rectangular garden 70 feet by 21 feet.

21. If the stripping in Exercise 19 costs \$3 per foot, find the total cost of the stripping.

22. If the fencing in Exercise 20 costs \$2 per foot, find the total cost of the fencing.

Find the perimeter of each figure. See Example 6.

23.

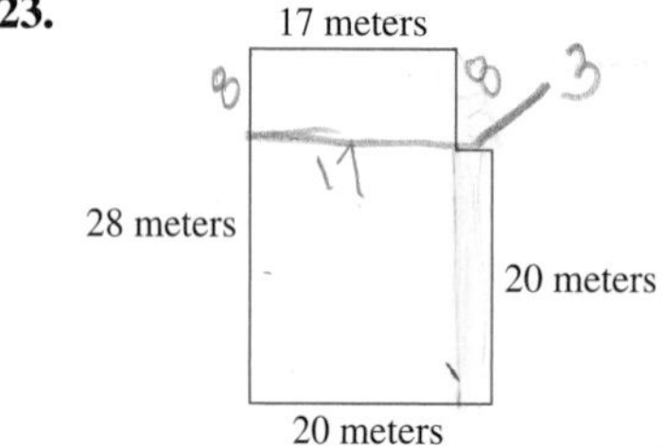

24.

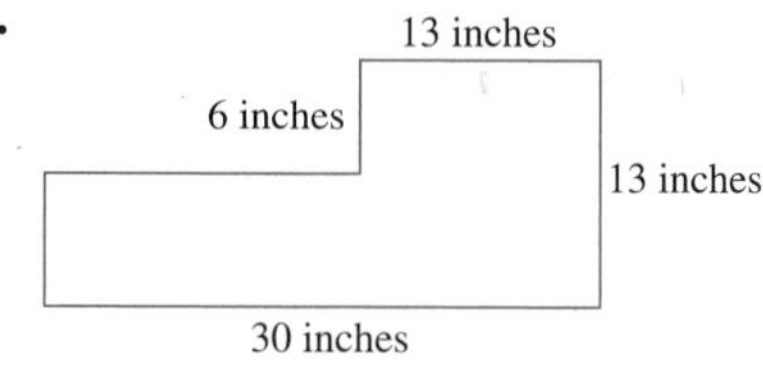

25.

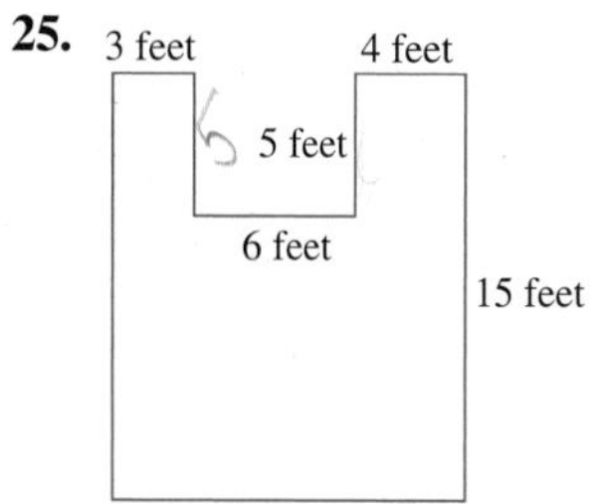

26.

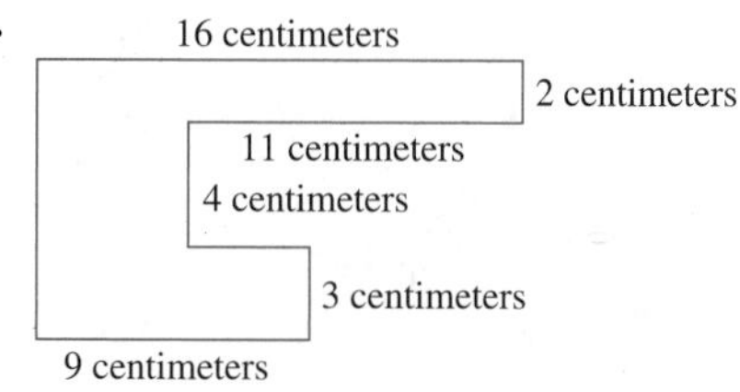

27.

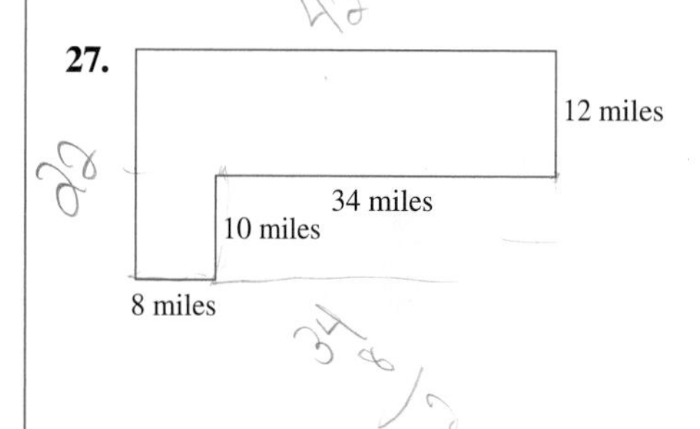

28.

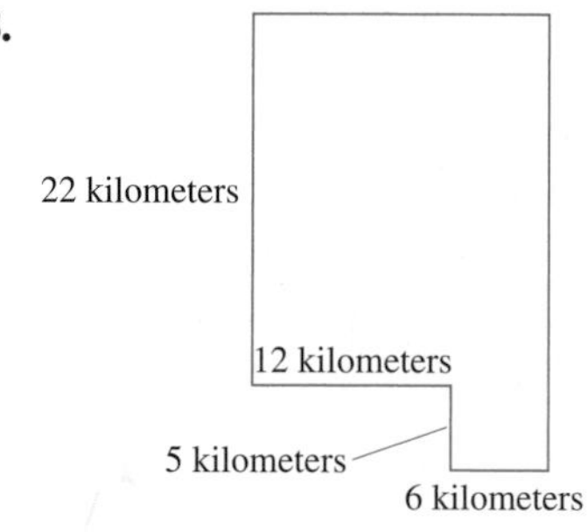

Solve. See Example 7.

29. The perimeter of a rectangular garden is 112 meters. If the garden is three times longer than it is wide, find the length of the garden.

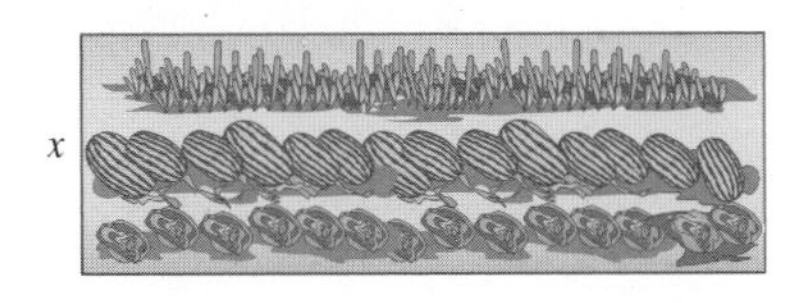

30. The perimeter of a rectangular picture frame is 84 inches. If the length of the picture frame is five times its width, find the length and the width of the frame.

31. The perimeter of a square compact disc box is 28 inches. What is the length of just one side?

32. Find the perimeter of a square ceramic tile with a side of length 5 inches.

33. The length of a rectangular plot of land is 23 yards longer than its width. If the perimeter of the land is 1562 yards, find its length.

34. The width of a rectangle is 8 centimeters less than its length. If the perimeter is 72 centimeters, find its width.

35. Two sides of a triangle are the same length. The third side is 18 kilometers. If the perimeter of the triangle is 60 kilometers, find the length of each equal side.

36. The shortest side of a triangle is x feet. A second side of the triangle is twice the shortest side and the third side of the triangle is 14 feet. If the perimeter of the triangle is 44 feet, find the length of each side of the triangle.

Given the following situations, tell whether you are more likely to be concerned with area or perimeter.

37. ordering fencing to fence a yard

38. ordering grass seed to plant in a yard

39. buying carpet to install in a room

40. buying gutters to install on a house

41. ordering paint to paint a wall

42. ordering baseboards to install in a room

43. buying a wallpaper border to go on the walls around a room

44. buying fertilizer for your yard

Review Exercises

Simplify. See Section 1.7.

45. $5 + 6 \cdot 3$

46. $25 - 3 \cdot 7$

47. $(20 - 16) \div 4$

48. $6 \cdot (8 + 2)$

49. $(18 + 8) - (12 + 4)$

50. $72 \div (2 \cdot 6)$

51. $(72 \div 2) \cdot 6$

52. $4^1 \cdot (2^3 - 8)$

ANSWERS

29. ____
30. ____
31. ____
32. ____
33. ____
34. ____
35. ____
36. ____
37. ____
38. ____
39. ____
40. ____
41. ____
42. ____
43. ____
44. ____
45. ____
46. ____
47. ____
48. ____
49. ____
50. ____
51. ____
52. ____

GROUP ACTIVITY

ANALYZING OPTIONS

Fritz and Maria's daughter is getting married and they have agreed to pay for their daughter's wedding reception. They have decided that the reception will be held at a local hotel that charges a flat fee of $700 for renting their large reception hall. This cost does not include music, decorations, buffet, or wedding cake. The costs of the wedding cake and buffet will depend on the number of guests, while the wedding decorations and the music will not.

1. Suppose that the hotel offers several different buffet and cake packages, as outlined in the table. Translate each option into an algebraic expression that describes the cost of the package in terms of the number of guests x who are invited to the reception.

OPTION	DESCRIPTION	COST OF BUFFET PER PERSON	COST OF CAKE AND BEVERAGES PER PERSON
A	Cold cuts, white cake, punch	$18	$2
B	Hot buffet, white cake, punch	$23	$2
C	Hot buffet, salad bar, white cake, mints, nuts, punch, coffee	$28	$3

2. Suppose that decorations for the reception will cost $300 and that music will cost $800. Fritz and Maria decide that they will spend $8000 on the reception. For each wedding reception option, write an equation for the total cost of the reception (including both fixed and variable costs) in terms of the number of guests x who are invited to the reception.

3. If Fritz and Maria spend only a total of $8000 on the reception, how many guests can be invited under each option?

4. If the bride and groom prefer Option C but hope to invite a total of 240 people, what additional amount would they have to contribute above Fritz and Maria's $8000 to cover all of the reception expenses?

CHAPTER 3 HIGHLIGHTS

DEFINITIONS AND CONCEPTS	EXAMPLES
SECTION 3.1 SIMPLIFYING ALGEBRAIC EXPRESSIONS	
The addends of an algebraic expression are called the **terms** of the expression.	$5x^2 + (-4x) + (-2)$ — 3 terms
The number factor of a variable term is called the **numerical coefficient.**	TERM / NUMERICAL COEFFICIENT: $7x$ — 7; $-6y$ — -6; x or $1x$ — 1
Terms that are exactly the same, except that they may have different numerical coefficients, are called **like terms.**	$5x + 11x = (5 + 11)x = 16x$ (like terms) $y - 6y = (1 - 6)y = -5y$ (like terms)
An algebraic expression is **simplified** when all like terms have been **combined.** Use the distributive property to multiply an algebraic expression by a term.	Simplify. $-4(x + 2) + 3(5x - 7)$ $= -4(x) + (-4)(2) + 3(5x) + 3(-7)$ $= -4x + (-8) + 15x + (-21)$ $= 11x + (-29)$ or $11x - 29$
SECTION 3.2 SOLVING EQUATIONS: THE ADDITION PROPERTY OF EQUALITY	
ADDITION PROPERTY OF EQUALITY	
Let a, b, and c represent numbers. If $a = b$, then $a + c = b + c$. In other words, the same number may be added to both sides of an equation without changing the solution of the equation.	Solve for x: $x + 8 = 2 + (-1)$ $x + 8 = 1$ $x + 8 + (-8) = 1 + (-8)$ Add -8. $x = -7$ Simplify. The solution is -7.
SECTION 3.3 SOLVING EQUATIONS: THE DIVISION PROPERTY OF EQUALITY	
DIVISION PROPERTY OF EQUALITY	
If $a = b$, then $\frac{a}{c} = \frac{b}{c}$ as long as c is not 0. In other words, both sides of an equation may be divided by the same nonzero number without changing the solution of the equation.	Solve for y: $y - 7y = 30$ $-6y = 30$ Combine like terms. $\frac{-6y}{-6} = \frac{30}{-6}$ Divide by -6. $y = -5$ Simplify. The solution is -5.

DEFINITIONS AND CONCEPTS	EXAMPLES

SECTION 3.4 SOLVING LINEAR EQUATIONS IN ONE VARIABLE

STEPS FOR SOLVING AN EQUATION IN x

Step 1. If parentheses are present, use the distributive property.

Step 2. Combine any like terms on each side of the equation.

Step 3. Use the addition property of equality to rewrite the equation so that variable terms are on one side of the equation and constant terms are on the other side.

Step 4. Use the division property of equality to divide both sides by the numerical coefficient of x to solve.

Step 5. Check the solution in the **original equation.**

Solve for x:

$$5(3x - 1) + 15 = -5$$

Step 1. $15x - 5 + 15 = -5$ Apply the distributive property.

Step 2. $15x + 10 = -5$ Combine like terms.

Step 3. $15x + 10 + (-10) = -5 + (-10)$ Add -10.

$$15x = -15$$

Step 4. $\frac{15x}{15} = \frac{-15}{15}$ Divide by 15.

$$x = -1$$

Step 5. Check to verify that -1 is the solution.

SECTION 3.5 LINEAR EQUATIONS IN ONE VARIABLE AND PROBLEM SOLVING

PROBLEM-SOLVING STEPS

1. UNDERSTAND the problem. During this step, don't work with variables, but simply become comfortable with the problem. Some ways of accomplishing this are to
   *Read and reread the problem.
   *Construct a drawing.
2. ASSIGN a variable to an unknown in the problem. Use this variable to represent any other unknown quantities.
3. TRANSLATE the problem into an equation.
4. SOLVE the equation.
5. CHECK the proposed solution in the stated problem.
6. STATE your conclusion.

The incubation period for a golden eagle is three times the incubation period for a hummingbird. If the total of their incubation periods is 60 days, find the incubation period for each bird. (*Source: Wildlife Fact File,* International Masters Publishers)

1. UNDERSTAND the problem.
2. ASSIGN a variable. Let x = incubation period of a hummingbird. Then $3x$ = incubation period of a golden eagle.
3. TRANSLATE

Incubation of hummingbird	+	incubation of golden eagle	is	60
x	+	$3x$	=	60

4. SOLVE.

$$x + 3x = 60$$
$$4x = 60$$
$$\frac{4x}{4} = \frac{60}{4}$$
$$x = 15$$

5. CHECK. The incubation period for the hummingbird is 15 days. The incubation period for the golden eagle is $3x = 3 \cdot 15 = 45$ days. Since 15 days + 45 days = 60 days and 45 is 3(15), the solution checks.
6. STATE. The incubation period for the hummingbird is 15 days. The incubation period for the golden eagle is 45 days.

DEFINITIONS AND CONCEPTS	EXAMPLES

SECTION 3.6 PERIMETER AND PROBLEM SOLVING

Perimeter of a Rectangle

$P = 2l + 2w$

Perimeter of a Square

$P = 4s$

Perimeter of a Triangle

$P = a + b + c$

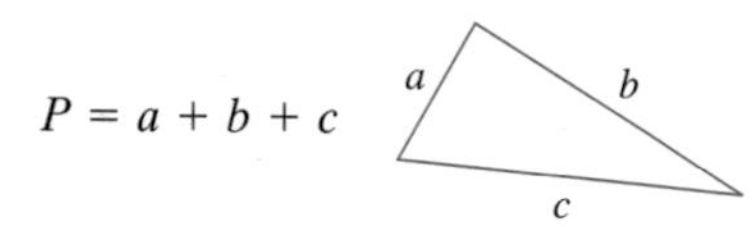

Find the perimeter of an 8-inch by 10-inch photograph.

$$P = 2l + 2w \quad \text{Perimeter of a rectangle.}$$
$$= 2(10 \text{ in.}) + 2(8 \text{ in.}) \quad \text{Let } l = 10 \text{ in. and } w = 8 \text{ in.}$$
$$= 20 \text{ in.} + 16 \text{ in.} \quad \text{Simplify.}$$
$$= 36 \text{ inches}$$

CHAPTER 3 REVIEW

(3.1) *Simplify the following by combining like terms.*

1. $3y + 7y - 15$

2. $2y - 10 - 8y$

3. $8a + a - 7 - 15a$

4. $y + 3 - 9y - 1$

Multiply.

5. $-2(x + 5)$

6. $-3(y + 8)$

Simplify.

7. $7x + 3(x - 4) + x$

8. $10 - 2(m - 3) - m$

9. $3(5a - 2) - 20a + 10$

10. $6y + 3 + 2(3y - 6)$

Find the perimeter of each figure.

11.

(2x – 1) yards

3 yards

Rectangle

12.

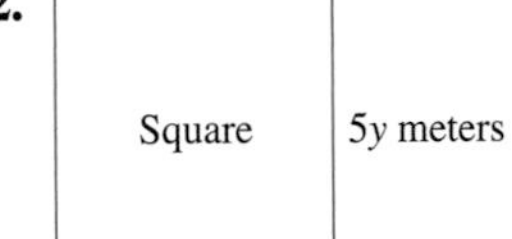

Find the area of each figure.

13.

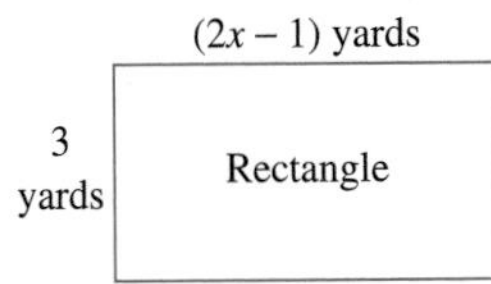

14.

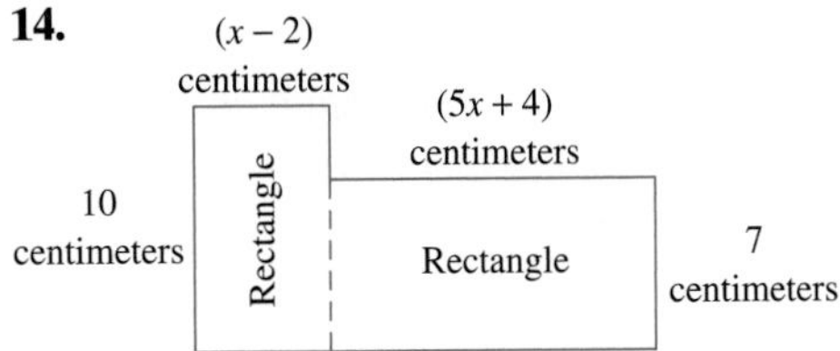

Simplify.

15. $85(7068x - 108) + 42x$

16. $-4268y + 120(63y - 32)$

(3.2)

17. Is 4 a solution of $5(2 - x) = -10$?

18. Is 0 a solution of $(6y + 2) = 23 + 4y$?

Solve each of the following.

19. $z - 5 = -7$

20. $x + 1 = 8$

21. $n + 18 = 10 - (-2)$

22. $c - 5 = -13 + 7$

23. $7x + 5 - 6x = -20$

24. $15 = 8x + 35 - 7x$

(3.3) *Solve each of the following.*

25. $-3y = -21$

26. $-8x = 72$

27. $-5n = -5$

28. $-3a = 15$

ANSWERS

1. ______
2. ______
3. ______
4. ______
5. ______
6. ______
7. ______
8. ______
9. ______
10. ______
11. ______
12. ______
13. ______
14. ______
15. ______
16. ______
17. ______
18. ______
19. ______
20. ______
21. ______
22. ______
23. ______
24. ______
25. ______
26. ______
27. ______
28. ______

ANSWERS

29.
30.
31.
32.
33.
34.
35.
36.
37.
38.
39.
40.
41.
42.
43.
44.
45.
46.
47.
48.
49.
50.
51.
52.
53.
54.

29. $0 = 12x$ **30.** $-14 = 7x$ **31.** $-5t + 32 + 4t = 32$ **32.** $3z + 72 - 2z = -56$

Translate each phrase to an algebraic expression.

33. Eleven added to twice a number.

34. The product of -5 and a number, decreased by 50.

35. The quotient of 70 and the sum of a number and 6.

36. Twice the difference of a number and 13.

(3.4) *Solve each of the following.*

37. $3x - 4 = 11$ **38.** $6y + 1 = 73$ **39.** $14 - y = -3$ **40.** $7 - z = 0$

41. $4z - z = -6$ **42.** $t - 9t = -64$ **43.** $5(n - 3) = 7 + 3n$

44. $7(2 + x) = 4x - 1$ **45.** $2x + 7 = 6x - 1$ **46.** $5x - 18 = -4x$

Write each sentence as an equation.

47. The difference of 20 and -8 is 28.

48. The product of 5 and the sum of 2 and -6 yields -20.

49. The quotient of -75 and the sum of 5 and 20 is equal to -3.

50. Nineteen subtracted from -2 amounts to -21.

(3.5)

Translate each sentence into an equation using x as the variable.

51. Twice a number minus 8 is 40.

52. Twelve subtracted from the quotient of a number and 2 is 10.

53. The difference of a number and 3 is the quotient of the number and 4.

54. The product of some number and 6 is equal to the sum of the number and 2.

EXERCISE SET 4.1

Represent the shaded part of each geometric figure by a proper fraction.

1. 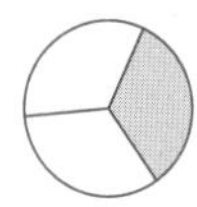

2. 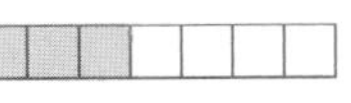

3. 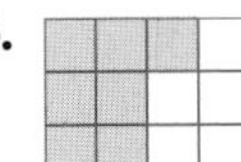

4. 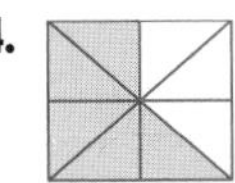

5. 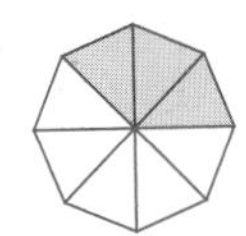

6. 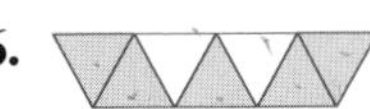

Represent the shaded part of the geometric figures by an improper fraction.

7. 

8. 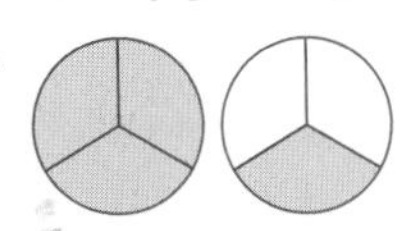

9. 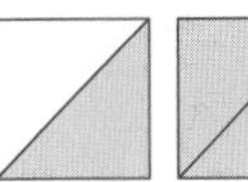 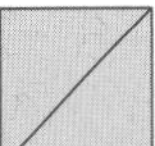

10. 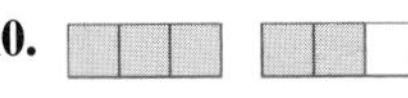

11. 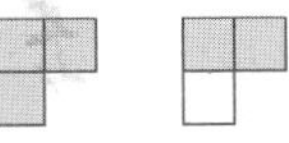

12. 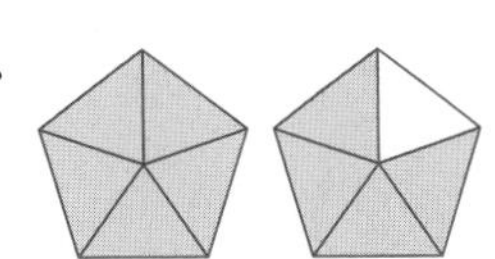

Graph the fraction on a number line. See Examples 1 and 2.

13. $\frac{1}{4}$ 0 1

14. $\frac{1}{3}$ 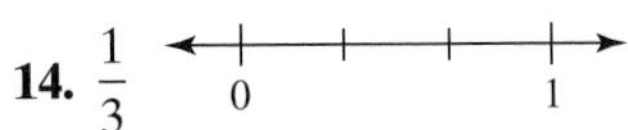

15. $\frac{4}{7}$ 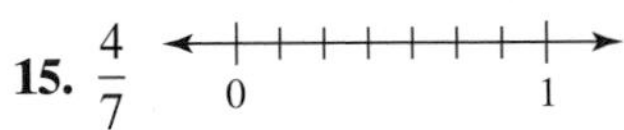

16. $\frac{5}{6}$ 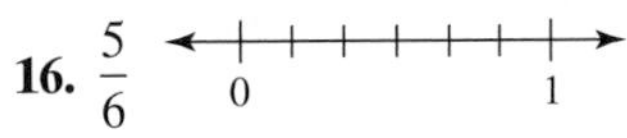

17. $\frac{8}{5}$ 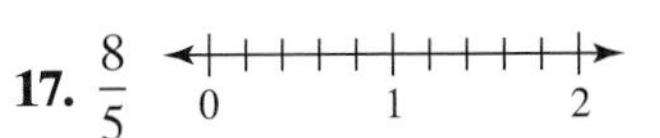

18. $\frac{9}{8}$ 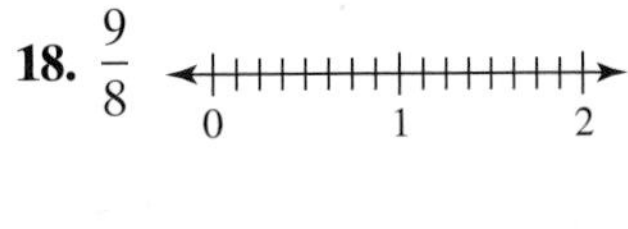

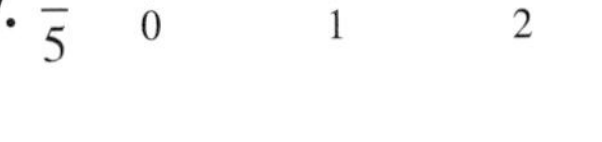

19. $\frac{7}{3}$ 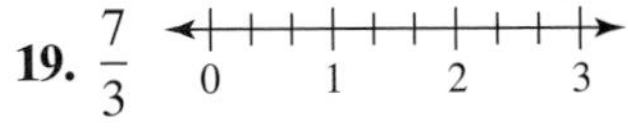

20. $\frac{15}{7}$ 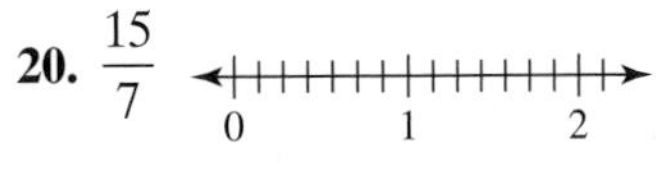

21. $\frac{3}{8}$ 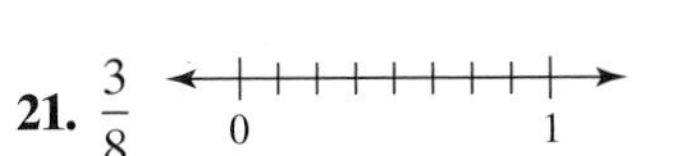

22. $\frac{8}{3}$ 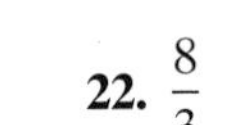 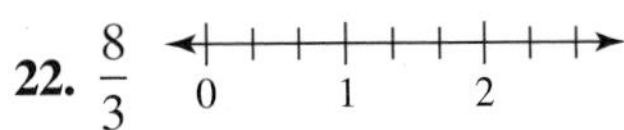

Write each fraction as an equivalent fraction with the given denominator. See Examples 3 through 6.

23. $\frac{4}{7}$; denominator of 35 **24.** $\frac{3}{5}$; denominator of 20 **25.** $\frac{2}{3}$; denominator of 21

26. $\frac{1}{6}$; denominator of 24 **27.** $\frac{2y}{5}$; denominator of 25 **28.** $\frac{9a}{10}$; denominator of 70

ANSWERS

1. ______
2. ______
3. ______
4. ______
5. ______
6. ______
7. ______
8. ______
9. ______
10. ______
11. ______
12. ______
23. ______
24. ______
25. ______
26. ______
27. ______
28. ______

ANSWERS

29. ______
30. ______
31. ______
32. ______
33. ______
34. ______
35. ______
36. ______
37. ______
38. ______
39. ______
40. ______
41. ______
42. ______
43. ______
44. ______
45. ______
46. ______
47. ______
48. ______
49. ______
50. ______
51. ______
52. ______
53. ______
54. ______
55. ______
56. ______

29. $\frac{1}{2}$; denominator of 30 **30.** $\frac{1}{3}$; denominator of 30 **31.** $\frac{10}{7x}$; denominator of $21x$

32. $\frac{5}{3b}$; denominator of $21b$ **33.** 2; denominator of 5 **34.** 5; denominator of 8

Simplify by dividing. See Example 7.

35. $\frac{12}{12}$ **36.** $\frac{-3}{-3}$ **37.** $\frac{-5}{1}$ **38.** $\frac{10}{10}$

39. $\frac{36}{3}$ **40.** $\frac{-12}{3}$ **41.** $\frac{-8}{-8}$ **42.** $\frac{7}{1}$

43. $\frac{-14}{2}$ **44.** $\frac{-8}{4}$

Write each fraction as an equivalent fraction whose denominator is 12.

45. $\frac{3}{4}$ **46.** $\frac{4}{6}$ **47.** $\frac{2y}{3}$

48. $\frac{2}{3}$ **49.** $\frac{1}{2}$ **50.** $\frac{3x}{2}$

Write each fraction as an equivalent fraction whose denominator is 36x.

51. $\frac{4}{3}$ **52.** $\frac{3}{4}$ **53.** $\frac{5}{9}$

54. $\frac{7}{6}$ **55.** 1 **56.** 2

57. There are 12 inches in a foot. What fractional part of a foot does 5 inches represent?

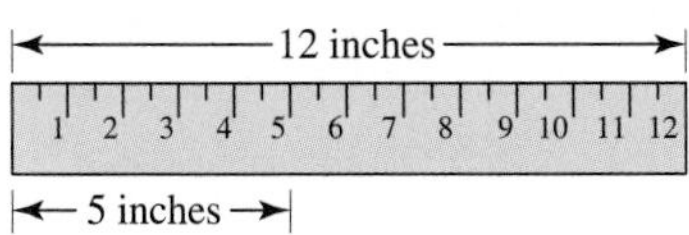

58. A 3M high-density disk holds 2 megabytes of memory. What fractional part of the memory on this disk does 1 megabyte represent?

59. There are 24 hours in a day. What fraction of a day does 11 hours represent?

60. There are 60 minutes in an hour. What fraction of an hour does 37 minutes represent?

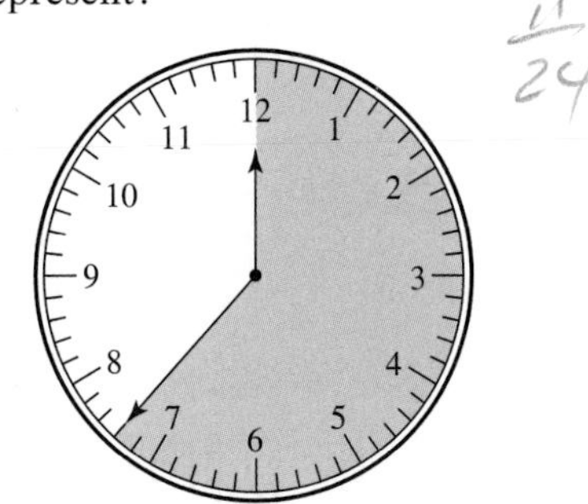

61. In a family with 11 children, there are 4 boys and 7 girls. What fraction of the children are girls?

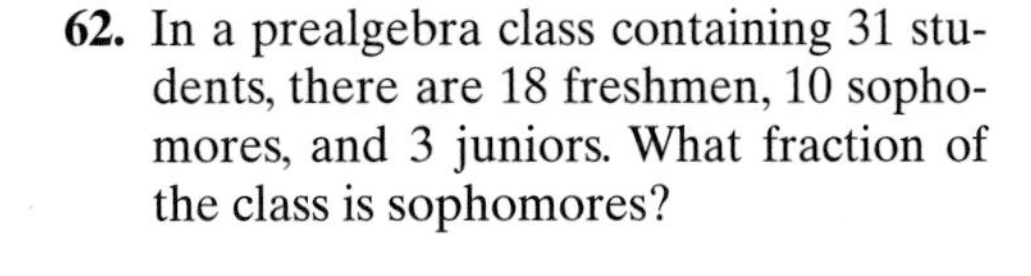

62. In a prealgebra class containing 31 students, there are 18 freshmen, 10 sophomores, and 3 juniors. What fraction of the class is sophomores?

63. Consumer fireworks are legal in 32 states in the United States.

a. In what fraction of the states are consumer fireworks legal?

b. In how many states are consumer fireworks illegal?

c. In what fraction of the states are consumer fireworks illegal? (*Source: USA Today,* July 3, 1995)

64. Thirty-three states in the United States contain federal Indian reservations.

a. What fraction of the states contain Indian reservations?

b. How many states do not contain federal Indian reservations?

c. What fraction of the states do not contain Indian reservations? (*Source:* Tiller Research, Inc., Albuquerque, NM)

65. Write $\frac{2}{9}$ as an equivalent fraction with a denominator of 2088.

66. Write $\frac{3}{11}$ as an equivalent fraction with a denominator of 6479.

Find the missing number so that the fractions are equivalent.

67. $\frac{4}{5} = \frac{?}{5105}$

68. $\frac{9}{12} = \frac{?}{3744}$

69. $\frac{29}{30} = \frac{?}{23{,}640}$

70. $\frac{2}{7} = \frac{?}{14{,}392}$

ANSWERS

57. ______
58. ______
59. ______
60. ______
61. ______
62. ______
63. a. b. c. ______
64. a. b. c. ______
65. ______
66. ______
67. ______
68. ______
69. ______
70. ______

ANSWERS

71. ______

72. ______

73. ______

74. ______

75. ______

76. ______

77. ______

78. ______

Review Exercises

Evaluate each expression using the given replacement numbers. *See Section 2.5.*

71. $y + z$, when $y = 10$ and $z = 32$

72. $12 + x$, when $x = -5$

73. $m + 100$, when $m = 62$

74. $x + y$, when $x = -18$ and $y = -32$

75. $x - y$, when $x = -20$ and $y = 0$

76. $z - x$, when $z = 32$ and $x = 20$

77. $x + y + z$, when $x = -3, y = -7$, and $z = 9$

78. $a + b + c$, when $a = -10, b = 20$, and $c = 30$

TAPE PA 4.2

4.2

FACTORS AND LOWEST TERMS

OBJECTIVES

1 Write a number as a product of primes.

2 Write a fraction in lowest terms.

1 Of all the equivalent ways to write a particular fraction, one special way is called **lowest terms.** To help us write a fraction in lowest terms, we first practice writing a number as a product of primes.

> A **prime number** is a whole number greater than 1 whose only divisors are 1 and itself. The first few prime numbers are 2, 3, 5, 7, 11, 13, 17, 19, 23, 29, . . .
> A **composite number** is a whole number greater than 1 that is not prime.

When a number is written as a product of primes, this product is called the **prime factorization** of the number. For example, the prime factorization of 12 is $2 \cdot 2 \cdot 3$ since $12 = 2 \cdot 2 \cdot 3$ and all the factors are prime. Because multiplication is commutative, the order of the factors is not important. We can write the factorization $2 \cdot 2 \cdot 3$ as $2 \cdot 3 \cdot 2$ or $3 \cdot 2 \cdot 2$. Any of these is called the prime factorization of 12.

> Every whole number greater than 1 has exactly one prime factorization.

Recall from Section 1.5 that since $12 = 2 \cdot 2 \cdot 3$, the numbers 2 and 3 are called **factors** of 12. A **factor** of 12 is any number that divides it evenly (with a remainder of 0).

EXAMPLE 1

Write the prime factorization of 45.

Solution:

We can begin by writing 45 as the product of two numbers, say 5 and 9.

$$45 = 5 \cdot 9$$

The number 5 is prime, but 9 is not. Write 9 as $3 \cdot 3$.

$$\begin{aligned} 45 &= 5 \cdot 9 \\ &\ \ \downarrow\ \ \downarrow\searrow \\ &= 5 \cdot 3 \cdot 3 \end{aligned}$$

Each factor is now a prime number, so the prime factorization of 45 is $3 \cdot 3 \cdot 5$ or $3^2 \cdot 5$.

PRACTICE PROBLEM 1

Find the prime factorization of 28.

Answer:

1. $28 = 2 \cdot 2 \cdot 7$ or $2^2 \cdot 7$

PRACTICE PROBLEM 2
Write the prime factorization of 60.

EXAMPLE 2

Write the prime factorization of 80.

Solution:

Write 80 as a product of two numbers. Continue this process until all factors are prime.

$$80 = 8 \cdot 10$$
$$4 \cdot 2 \cdot 2 \cdot 5$$
$$= 2 \cdot 2 \cdot 2 \cdot 2 \cdot 5$$

All factors are now prime, so the prime factorization of 80 is

$$2 \cdot 2 \cdot 2 \cdot 2 \cdot 5 \text{ or } 2^4 \cdot 5.$$

There are a few quick **divisibility tests** to determine if a number is divisible by the primes 2, 3, or 5.

A whole number is divisible by

- **2** if the ones digit is 0, 2, 4, 6, or 8.
  ↓
  132 is divisible by 2
- **3** if the sum of the digits is divisible by 3.
  144 is divisible by 3 since $1 + 4 + 4 = 9$ is divisible by 3.
- **5** if the ones digit is 0 or 5.
  ↓
  1,115 is divisible by 5

When finding the prime factorization of larger numbers, you may want to use the following procedure shown in Example 3.

PRACTICE PROBLEM 3
Write the prime factorization of 297.

EXAMPLE 3

Write the prime factorization of 252.

Solution:

Since the ones digit of 252 is 2, we know that 252 is divisible by 2.

$$2\overline{)252}\quad 126$$

126 is divisible by 2 also.

$$2\overline{)126}\quad 63$$
$$2\overline{)252}$$

63 is not divisible by 2 but is divisible by 3.

Answers:
2. $60 = 2 \cdot 2 \cdot 3 \cdot 5$ or $2^2 \cdot 3 \cdot 5$
3. $297 = 3 \cdot 3 \cdot 3 \cdot 11$ or $3^3 \cdot 11$

Divide 63 by 3 and continue in this same manner until the quotient is a prime number.

$$\begin{array}{r} 7 \\ 3\overline{)21} \\ 3\overline{)63} \\ 2\overline{)126} \\ 2\overline{)252} \end{array}$$

The prime factorization of 252 is $2 \cdot 2 \cdot 3 \cdot 3 \cdot 7$ or $2^2 \cdot 3^2 \cdot 7$.

2 We can use the prime factorization of a number to help us write a fraction in **lowest terms** or **simplest form.**

LOWEST TERMS

A fraction is in lowest terms, or simplest form, when the numerator and denominator have no common factors other than 1.

To write a fraction in lowest terms, write the prime factorization of the numerator and the denominator and then use the fundamental principle of fractions to divide both by all common factors.

For example,

$$\frac{6}{10} = \frac{2 \cdot 3}{2 \cdot 5} = \frac{2 \cdot 3 \div 2}{2 \cdot 5 \div 2} = \frac{3}{5}$$

The fraction $\frac{3}{5}$ is in lowest terms, since the numerator and the denominator have no common factors (other than 1).

In the future, we will use the following notation to show dividing the numerator and denominator by common factors.

$$\frac{6}{10} = \frac{2 \cdot 3}{2 \cdot 5} = \frac{2 \cdot 3}{2 \cdot 5} = \frac{3}{5}$$

EXAMPLE 4

Write $\frac{12}{20}$ in lowest terms.

Solution:

First, write the prime factorization of the numerator and the denominator.

$$\frac{12}{20} = \frac{2 \cdot 2 \cdot 3}{2 \cdot 2 \cdot 5}$$

Next, divide the numerator and the denominator by all common factors.

$$\frac{12}{20} = \frac{2 \cdot 2 \cdot 3}{2 \cdot 2 \cdot 5} = \frac{3}{5}$$

EXAMPLE 5

Write $\frac{42}{66}$ in lowest terms.

Solution:

$$\frac{42}{66} = \frac{2 \cdot 3 \cdot 7}{2 \cdot 3 \cdot 11} = \frac{7}{11}$$

PRACTICE PROBLEM 4

Write $\frac{30}{45}$ in lowest terms.

PRACTICE PROBLEM 5

Write $\frac{39}{51}$ in lowest terms.

Answers:

4. $\frac{2}{3}$

5. $\frac{13}{17}$

PRACTICE PROBLEM 6

Write $\frac{45}{105y}$ in lowest terms.

EXAMPLE 6

Write $\frac{84x}{90}$ in lowest terms.

Solution:

$$\frac{84x}{90} = \frac{2 \cdot 2 \cdot 3 \cdot 7 \cdot x}{2 \cdot 3 \cdot 3 \cdot 5} = \frac{14x}{15}$$

PRACTICE PROBLEM 7

Write $\frac{9a}{50a}$ in lowest terms.

EXAMPLE 7

Write $\frac{10y}{27y}$ in lowest terms.

Solution:

$$\frac{10y}{27y} = \frac{2 \cdot 5 \cdot y}{3 \cdot 3 \cdot 3 \cdot y} = \frac{10}{27}$$

PRACTICE PROBLEM 8

Write $\frac{38}{4}$ in lowest terms.

EXAMPLE 8

Write $\frac{72}{26}$ in lowest terms.

Solution:

$$\frac{72}{26} = \frac{2 \cdot 2 \cdot 2 \cdot 3 \cdot 3}{2 \cdot 13} = \frac{36}{13}$$

PRACTICE PROBLEM 9

Write $\frac{7a^3}{56a^2}$ in lowest terms.

EXAMPLE 9

Write $\frac{6x^2}{60x^3}$ in lowest terms.

Solution:

$$\frac{6x^2}{60x^3} = \frac{2 \cdot 3 \cdot x \cdot x}{2 \cdot 2 \cdot 3 \cdot 5 \cdot x \cdot x \cdot x} = \frac{1}{10x}$$

REMINDER Be careful when all factors of the numerator or denominator are divided out as in Example 9. In this example, notice that the numerator and denominator were both divided by $2 \cdot 3 \cdot x \cdot x = 6x^2$. When the numerator of $6x^2$ is divided by $6x^2$, the result is 1, not 0. In general, when all factors of the numerator or the denominator are divided out, the result is 1 in the appropriate numerator or denominator.

Answers:

6. $\frac{3}{7y}$
7. $\frac{9}{50}$
8. $\frac{19}{2}$
9. $\frac{a}{8}$

MENTAL MATH

1. Is 2430 divisible by 2? by 3? by 5?

Find the prime factorization of each number.

2. 15
3. 10
4. 6
5. 21
6. 4
7. 9
8. 14

EXERCISE SET 4.2

Write the prime factorization of each number. See Examples 1 through 3.

1. 20 **2.** 12 **3.** 48 **4.** 80

5. 64 **6.** 45 **7.** 240 **8.** 128

Write each fraction in lowest terms. See Examples 4 through 9.

9. $\frac{3}{12}$ **10.** $\frac{5}{20}$ **11.** $\frac{7x}{35}$ **12.** $\frac{9}{48z}$

13. $\frac{14}{16}$ **14.** $\frac{18}{4}$ **15.** $\frac{24a}{30a}$ **16.** $\frac{70y}{80xy}$

17. $\frac{35}{42}$ **18.** $\frac{25}{55}$ **19.** $\frac{30x^2}{36x}$ **20.** $\frac{45b}{80b^2}$

21. $\frac{16}{24}$ **22.** $\frac{18}{45}$ **23.** $\frac{45xz}{60z}$ **24.** $\frac{22a}{99ab}$

25. $\frac{26a^2}{39ab}$ **26.** $\frac{24x}{42x^2}$ **27.** $\frac{63}{72}$ **28.** $\frac{56}{64}$

ANSWERS

1. ______
2. ______
3. ______
4. ______
5. ______
6. ______
7. ______
8. ______
9. ______
10. ______
11. ______
12. ______
13. ______
14. ______
15. ______
16. ______
17. ______
18. ______
19. ______
20. ______
21. ______
22. ______
23. ______
24. ______
25. ______
26. ______
27. ______
28. ______

ANSWERS

29. ____
30. ____
31. ____
32. ____
33. ____
34. ____
35. ____
36. ____
37. ____
38. ____
39. ____
40. ____
41. ____
42. ____
43. ____
44. ____
45. ____
46. ____
47. ____
48. ____
49. ____
50. ____
51. ____
52. ____
53. ____
54. ____

29. $\frac{21}{49}$ **30.** $\frac{14}{35}$ **31.** $\frac{24y}{40}$ **32.** $\frac{36}{54x}$

33. $\frac{36z}{63z}$ **34.** $\frac{39b}{52b}$ **35.** $\frac{72x^3y^2}{90xy}$ **36.** $\frac{24a^2b}{54ab^3}$

37. $\frac{12}{15}$ **38.** $\frac{18}{24}$ **39.** $\frac{25x^2}{40x}$ **40.** $\frac{36y^2}{42y}$

41. $\frac{27xy}{90y}$ **42.** $\frac{60y}{150yz}$ **43.** $\frac{36a^3bc^2}{24ab^4c^2}$ **44.** $\frac{60x^2yz}{36x^3y^3z^3}$

45. $\frac{40xy}{64xyz}$ **46.** $\frac{28abc}{60ac}$

For Exercises 47 through 54, write all answers in lowest terms.

47. There are 5280 feet in a mile. What fraction of a mile is represented by 2640 feet?

48. A Macintosh computer contains 40 megabytes of memory. What fraction of memory is 10 megabytes?

49. A work shift for an employee at McDonald's consists of 8 hours. What fraction of the employee's work shift is represented by 6 hours?

50. Two thousand baseball caps were sold one year at the U.S. Open Golf Tournament. What fractional part of this total does 200 caps represent?

51. There are 35 students in a biology class. If 10 students made an A on the first test, what fraction of the students made an A?

52. Four out of 10 marbles are red. What fraction of marbles are *not red?*

53. There are 16,000 students at a local university. If 8800 are females, what fraction of the students are *male?*

54. There are 100 centimeters in 1 meter. What fraction of a meter is 20 centimeters?

The rule for multiplying fractions is the same if variables are involved.

EXAMPLE 6

Multiply: $\frac{3x}{4} \cdot \frac{8}{5x}$.

Solution:

$$\frac{3x}{4} \cdot \frac{8}{5x} = \frac{3 \cdot x \cdot 8}{4 \cdot 5 \cdot x} = \frac{3 \cdot 4 \cdot 2}{4 \cdot 5} = \frac{6}{5}$$

PRACTICE PROBLEM 6

Multiply: $\frac{2}{3} \cdot \frac{3y}{2}$.

REMINDER Recall that when the denominator of a fraction contains a variable, such as $\frac{8}{5x}$, we assume that the variable does not represent 0.

EXAMPLE 7

Multiply: $\frac{x^2}{y} \cdot \frac{y^3}{x}$.

Solution:

$$\frac{x^2}{y} \cdot \frac{y^3}{x} = \frac{x^2 \cdot y^3}{y \cdot x} = \frac{x \cdot x \cdot y \cdot y \cdot y}{y \cdot x} = \frac{x \cdot y \cdot y}{1} = xy^2$$

PRACTICE PROBLEM 7

Multiply: $\frac{a^3}{b^2} \cdot \frac{b}{a^2}$.

2 The base of an exponential expression can also be a fraction.

$$\left(\frac{1}{3}\right)^4 = \underbrace{\frac{1}{3} \cdot \frac{1}{3} \cdot \frac{1}{3} \cdot \frac{1}{3}}_{} = \frac{1 \cdot 1 \cdot 1 \cdot 1}{3 \cdot 3 \cdot 3 \cdot 3} = \frac{1}{81}$$

$\frac{1}{3}$ is a factor 4 times.

EXAMPLE 8

Evaluate:

a. $\left(\frac{2}{5}\right)^4$ b. $\left(-\frac{1}{4}\right)^2$

Solution:

a. $\left(\frac{2}{5}\right)^4 = \frac{2}{5} \cdot \frac{2}{5} \cdot \frac{2}{5} \cdot \frac{2}{5} = \frac{2 \cdot 2 \cdot 2 \cdot 2}{5 \cdot 5 \cdot 5 \cdot 5} = \frac{16}{625}$

b. $\left(-\frac{1}{4}\right)^2 = \left(-\frac{1}{4}\right) \cdot \left(-\frac{1}{4}\right) = \frac{1 \cdot 1}{4 \cdot 4} = \frac{1}{16}$

PRACTICE PROBLEM 8

Evaluate:

a. $\left(\frac{3}{4}\right)^3$ b. $\left(-\frac{2}{3}\right)^2$

Before we can divide fractions, we need to know how to find the reciprocal of a fraction.

RECIPROCAL OF A FRACTION

Two numbers are **reciprocals** of each other if their product is 1. The reciprocal of the fraction $\frac{a}{b}$ is $\frac{b}{a}$ because $\frac{a}{b} \cdot \frac{b}{a} = \frac{a \cdot b}{b \cdot a} = 1$.

Answers:

6. y

7. $\frac{a}{b}$

8a. $\frac{27}{64}$ **b.** $\frac{4}{9}$

The reciprocal of $\frac{2}{5}$ is $\frac{5}{2}$ because $\frac{2}{5} \cdot \frac{5}{2} = \frac{10}{10} = 1$.

The reciprocal of 5 is $\frac{1}{5}$ because $5 \cdot \frac{1}{5} = \frac{5}{1} \cdot \frac{1}{5} = \frac{5}{5} = 1$.

The reciprocal of $-\frac{7}{11}$ is $-\frac{11}{7}$ because $-\frac{7}{11} \cdot -\frac{11}{7} = \frac{77}{77} = 1$.

REMINDER Every number has a reciprocal except 0. The number 0 has no reciprocal because there is no number that when multiplied by 0 gives a result of 1.

We use reciprocals to divide fractions.

TO DIVIDE FRACTIONS

If b, c, and d are not 0, then $\frac{a}{b} \div \frac{c}{d} = \frac{a}{b} \cdot \frac{d}{c} = \frac{a \cdot d}{b \cdot c}$.

In other words, to divide fractions, multiply the first fraction by the reciprocal of the second fraction. For example

multiply by reciprocal

$$\frac{1}{4} \div \frac{2}{5} = \frac{1}{4} \cdot \frac{5}{2} = \frac{5}{8}$$

PRACTICE PROBLEM 9

Divide: $\frac{3}{2} \div \frac{14}{5}$.

EXAMPLE 9

Divide: $\frac{7}{8} \div \frac{2}{9}$.

Solution:

$$\frac{7}{8} \div \frac{2}{9} = \frac{7}{8} \cdot \frac{9}{2} = \frac{7 \cdot 9}{8 \cdot 2} = \frac{63}{16}$$

REMINDER When dividing by a fraction, do not look for common factors to divide out until you rewrite the division as multiplication.

Do not try to divide out these two 2s.

$$\frac{1}{\mathbf{2}} \div \frac{\mathbf{2}}{3} = \frac{1}{2} \cdot \frac{3}{2} = \frac{3}{4}$$

When performing operations on fractions, write all results in lowest terms.

PRACTICE PROBLEM 10

Divide: $\frac{10}{4} \div \frac{2}{9}$.

Answers:

9. $\frac{15}{28}$

10. $\frac{45}{4}$

EXAMPLE 10

Divide: $-\frac{5}{16} \div -\frac{3}{4}$.

Solution:

Recall that the quotient (or product) of two negative numbers is a positive number.

$$-\frac{5}{16} \div -\frac{3}{4} = -\frac{5}{16} \cdot -\frac{4}{3} = \frac{5 \cdot 4}{4 \cdot 4 \cdot 3} = \frac{5}{12}$$

EXAMPLE 11

Divide: $\frac{2x}{3} \div 3x^2$.

Solution:

$$\frac{2x}{3} \div 3x^2 = \frac{2x}{3} \div \frac{3x^2}{1} = \frac{2x}{3} \cdot \frac{1}{3x^2} = \frac{2 \cdot x \cdot 1}{3 \cdot 3 \cdot x \cdot x} = \frac{2}{9x}$$

PRACTICE PROBLEM 11

Divide: $\frac{3y}{4} \div 5y^3$.

EXAMPLE 12

Simplify $\left(\frac{4}{7} \cdot \frac{3}{8}\right) \div -\frac{3}{4}$.

Solution:

Remember to perform the operations inside ( ) first.

$$\left(\frac{4}{7} \cdot \frac{3}{8}\right) \div -\frac{3}{4} = \left(\frac{4 \cdot 3}{7 \cdot 2 \cdot 4}\right) \div -\frac{3}{4} = \frac{3}{14} \div -\frac{3}{4}$$

Now divide.

$$\frac{3}{14} \div -\frac{3}{4} = \frac{3}{14} \cdot -\frac{4}{3} = -\frac{3 \cdot 2 \cdot 2}{2 \cdot 7 \cdot 3} = -\frac{2}{7}$$

PRACTICE PROBLEM 12

Evaluate $\left(-\frac{2}{3} \cdot \frac{9}{14}\right) \div \frac{7}{15}$.

EXAMPLE 13

If $x = \frac{7}{8}$ and $y = -\frac{1}{3}$, evaluate (a) xy and (b) $x \div y$.

Solution:

Replace x with $\frac{7}{8}$ and y with $-\frac{1}{3}$.

a. $xy = \frac{7}{8} \cdot -\frac{1}{3}$

$= -\frac{7 \cdot 1}{8 \cdot 3}$

$= -\frac{7}{24}$

b. $x \div y = \frac{7}{8} \div -\frac{1}{3}$

$= \frac{7}{8} \cdot -\frac{3}{1}$

$= -\frac{7 \cdot 3}{8 \cdot 1}$

$= -\frac{21}{8}$

PRACTICE PROBLEM 13

If $x = -\frac{3}{4}$ and $y = \frac{9}{2}$, evaluate (**a**) xy, and (**b**) $x \div y$.

EXAMPLE 14

Is $-\frac{2}{3}$ a solution of the equation $-\frac{1}{2}x = \frac{1}{3}$?

Solution:

To check whether a number is a solution of an equation, recall that we replace the variable with the given number and see if a true statement results.

PRACTICE PROBLEM 14

Is $-\frac{9}{8}$ a solution of the equation $2x = -\frac{9}{4}$?

Answers:

11. $\frac{3}{20y^2}$

12. $-\frac{45}{49}$

13a. $-\frac{27}{8}$ **b.** $-\frac{1}{6}$

14. yes

$$-\frac{1}{2} \cdot x = \frac{1}{3} \qquad \text{Recall that } -\frac{1}{2}x \text{ means } -\frac{1}{2} \cdot x$$

$$-\frac{1}{2} \cdot -\frac{2}{3} = \frac{1}{3} \qquad \text{Replace } x \text{ with } -\frac{2}{3}.$$

$$\frac{1 \cdot 2}{2 \cdot 3} = \frac{1}{3} \qquad \text{The product of two negative numbers is a positive number.}$$

$$\frac{1}{3} = \frac{1}{3} \qquad \text{True.}$$

Since we have a true statement, $-\frac{2}{3}$ is a solution.

PRACTICE PROBLEM 15

In Example 15, Ms. Thompson kept $\frac{1}{4}$ of her $\frac{1}{3}$ interest in the oil drilling company. What part of the company does she still own?

EXAMPLE 15

Maria Thompson owns $\frac{1}{3}$ interest in an oil drilling company. She sold $\frac{3}{4}$ of her interest to another partner. Find the fractional part of the company she sold.

Solution:

1. UNDERSTAND the problem. To do so, read and reread the problem.
2. ASSIGN a variable to an unknown.

   Let x = the fractional part of the company sold.
3. TRANSLATE the problem into an equation.

   In words: (fraction of company sold) = (fraction of interest sold) · (interest owned)

   Translate: $x = \frac{3}{4} \cdot \frac{1}{3}$
4. SOLVE the equation.

$$x = \frac{3}{4} \cdot \frac{1}{3}$$

$$x = \frac{3 \cdot 1}{4 \cdot 3}$$

$$x = \frac{1}{4}$$

5. CHECK the proposed solution.
6. STATE your conclusion. Maria sold $\frac{1}{4}$ of the company.

PRACTICE PROBLEM 16

The Nothing But Fruit Jelly Company markets jellies in $\frac{3}{4}$-pound jars. How many jars can be filled from 15 pounds of jelly?

15 pounds

$\frac{3}{4}$ pound

EXAMPLE 16

The We-Brake-For-U brake repair service estimates that it takes one person $\frac{2}{3}$ of an hour to replace the brake pads on a car. On how many cars can one person replace brake pads in an 8-hour work day?

Solution:

1. UNDERSTAND the problem. To do so, read and reread the problem.

Answers:

15. $\frac{1}{12}$ of the company

16. 20 jars

55. $7 \div \frac{2}{11}$

56. $-100 \div \frac{1}{2}$

57. $-3x \div \frac{x^2}{12}$

58. $7x \div \frac{14x}{3}$

59. $\left(\frac{2}{7} \div \frac{7}{2}\right) \cdot \frac{3}{4}$

60. $\frac{1}{2} \cdot \left(\frac{5}{6} \div \frac{1}{12}\right)$

61. $-\frac{19}{63y} \cdot 9y^2$

62. $16a^2 \cdot -\frac{31}{24a}$

63. $-\frac{2}{3} \cdot -\frac{6}{11}$

64. $-\frac{1}{5} \cdot -\frac{6}{7}$

65. $\frac{4}{8} \div \frac{3}{16}$

66. $\frac{9}{2} \div \frac{16}{15}$

67. $\frac{21x^2}{10y} \div \frac{14x}{25y}$

68. $\frac{17y^2}{24x} \div \frac{13y}{18x}$

69. $\left(1 \div \frac{3}{4}\right) \cdot \frac{2}{3}$

70. $\left(33 \div \frac{2}{11}\right) \cdot \frac{5}{9}$

71. $\frac{a^3}{2} \div 30a^3$

72. $15c^3 \div \frac{3c^2}{5}$

73. $\frac{ab^2}{c} \cdot \frac{c}{ab}$

74. $\frac{ac}{b} \cdot \frac{b^3}{a^2c}$

Given the following replacement values, evaluate (a) xy and (b) x ÷ y.

75. $y = \frac{6}{14}$ and $x = 7$

76. $x = \frac{4}{3}$ and $y = 3$

77. $y = -\frac{1}{2}$ and $x = 5$

78. $y = -\frac{2}{3}$ and $x = 2$

Solve. See Examples 15 and 16.

79. A veterinarian's dipping vat holds 36 gallons of liquid. She normally fills it $\frac{5}{6}$ full of a medicated flea dip solution. Find how many gallons of solution are normally in the vat.

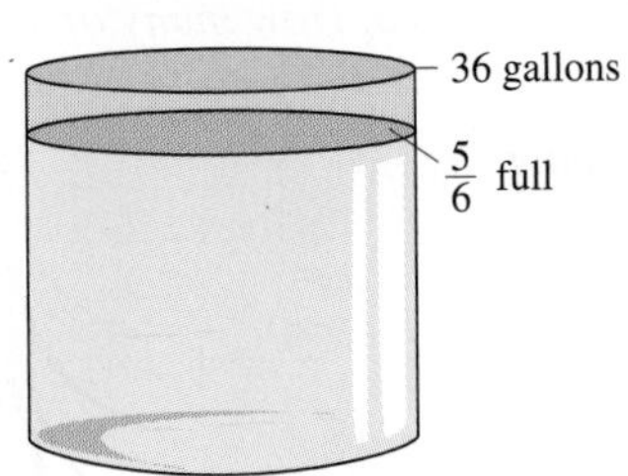

80. The winnings of a race horse are distributed among the owners of the horse according to the amount of ownership each has in the horse. Mr. Gamble has a $\frac{3}{25}$ interest in the horse, Rainbow, which won $285,000 in the Rye Grass Stakes. Find Mr. Gamble's winnings.

ANSWERS

55. ______
56. ______
57. ______
58. ______
59. ______
60. ______
61. ______
62. ______
63. ______
64. ______
65. ______
66. ______
67. ______
68. ______
69. ______
70. ______
71. ______
72. ______
73. ______
74. ______
75. ______
76. ______
77. ______
78. ______
79. ______
80. ______

ANSWERS

81. ______

82. ______

83. ______

84. ______

85. ______

86. ______

87. ______

88. ______

89. ______

90. ______

81. Each turn of a screw sinks it $\frac{3}{16}$ of an inch deeper into a piece of wood. Find how deep the screw is after 8 turns.

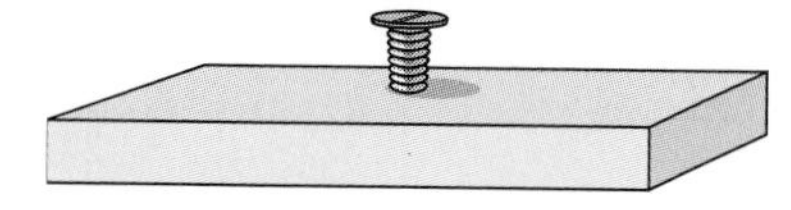

82. The O'Neill family has a weekly income of \$450, $\frac{2}{15}$ of which must be budgeted for utilities. Find how many dollars are budgeted each week to pay the utilities.

83. Campbell Soup Company ships its soup in boxes containing 24 cans. If each can weighs $\frac{5}{4}$ pounds, find how much the contents of a box weighs.

84. An estimate for the measure of an adult's wrist is $\frac{1}{4}$ of the waist size. If Jorge has a 34-inch waist, estimate the size of his wrist.

85. An estimate for an adult's waist measurement is found by dividing the neck size (in inches) by $\frac{1}{2}$. Jock's neck measures 18 inches. Estimate his waist measurement.

86. The market value of a house is $\frac{3}{4}$ of its appraised value. If the appraised value of a house is \$60,000, find its market value.

The following circle graph first appeared in Section 4.2. Recall that it shows the fractional part of a car's total mileage in each category. Use this graph for Exercises 87 through 90.

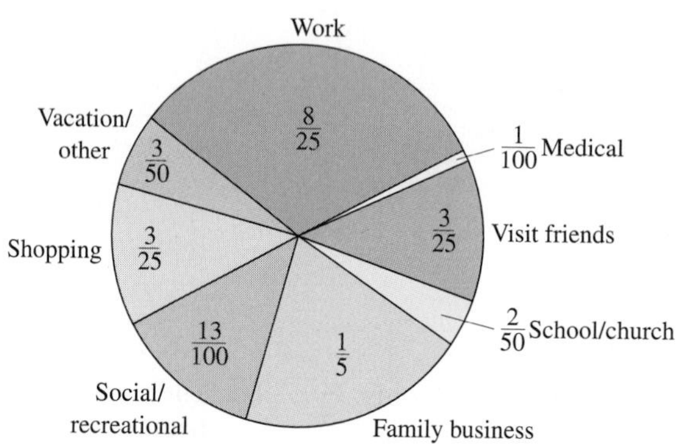

(*Source:* The American Automobile Manufacturers Assn. and The National Automobile Dealers Assn.)

In one year, the Rodriguez family drove 12,000 miles in the family car. How many of these miles might we expect to fall in each category below?

87. Work

88. Shopping

89. Family business

90. Medical

When setting a post for building a deck or fence, it is recommended that $\frac{1}{3}$ of the total length of the post be buried in the ground. Find the amount of post to be buried in the ground for the given post lengths.

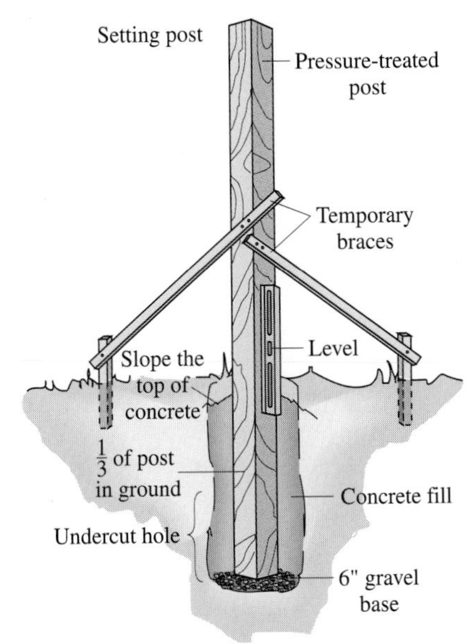

(*Source: Southern Living*, May, 1995.)

91. a 9-foot post

92. a 12-foot post

93. In 1995, approximately $\frac{7}{10}$ of households that have on-line service in the United States made on-line purchases. If 10,300,000 households have on-line services in the U.S., how many of these made on-line purchases? (*Source:* Inteco Corp.)

94. Approximately $\frac{3}{25}$ of the United States population is in the state of California. If the U.S. population is 260,341,000, find the population of California. (*Source:* U.S. Bureau of the Census, 1994)

95. In your own words, describe how to multiply fractions.

96. In your own words, describe how to divide fractions.

Review Exercises

Write the prime factorization of each number. See Section 4.2.

97. 90 **98.** 42 **99.** 65

100. 72 **101.** 126 **102.** 112

ANSWERS

91. ______
92. ______
93. ______
94. ______
95. ______
96. ______
97. ______
98. ______
99. ______
100. ______
101. ______
102. ______

EXAMPLE 8

Find the LCD of $\frac{1}{18}$ and $\frac{5}{24}$.

Solution:

First, write each denominator as a product of primes.

$$18 = 2 \cdot 3 \cdot 3$$
$$24 = 2 \cdot 2 \cdot 2 \cdot 3$$

Write each factor the greatest number of times that it appears in any **one** prime factorization.

The greatest number of times that 2 appears is **3** times.

The greatest number of times that 3 appears is **2** times.

$$\text{LCD} = \underbrace{2 \cdot 2 \cdot 2}_{\substack{\text{2 is a factor} \\ \text{3 times}}} \cdot \underbrace{3 \cdot 3}_{\substack{\text{3 is a factor} \\ \text{2 times}}} = 72$$

Notice that 72 is the smallest positive number that is divisible by both 18 and 24.

PRACTICE PROBLEM 8

Find the LCD of $\frac{9}{14}$ and $\frac{11}{35}$.

EXAMPLE 9

Find the LCD of $-\frac{2}{5}, \frac{1}{6}$, and $\frac{5}{12}$.

Solution:

$$5 = 5$$
$$6 = 2 \cdot 3$$
$$12 = 2 \cdot 2 \cdot 3$$
$$\text{LCD} = 2 \cdot 2 \cdot 3 \cdot 5 = 60$$

PRACTICE PROBLEM 9

Find the LCD of $\frac{7}{4}, \frac{7}{15}$, and $\frac{3}{10}$.

EXAMPLE 10

Find the LCD of $\frac{3}{5}$ and $\frac{2}{x}$.

Solution:

$$5 = 5$$
$$x = x$$
$$\text{LCD} = 5 \cdot x = 5x$$

PRACTICE PROBLEM 10

Find the LCD of $\frac{7}{y}$ and $\frac{6}{11}$.

Answers:

8. 70

9. 60

10. $11y$

EXERCISE SET 4.4

Add. See Examples 1 and 2.

1. $\frac{3}{7} + \frac{2}{7}$

2. $\frac{5}{9} + \frac{2}{9}$

3. $-\frac{1}{2} + \frac{1}{2}$

4. $-\frac{3}{x} + \frac{1}{x}$

5. $\frac{2}{9x} + \frac{4}{9x}$

6. $\frac{3}{10y} + \frac{2}{10y}$

7. $-\frac{4}{13} + \frac{2}{13} + \frac{1}{13}$

8. $-\frac{5}{11} + \frac{1}{11} + \frac{2}{11}$

9. $\frac{7}{18} + \frac{3}{18} + \frac{2}{18}$

10. $\frac{2}{15} + \frac{4}{15} + \frac{9}{15}$

Subtract. See Examples 3 and 4.

11. $\frac{10}{11} - \frac{4}{11}$

12. $\frac{9}{13} - \frac{5}{13}$

13. $\frac{1}{y} - \frac{4}{y}$

14. $\frac{4}{z} - \frac{7}{z}$

15. $\frac{7a}{4} - \frac{3}{4}$

16. $\frac{18b}{5} - \frac{3}{5}$

17. $\frac{1}{8} - \frac{7}{8}$

18. $\frac{1}{6} - \frac{5}{6}$

19. $\frac{20}{21} - \frac{10}{21} - \frac{17}{21}$

20. $\frac{27}{28} - \frac{5}{28} - \frac{28}{28}$

Evaluate each expression given the replacement values. See Example 6.

21. $x + y; x = \frac{3}{4}, y = \frac{2}{4}$

22. $x - y; x = \frac{7}{8}, y = \frac{9}{8}$

23. $x - y; x = -\frac{1}{5}, y = \frac{3}{5}$

24. $x + y; x = -\frac{1}{6}, y = \frac{5}{6}$

25. $x - y + z; x = \frac{3}{12}, y = \frac{5}{12}, z = -\frac{7}{12}$

26. $x + y - z; x = \frac{2}{14}, y = \frac{3}{14}, z = \frac{8}{14}$

ANSWERS

1. ______
2. ______
3. ______
4. ______
5. ______
6. ______
7. ______
8. ______
9. ______
10. ______
11. ______
12. ______
13. ______
14. ______
15. ______
16. ______
17. ______
18. ______
19. ______
20. ______
21. ______
22. ______
23. ______
24. ______
25. ______
26. ______

ANSWERS

27. ______
28. ______
29. ______
30. ______
31. ______
32. ______
33. ______
34. ______
35. ______
36. ______
37. ______
38. ______
39. ______
40. ______
41. ______
42. ______
43. ______
44. ______
45. ______
46. ______
47. ______
48. ______
49. ______
50. ______
51. ______
52. ______
53. ______
54. ______

Determine whether the given value is a solution of the given equation. See Example 7.

27. Is $-\frac{1}{8}$ a solution to $x + \frac{7}{8} = \frac{3}{4}$?

28. Is $\frac{2}{10}$ a solution to $\frac{7}{10} - y = \frac{1}{2}$?

29. Is $\frac{7}{10}$ a solution to $y - \frac{2}{10} = -\frac{1}{2}$?

30. Is $-\frac{1}{8}$ a solution to $x + \frac{9}{8} = 1$?

31. Is $\frac{1}{3}$ a solution to $-\frac{2}{3} + x = -\frac{1}{3}$?

32. Is $\frac{1}{16}$ a solution to $z - \frac{11}{16} = -\frac{5}{8}$?

Find the LCD of the list of fractions. See Examples 8 and 9.

33. $\frac{1}{3}, \frac{3}{4}$

34. $\frac{1}{4}, \frac{5}{6}$

35. $-\frac{2}{9}, \frac{6}{15}$

36. $-\frac{7}{12}, \frac{3}{20}$

37. $\frac{5}{12}, \frac{5}{18}$

38. $\frac{7}{12}, \frac{7}{15}$

39. $-\frac{7}{24}, -\frac{5}{x}$

40. $-\frac{11}{y}, -\frac{13}{70}$

41. $\frac{2}{25}, \frac{3}{15}, \frac{5}{6}$

42. $\frac{3}{4}, \frac{1}{6}, \frac{13}{18}$

Add or subtract the following.

43. $\frac{5}{11} + \frac{2}{11}$

44. $\frac{4}{7} + \frac{2}{7}$

45. $-\frac{1}{15} + \frac{9}{15}$

46. $\frac{3}{15} - \frac{1}{15}$

47. $\frac{9x}{15} + \frac{1x}{15}$

48. $\frac{2x}{15} - \frac{7}{15}$

49. $\frac{7x}{16} - \frac{15x}{16}$

50. $\frac{15b}{16} + \frac{7b}{16}$

51. $\frac{15}{16z} - \frac{3}{16z}$

52. $\frac{7}{16a} + \frac{15}{16a}$

53. $\frac{3}{10} - \frac{6}{10}$

54. $-\frac{6}{10} + \frac{3}{10}$

4.6

COMPLEX FRACTIONS AND REVIEW OF OPERATIONS

OBJECTIVES

1. Simplify complex fractions.
2. Review addition, subtraction, multiplication, and division of fractions.

TAPE PA 4.6

1 Thus far, we have studied operations on fractions. We now practice simplifying fractions whose numerators or denominators themselves contain fractions. These fractions are called **complex fractions.**

COMPLEX FRACTION

A fraction whose numerator or denominator or both numerator and denominator contain fractions is called a complex fraction.

Examples of complex fractions are

$$\frac{\frac{x}{4}}{\frac{3}{2}} \qquad \frac{\frac{1}{2}+\frac{3}{8}}{\frac{3}{4}-\frac{1}{6}} \qquad \frac{\frac{y}{5}-2}{\frac{3}{10}}$$

Two methods are presented to simplify complex fractions. The first method makes use of the fact that a fraction bar means division.

EXAMPLE 1

Simplify $\dfrac{\frac{x}{4}}{\frac{3}{2}}$.

Solution:

Since a fraction bar means division, the complex fraction $\dfrac{\frac{x}{4}}{\frac{3}{2}}$ can be written as $\frac{x}{4} \div \frac{3}{2}$. Then divide as usual to simplify.

$$\frac{x}{4} \div \frac{3}{2} = \frac{x}{4} \cdot \frac{2}{3} \quad \text{Multiply by the reciprocal.}$$

$$= \frac{2 \cdot x}{2 \cdot 2 \cdot 3}$$

$$= \frac{x}{6}$$

PRACTICE PROBLEM 1

Simplify $\dfrac{\frac{7y}{10}}{\frac{1}{5}}$.

Answer:

1. $\frac{7y}{2}$

PRACTICE PROBLEM 2

Simplify $\dfrac{\dfrac{1}{2}+\dfrac{1}{6}}{\dfrac{3}{4}-\dfrac{2}{3}}$.

Answer:

2. $\dfrac{8}{1}$ or 8

EXAMPLE 2

Simplify $\dfrac{\dfrac{1}{2}+\dfrac{3}{8}}{\dfrac{3}{4}-\dfrac{1}{6}}$.

Solution:

Recall the order of operations. Before performing the division we simplify the numerator and the denominator of the complex fraction separately.

$$\dfrac{\dfrac{1}{2}+\dfrac{3}{8}}{\dfrac{3}{4}-\dfrac{1}{6}} = \dfrac{\dfrac{1\cdot 4}{2\cdot 4}+\dfrac{3}{8}}{\dfrac{3\cdot 3}{4\cdot 3}-\dfrac{1\cdot 2}{6\cdot 2}} = \dfrac{\dfrac{4}{8}+\dfrac{3}{8}}{\dfrac{9}{12}-\dfrac{2}{12}} = \dfrac{\dfrac{7}{8}}{\dfrac{7}{12}}$$

Thus,

$$\dfrac{\dfrac{1}{2}+\dfrac{3}{8}}{\dfrac{3}{4}-\dfrac{1}{6}} = \dfrac{\dfrac{7}{8}}{\dfrac{7}{12}}$$

$= \dfrac{7}{8} \div \dfrac{7}{12}$ Rewrite the quotient using the ÷ sign.

$= \dfrac{7}{8} \cdot \dfrac{12}{7}$ Multiply by the reciprocal.

$= \dfrac{7\cdot 3\cdot 4}{2\cdot 4\cdot 7}$ Multiply.

$= \dfrac{3}{2}$ Simplify.

The second method for simplifying complex fractions is to multiply the numerator and the denominator of the complex fraction by the LCD of all the fractions in its numerator and its denominator. Since this LCD is divisible by all denominators, this has the effect of leaving sums and differences of integers in the numerator and the denominator. Let's use this second method to simplify the complex fraction in Example 2 again.

The complex fraction $\dfrac{\dfrac{1}{2}+\dfrac{3}{8}}{\dfrac{3}{4}-\dfrac{1}{6}}$ contains fractions with denominators 2, 8, 4, and 6. The LCD is 24. By the fundamental property of fractions, we can multiply the numerator and the denominator of the complex fraction by 24.

$$\dfrac{\dfrac{1}{2}+\dfrac{3}{8}}{\dfrac{3}{4}-\dfrac{1}{6}} = \dfrac{24\left(\dfrac{1}{2}+\dfrac{3}{8}\right)}{24\left(\dfrac{3}{4}-\dfrac{1}{6}\right)}$$

$$= \frac{\left(24 \cdot \frac{1}{2}\right) + \left(24 \cdot \frac{3}{8}\right)}{\left(24 \cdot \frac{3}{4}\right) - \left(24 \cdot \frac{1}{6}\right)}$$ Apply the distributive property.

$$= \frac{12 + 9}{18 - 4}$$ Multiply.

$$= \frac{21}{14}$$

$$= \frac{7 \cdot 3}{7 \cdot 2} = \frac{3}{2}$$ Simplify.

The simplified result is the same, of course, no matter which method is used.

EXAMPLE 3

Simplify $\dfrac{\frac{y}{5} - 2}{\frac{3}{10}}$.

Solution:

We use the second method and multiply the numerator and the denominator of the complex fraction by the LCD of all fractions. Recall that $2 = \frac{2}{1}$. The LCD of denominators 5, 1, and 10 is 10.

$$\frac{\frac{y}{5} - \frac{2}{1}}{\frac{3}{10}} = \frac{10\left(\frac{y}{5} - \frac{2}{1}\right)}{10\left(\frac{3}{10}\right)}$$ Multiply the numerator and denominator by 10.

$$= \frac{\left(10 \cdot \frac{y}{5}\right) - (10 \cdot 2)}{10 \cdot \frac{3}{10}}$$ Apply the distributive property.

$$= \frac{2y - 20}{3}$$ Multiply.

PRACTICE PROBLEM 3

Simplify $\dfrac{\frac{x}{4}}{\frac{2}{5} - 1}$.

2 At this time, it is probably a good idea to review operations on fractions and order of operations.

OPERATIONS ON FRACTIONS

Let a, b, c, and d be integers.

Addition: $\frac{a}{b} + \frac{c}{b} = \frac{a + c}{b}$ $(b \neq 0)$ ↑ ↑ **common denominators**

Subtraction: $\frac{a}{b} - \frac{c}{b} = \frac{a - c}{b}$ $(b \neq 0)$ ↑ ↑ **common denominators**

Multiplication: $\frac{a}{b} \cdot \frac{c}{d} = \frac{a \cdot c}{b \cdot d}$ $(b \neq 0, d \neq 0)$

Division: $\frac{a}{b} \div \frac{c}{d} = \frac{a}{b} \cdot \frac{d}{c} = \frac{a \cdot d}{b \cdot c}$ $(b \neq 0, d \neq 0, c \neq 0)$

Answer:

3. $\frac{5x}{-12}$ or $-\frac{5x}{12}$

ORDER OF OPERATIONS

Simplify expressions using the following order. If grouping symbols such as parentheses or brackets are present, simplify expressions within those first, starting with the innermost set. If fraction bars are present, simplify above and below the fraction bar separately.

1. Simplify any expressions with exponents.
2. Perform multiplications or divisions in order from left to right.
3. Perform additions or subtractions in order from left to right.

PRACTICE PROBLEM 4

Simplify $\left(2 - \frac{2}{3}\right)^3$.

EXAMPLE 4

Simplify $\left(\frac{4}{5}\right)^2 - 1$.

Solution:

According to the order of operations, first evaluate $\left(\frac{4}{5}\right)^2$.

$$\left(\frac{4}{5}\right)^2 - 1 = \frac{16}{25} - 1$$

Next, combine the fractions. The LCD of 25 and 1 is 25.

$$\frac{16}{25} - 1 = \frac{16}{25} - \frac{25}{25}$$
$$= \frac{-9}{25} \text{ or } -\frac{9}{25}$$

PRACTICE PROBLEM 5

Simplify $\left(-\frac{1}{2} + \frac{1}{5}\right)\left(\frac{7}{8} + \frac{1}{8}\right)$.

EXAMPLE 5

Simplify $\left(\frac{1}{4} + \frac{2}{3}\right)\left(\frac{11}{12} + \frac{1}{4}\right)$.

Solution:

First, perform operations inside parentheses. Then multiply.

$$\left(\frac{1}{4} + \frac{2}{3}\right)\left(\frac{11}{12} + \frac{1}{4}\right) = \left(\frac{1 \cdot 3}{4 \cdot 3} + \frac{2 \cdot 4}{3 \cdot 4}\right)\left(\frac{11}{12} + \frac{1 \cdot 3}{4 \cdot 3}\right) \quad \text{The LCD is 12.}$$
$$= \left(\frac{3}{12} + \frac{8}{12}\right)\left(\frac{11}{12} + \frac{3}{12}\right)$$
$$= \left(\frac{11}{12}\right)\left(\frac{14}{12}\right) \quad \text{Add.}$$
$$= \frac{11 \cdot 2 \cdot 7}{2 \cdot 6 \cdot 12} \quad \text{Multiply.}$$
$$= \frac{77}{72} \quad \text{Simplify.}$$

Answers:

4. $\frac{64}{27}$

5. $-\frac{3}{10}$

EXAMPLE 4

Solve for y: $3y = -\frac{2}{11}$.

Solution:

We can either divide both sides by 3 or multiply both sides by the reciprocal of 3.

$$3y = -\frac{2}{11}$$

$$\frac{1}{3} \cdot 3y = \frac{1}{3} \cdot -\frac{2}{11} \quad \text{Multiply both sides by } \frac{1}{3}.$$

$$1y = -\frac{1 \cdot 2}{3 \cdot 11} \quad \text{Multiply.}$$

$$y = -\frac{2}{33} \quad \text{Simplify.}$$

Check to see that the solution is $-\frac{2}{33}$.

Solving equations with fractions can be tedious. If an equation contains fractions, it is often helpful to first multiply both sides of the equation by the LCD of the fractions. This has the effect of eliminating the fractions in the equation, as shown in the next example.

EXAMPLE 5

Solve for x: $\frac{x}{6} + 1 = \frac{4}{3}$.

Solution:

First, multiply both sides of the equation by the LCD of the fractions. The LCD for denominators 6 and 3 is 6.

$$\frac{x}{6} + 1 = \frac{4}{3}$$

$$6\left(\frac{x}{6} + 1\right) = 6\left(\frac{4}{3}\right) \quad \text{Multiply both sides by 6.}$$

$$6\left(\frac{x}{6}\right) + 6(1) = 6\left(\frac{4}{3}\right) \quad \text{Apply the distributive property.}$$

$$x + 6 = 8 \quad \text{Simplify.}$$

$$x + 6 + (-6) = 8 + (-6) \quad \text{Add } -6 \text{ to both sides.}$$

$$x = 2 \quad \text{Simplify.}$$

To check, replace x with 2 in the original equation.

$$\frac{x}{6} + 1 = \frac{4}{3} \quad \text{Original equation.}$$

$$\frac{2}{6} + 1 = \frac{4}{3} \quad \text{Replace } x \text{ with 2.}$$

$$\frac{1}{3} + \frac{3}{3} = \frac{4}{3} \quad \text{Simplify } \frac{2}{6}\text{. The LCD for 3 and 1 is 3.}$$

$$\frac{4}{3} = \frac{4}{3} \quad \text{True.}$$

Since we arrived at a true statement, 2 is the solution of $\frac{x}{6} + 1 = \frac{4}{3}$.

PRACTICE PROBLEM 4

Solve for x: $5x = -\frac{3}{4}$.

PRACTICE PROBLEM 5

Solve for y: $\frac{y}{8} + \frac{3}{4} = 2$.

Answers:

4. $-\frac{3}{20}$

5. 10

Let's review the steps for solving equations in x. An extra step is now included to handle equations containing fractions.

TO SOLVE AN EQUATION IN X

Step 1. If fractions are present, multiply both sides of the equation by the LCD of the fractions.

Step 2. If parentheses are present, use the distributive property.

Step 3. Combine any like terms on each side of the equation.

Step 4. Use the addition property of equality to rewrite the equation so that variable terms are on one side of the equation and constant terms are on the other side.

Step 5. Divide both sides of the equation by the numerical coefficient of x to solve.

Step 6. Check the answer in the **original equation.**

PRACTICE PROBLEM 6

Solve for x: $\frac{x}{5} - x = \frac{1}{5}$.

EXAMPLE 6

Solve for z: $\frac{z}{5} - \frac{z}{3} = 6$.

Solution:

$$\frac{z}{5} - \frac{z}{3} = 6$$

$$15\left(\frac{z}{5} - \frac{z}{3}\right) = 15(6) \quad \text{Multiply both sides by the LCD 15.}$$

$$15\left(\frac{z}{5}\right) - 15\left(\frac{z}{3}\right) = 15(6) \quad \text{Apply the distributive property.}$$

$$3z - 5z = 90 \quad \text{Simplify.}$$

$$-2z = 90 \quad \text{Combine like terms.}$$

$$\frac{-2z}{-2} = \frac{90}{-2} \quad \text{Divide both sides by } -2\text{, the coefficient of } x.$$

$$z = -45 \quad \text{Simplify.}$$

To check, replace z with -45 in the **original equation** to see that a true statement results.

PRACTICE PROBLEM 7

Solve $\frac{y}{2} = \frac{y}{5} + \frac{3}{2}$.

EXAMPLE 7

Solve $\frac{x}{2} = \frac{x}{3} + \frac{1}{2}$.

Solution:

First, multiply both sides by the LCD 6.

$$\frac{x}{2} = \frac{x}{3} + \frac{1}{2}$$

$$6\left(\frac{x}{2}\right) = 6\left(\frac{x}{3} + \frac{1}{2}\right) \quad \text{Multiply both sides by 6.}$$

Answers:

6. $-\frac{1}{4}$

7. 5

$$6\left(\frac{x}{2}\right) = 6\left(\frac{x}{3}\right) + 6\left(\frac{1}{2}\right)$$ Apply the distributive property.

$$3x = 2x + 3$$ Simplify.

$$3x + (-2x) = 2x + 3 + (-2x)$$ Add $(-2x)$ to both sides.

$$x = 3$$ Simplify.

To check, replace x with 3 in the original equation to see that a true statement results.

Make sure you understand the difference between **solving an equation** containing fractions and **adding or subtracting two fractions.** To solve an equation containing fractions, we use the multiplication property of equality and multiply both sides by the LCD of the fractions, thus eliminating the fractions. This method does not apply to adding or subtracting fractions. The multiplication property of equality applies only to equations. To add or subtract unlike fractions, we write each fraction as an equivalent fraction using the LCD of the fractions as the denominator. See the next example for a review.

EXAMPLE 8

Add: $\frac{x}{3} + \frac{2}{5}$.

Solution:

This expression is not an equation. Here, we are adding two unlike fractions. To add unlike fractions, we need to find the LCD. The LCD for denominators 3 and 5 is 15. Write each fraction as an equivalent fraction with a denominator of 15.

$$\frac{x}{3} + \frac{2}{5} = \frac{x \cdot 5}{3 \cdot 5} + \frac{2 \cdot 3}{5 \cdot 3}$$

$$= \frac{5x}{15} + \frac{6}{15}$$

$$= \frac{5x + 6}{15}$$

PRACTICE PROBLEM 8

Subtract: $\frac{9}{10} - \frac{y}{3}$.

Answer:

8. $\frac{27 - 10y}{30}$

EXERCISE SET 4.8

Write each mixed number as an improper fraction. See Examples 1 and 2.

1. $2\frac{1}{3}$ **2.** $6\frac{3}{4}$ **3.** $3\frac{3}{8}$

4. $2\frac{5}{8}$ **5.** $11\frac{6}{7}$ **6.** $8\frac{9}{10}$

Write each improper fraction as a whole number or a mixed number. See Example 3.

7. $\frac{13}{7}$ **8.** $\frac{42}{13}$ **9.** $\frac{47}{15}$

10. $\frac{65}{12}$ **11.** $\frac{37}{8}$ **12.** $\frac{42}{6}$

Multiply or divide. See Examples 4 and 5.

13. $2\frac{2}{3} \cdot \frac{1}{7}$ **14.** $\frac{5}{9} \cdot 4\frac{1}{5}$ **15.** $8 \div 1\frac{5}{7}$ **16.** $5 \div 3\frac{3}{4}$

17. $3\frac{2}{3} \cdot 1\frac{1}{2}$ **18.** $2\frac{4}{5} \cdot 2\frac{5}{8}$ **19.** $2\frac{2}{3} \div \frac{1}{7}$ **20.** $\frac{5}{9} \div 4\frac{1}{5}$

Add. See Examples 6 through 8.

21. $4\frac{7}{10} + 2\frac{1}{10}$ **22.** $7\frac{4}{9} + 3\frac{2}{9}$ **23.** $15\frac{4}{7} + 9\frac{11}{14}$

24. $23\frac{3}{5} + 8\frac{8}{15}$ **25.** $3\frac{5}{8} + 2\frac{1}{6} + 7\frac{3}{4}$ **26.** $4\frac{1}{3} + 9\frac{2}{5} + 3\frac{1}{6}$

ANSWERS

1. ______
2. ______
3. ______
4. ______
5. ______
6. ______
7. ______
8. ______
9. ______
10. ______
11. ______
12. ______
13. ______
14. ______
15. ______
16. ______
17. ______
18. ______
19. ______
20. ______
21. ______
22. ______
23. ______
24. ______
25. ______
26. ______

ANSWERS

27. ______
28. ______
29. ______
30. ______
31. ______
32. ______
33. ______
34. ______
35. ______
36. ______
37. ______
38. ______
39. ______
40. ______
41. ______
42. ______
43. ______
44. ______
45. ______
46. ______
47. ______
48. ______
49. ______
50. ______
51. ______
52. ______

Subtract. See Examples 9 through 11.

27. $4\frac{7}{10} - 2\frac{1}{10}$

28. $7\frac{4}{9} - 3\frac{2}{9}$

29. $10\frac{13}{14} - 3\frac{4}{7}$

30. $12\frac{5}{12} - 4\frac{1}{6}$

31. $9\frac{1}{5} - 8\frac{6}{25}$

32. $6\frac{2}{13} - 4\frac{7}{26}$

Perform the indicated operations.

33. $2\frac{3}{4} + 1\frac{1}{4}$

34. $5\frac{5}{8} + 2\frac{3}{8}$

35. $15\frac{4}{7} - 9\frac{11}{14}$

36. $23\frac{3}{5} - 8\frac{8}{15}$

37. $3\frac{1}{9} \cdot 2$

38. $4\frac{1}{2} \cdot 3$

39. $1\frac{2}{3} \div 2\frac{1}{5}$

40. $5\frac{1}{5} \div 3\frac{1}{4}$

41. $22\frac{4}{9} + 13\frac{5}{18}$

42. $15\frac{3}{25} - 5\frac{2}{5}$

43. $5\frac{2}{3} - 3\frac{1}{6}$

44. $5\frac{3}{8} - 2\frac{13}{16}$

45. $15\frac{1}{5} + 20\frac{3}{10} + 37\frac{2}{15}$

46. $7\frac{3}{7} + 15 + 20\frac{1}{2}$

47. $6\frac{4}{7} - 5\frac{11}{14}$

48. $47\frac{5}{12} - 23\frac{19}{24}$

49. $4\frac{2}{7} \cdot 1\frac{3}{10}$

50. $6\frac{2}{3} \cdot 2\frac{3}{4}$

51. $6\frac{2}{11} + 3 + 4\frac{10}{33}$

52. $3\frac{7}{16} + 6\frac{1}{2} + 9\frac{3}{8}$

DEFINITIONS AND CONCEPTS	EXAMPLES
SECTION 4.4 ADDING AND SUBTRACTING LIKE FRACTIONS AND LEAST COMMON DENOMINATOR	
Fractions that have a common denominator are called **like fractions.**	Like fractions: $-\frac{1}{3}$ and $\frac{2}{3}$; $\frac{5x}{7}$ and $\frac{6}{7}$
To add or subtract like fractions, $\frac{a}{b}+\frac{c}{b}=\frac{a+c}{b}$, $\frac{a}{b}-\frac{c}{b}=\frac{a-c}{b}$, $b \neq 0$	Find $\frac{10}{11}+\frac{3}{11}-\frac{8}{11}=\frac{10+3-8}{11}=\frac{5}{11}$.
The **least common denominator (LCD)** of a list of fractions is the smallest positive number divisible by all the denominators in the list.	Find the LCD of $\frac{7}{30}$ and $\frac{1}{12}$. $30 = 2 \cdot 3 \cdot 5$ $12 = 2 \cdot 2 \cdot 3$ Notice that the greatest number of times that 2 appears is 2 times. $\text{LCD} = 2 \cdot 2 \cdot 3 \cdot 5 = 60$
SECTION 4.5 ADDING AND SUBTRACTING FRACTIONS WITH UNLIKE DENOMINATORS	
To add or subtract fractions with unlike denominators,	Add: $\frac{3}{20}+\frac{2}{5}$.
Step 1. Find the LCD.	*Step 1.* The LCD is 20.
Step 2. Write equivalent fractions with the LCD as denominator.	*Step 2.* $\frac{2}{5}=\frac{2 \cdot 4}{5 \cdot 4}=\frac{8}{20}$.
Step 3. Add or subtract the like fractions.	*Step 3.* $\frac{3}{20}+\frac{2}{5}=\frac{3}{20}+\frac{8}{20}=\frac{11}{20}$.
Step 4. Write the result in lowest terms.	*Step 4.* $\frac{11}{20}$ is in lowest terms.
SECTION 4.6 COMPLEX FRACTIONS AND REVIEW OF OPERATIONS	
A fraction whose numerator or denominator or both contain fractions is called a **complex fraction.**	Complex Fractions $\frac{\frac{x}{4}}{\frac{7}{10}}$ $\frac{\frac{y}{6}-11}{\frac{4}{3}}$
One method for simplifying complex fractions is to multiply the numerator and the denominator of the complex fraction by the LCD of all fractions in its numerator and its denominator.	$\frac{\frac{y}{6}-11}{\frac{4}{3}}=\frac{6\left(\frac{y}{6}-11\right)}{6\left(\frac{4}{3}\right)}=\frac{6\left(\frac{y}{6}\right)-6(11)}{6\left(\frac{4}{3}\right)}$ $=\frac{y-66}{8}$
SECTION 4.7 SOLVING EQUATIONS CONTAINING FRACTIONS	
MULTIPLICATION AND DIVISION PROPERTIES OF EQUALITY Let a, b, and c be numbers, and $c \neq 0$. If $a = b$, then $a \cdot c = b \cdot c$ and $\frac{a}{c}=\frac{b}{c}$	

(continued)

DEFINITIONS AND CONCEPTS	EXAMPLES
SECTION 4.7 SOLVING EQUATIONS CONTAINING FRACTIONS	
To Solve an Equation in x	
Step 1. If fractions are present, multiply both sides of the equation by the LCD of the fractions.	Solve $\frac{x}{15} + 2 = \frac{7}{3}$. $15\left(\frac{x}{15} + 2\right) = 15\left(\frac{7}{3}\right)$ Multiply by the LCD 15.
Step 2. If parentheses are present, use the distributive property.	$15\left(\frac{x}{15}\right) + 15 \cdot 2 = 15\left(\frac{7}{3}\right)$
Step 3. Combine any like terms on each side of the equation.	$x + 30 = 35$
Step 4. Use the addition property of equality to rewrite the equation so that variable terms are on one side of the equation and constant terms are on the other side.	$x + 30 + (-30) = 35 + (-30)$ $x = 5$
Step 5. Divide both sides by the numerical coefficient of x to solve.	
Step 6. Check the answer in the **original equation.**	Check to see that 5 is the solution.
SECTION 4.8 OPERATIONS ON MIXED NUMBERS	
A **mixed number** is the sum of a whole number and a proper fraction.	Mixed Numbers $1\frac{2}{5}, 4\frac{7}{8}$
TO WRITE A MIXED NUMBER AS AN IMPROPER FRACTION	$4\frac{7}{8} = \frac{8 \cdot 4 + 7}{8} = \frac{39}{8}$
Step 1. Multiply the denominator of the fractional part by the whole-number part. Add the numerator of the fraction to this product.	
Step 2. Write this sum as the numerator of the improper fraction over the original denominator.	
TO WRITE AN IMPROPER FRACTION AS A MIXED NUMBER OR WHOLE NUMBER	$\frac{29}{7} = 4\frac{1}{7}$
Divide the denominator into the numerator. The whole-number part of the quotient is the whole-number part of the mixed number and the $\frac{\text{remainder}}{\text{divisor}}$ is the fractional part.	$7\overline{)29}$ quotient $4\frac{1}{7}$ $\underline{28}$ 1
To perform operations on mixed numbers, first write each mixed number as an improper fraction.	Multiply: $1\frac{3}{4} \cdot 2\frac{1}{5}$. $\frac{7}{4} \cdot \frac{11}{5} = \frac{77}{20} = 3\frac{17}{20}$
Addition and subtraction of mixed numbers can be performed using a vertical format.	Add: $2\frac{1}{2} + 5\frac{7}{8}$. $2\frac{1}{2} = 2\frac{4}{8}$ $+ 5\frac{7}{8} = 5\frac{7}{8}$ $7\frac{11}{8} = 7 + 1\frac{3}{8} = 8\frac{3}{8}$

112. $4\frac{1}{6} \cdot 2\frac{2}{5}$

113. $5\frac{2}{3} \cdot 2\frac{1}{4}$

114. $6\frac{3}{4} \div 1\frac{2}{7}$

115. $5\frac{1}{2} \div 2\frac{1}{11}$

116. $\frac{7}{2} \div 1\frac{1}{2}$

117. $1\frac{3}{5} \div \frac{1}{4}$

118. Two packages of soup bones weigh $3\frac{3}{4}$ pounds and $2\frac{3}{5}$ pounds. Find their combined weight.

119. A ribbon $5\frac{1}{2}$ yards long is cut from a reel of ribbon with 50 yards on it. Find the length of the piece remaining on the reel.

120. The average annual snowfall at a certain ski resort is $62\frac{3}{10}$ inches. Last year it had $54\frac{1}{2}$ inches. Find how many inches below average last year's annual snowfall was.

121. Find the area of a rectangular sheet of gift wrap that is $2\frac{1}{4}$ feet by $3\frac{1}{3}$ feet.

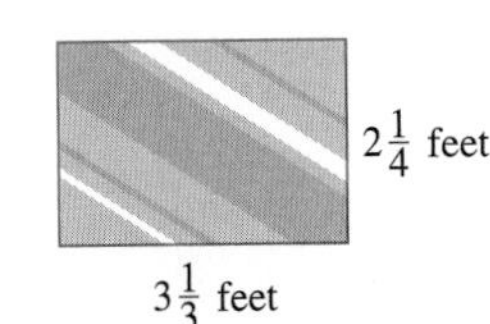

122. Find the perimeter of a sheet of shelf paper needed to fit exactly a square drawer $1\frac{1}{4}$ feet long on each side.

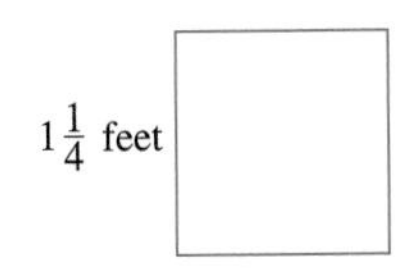

123. A can of fruit contains $15\frac{3}{5}$ ounces. A can of a different brand contains $15\frac{5}{8}$ ounces of sauce. Determine which brand contains more and by how much.

124. A flower gardener has a square flower bed $\frac{2}{3}$ meter on a side and a rectangular one that is $\frac{7}{8}$ by $\frac{1}{3}$ meters. Find the total perimeter of his flower beds.

125. There are 58 calories in 1 ounce of turkey. Find how many calories there are in a $3\frac{1}{2}$-ounce serving of turkey.

126. There are $3\frac{1}{3}$ grams of fat in each ounce of lean hamburger. Find how many grams of fat are in a 4-ounce hamburger.

127. Mr. Heltznutt walks at a pace of $2\frac{1}{2}$ miles per hour. Find how long it takes him to walk 7 miles at this pace.

ANSWERS

112. ______

113. ______

114. ______

115. ______

116. ______

117. ______

118. ______

119. ______

120. ______

121. ______

122. ______

123. ______

124. ______

125. ______

126. ______

127. ______

59. Write the numbers in order from smallest to largest. 9.672 9.68 9.67 9.682

60. Write the numbers in order from smallest to largest. 0.83 0.083 0.726 0.72

Round each decimal to the given place value.

61. 17.667, nearest hundredth

62. 0.766, nearest hundredth

63. 0.501, nearest tenth

64. 0.602, nearest tenth

65. 0.1295, nearest thousandth

66. 0.8295, nearest thousandth

67. 3829.34, nearest ten

68. 4520.876, nearest hundred

69. Which number(s) rounds to 0.26? 0.26559 0.26499 0.25786 0.25186

70. Which number(s) rounds to 0.06? 0.0612 0.066 0.0586 0.0506

Solve.

71. The price of a new jacket is twenty-nine dollars and sixty-two cents. Write this decimal in standard form.

72. The stock market average rose by twelve and six-tenths one day. Write this in standard decimal form.

73. On September 18, gasoline was nine hundred sixty-nine thousandths dollars per gallon in Podunk, Illinois. Write this decimal in standard form.

74. A used office desk is advertised at \$19.95 by Drawley's Office Furniture. Round this price to the nearest dollar.

75. The attendance at a Mets baseball game was reported to be 39,867. Round this number to the nearest thousand.

76. The weekly salaries of the three clerks in the Lease-A-Limo car rental agency were \$467.72 (José), \$456.83 (Sarah), and \$467.83 (Leona). List the clerks in order of salary, from highest to lowest.

77. During the 1995 Boston Marathon, Uta Pippig of Germany was the first woman to cross the finish line. Her time was 2.41972 hours. Round this time to the nearest hundredth. (*Source: 1996 World Almanac*)

78. The population density of the state of Ohio is roughly 271.0961 people per square mile. Round this population density to the nearest tenth. (*Source:* U.S. Bureau of the Census)

ANSWERS

59. ______
60. ______
61. ______
62. ______
63. ______
64. ______
65. ______
66. ______
67. ______
68. ______
69. ______
70. ______
71. ______
72. ______
73. ______
74. ______
75. ______
76. ______
77. ______
78. ______

ANSWERS

79. ______

80. ______

81. ______

82. ______

83. ______

84. ______

85. ______

86. ______

87. ______

88. ______

89. ______

90. ______

91. ______

92. ______

❑ **79.** Write a 5-digit number that rounds to 1.7.

❑ **80.** Write a 6-digit number that rounds to 28.60.

❑ **81.** Explain how to identify the value of the 9 in the decimal 486.3297.

❑ **82.** In your own words write a step-by-step procedure telling how to write a decimal number as a mixed number.

❑ **83.** Explain how to compare the size of two decimals.

❑ **84.** Check to see whether the procedure you wrote in Exercise 83 works for negative decimals, for example −6.28 and −6.23.

Review Exercises

Perform the indicated operation. See Sections 1.2 and 1.3.

85. 45 + 23

86. 57 + 33

87. 3452 + 2314

88. 8945 + 4536

89. 94 − 23

90. 82 − 47

91. 482 − 239

92. 4002 − 3897

57. $\begin{array}{r} 100.009 \\ 6.08 \\ +\ 9.034 \\ \hline \end{array}$

58. $\begin{array}{r} 200.89 \\ 7.49 \\ +\ 62.83 \\ \hline \end{array}$

59. $\begin{array}{r} 1000 \\ -\ 123.4 \\ \hline \end{array}$

60. $\begin{array}{r} 2000 \\ -\ 327.47 \\ \hline \end{array}$

61. $-0.003 + 0.091$

62. $-0.004 + 0.085$

63. $500 - 34.098$

64. $300 - 98.345$

Find the following when $x = 3.6$, $y = 5$, *and* $z = 0.21$.

65. $x + z$

66. $y + x$

67. $x - z$

68. $y - z$

69. $y - x$

70. $x + y + z$

Check to see if the given values are solutions to the given equations.

71. Is 1.4 a solution to $6 + x = 7.4$?

72. Is 2.8 a solution to $3 + y = 5.8$?

73. Is 0.8 a solution to $5.7 - y = 4.9$?

74. Is 2.9 a solution to $x + 0.9 = 3.8$?

75. Is 1 a solution to $2.3 + x = 5.3 - x$?

76. Is 0.9 a solution to $1.9 - x = x + 0.1$?

Solve.

77. The Black Bart Mutual Fund sold for \$29.34 per share on Wednesday and for \$31.03 on Thursday. How much did it increase per share from Wednesday to Thursday?

78. The price of oil was \$19.32 a barrel on May 1 and increased by \$0.97 on May 2. Find its price on May 2.

79. Tim Larson is building a horse corral shaped like a rectangle with dimensions 24.28 meters by 15.675 meters. (A meter is a unit of length in the metric system.) He plans to make a four-wire fence; that is, he will string four wires around the corral. How much wire will he need?

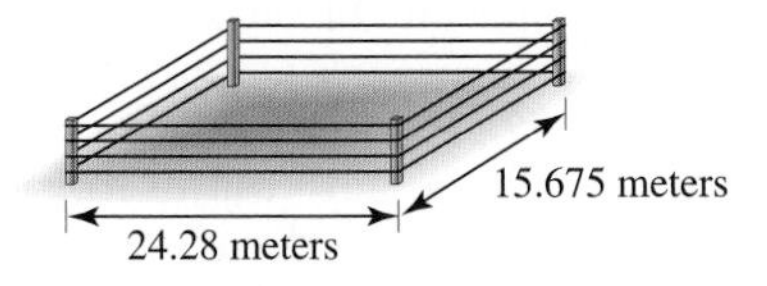

80. Laser beams can be used to measure the distance to the moon. One measurement showed the distance to the moon to be 256,435.235 miles. A later measurement showed that the distance is 256,436.012 miles. Find how much farther away the moon is in the second measurement compared to the first.

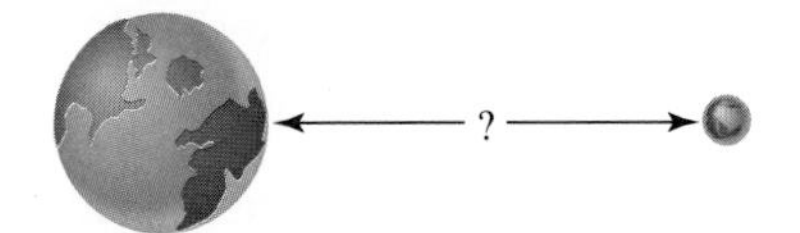

81. Gasoline was \$1.039 per gallon on one day and \$0.979 per gallon the next day. Find by how much the price changed.

82. A pair of eyeglasses costs \$347.89. The frames of the glasses are \$97.23. Find how much the lenses of the eyeglasses cost.

ANSWERS

57. ______
58. ______
59. ______
60. ______
61. ______
62. ______
63. ______
64. ______
65. ______
66. ______
67. ______
68. ______
69. ______
70. ______
71. ______
72. ______
73. ______
74. ______
75. ______
76. ______
77. ______
78. ______
79. ______
80. ______
81. ______
82. ______

ANSWERS

83. ______

84. ______

85. ______

86. ______

87. ______

88. ______

89. ______

90. ______

83. A landscape architect is planning a border for a flower garden shaped like a triangle. The sides of the garden measure 12.4 feet, 29.34 feet, and 25.7 feet. Find the length of border material needed.

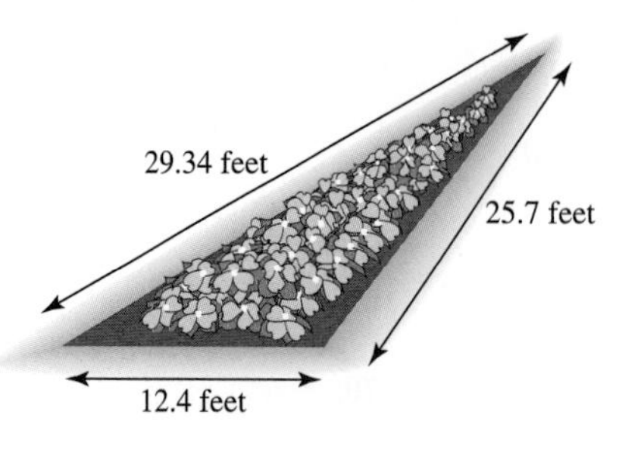

84. Find the perimeter of the deck.

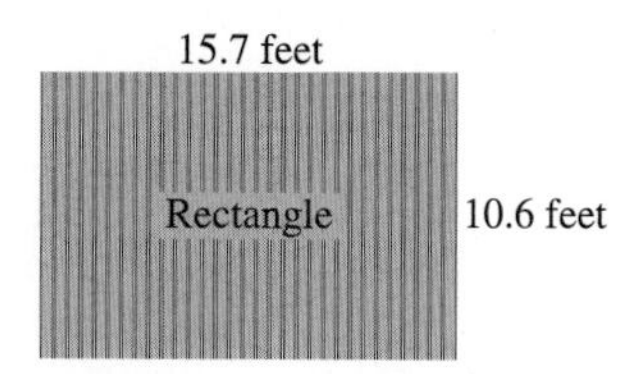

Use the tables given to answer the questions.

INDIANAPOLIS 500 WINNERS		
YEAR	WINNER	AVERAGE SPEED
1995	Jacques Villeneuve	153.616 mph
1950	Johnny Parsons	124.002 mph
1911	Ray Harroun	74.602 mph
Source: Indianapolis Motor Speedway		

85. How much faster was the average Indianapolis 500 speed in 1995 than in 1950?

86. How much faster was the average Indianapolis 500 speed in 1995 than in 1911?

SPACEFLIGHTS OF JAMES A. LOVELL		
YEAR	MISSION	DURATION (IN HOURS)
1965	Gemini 6	330.583
1966	Gemini 12	94.567
1968	Apollo 8	147.0
1970	Apollo 13	142.9
Source: NASA		

87. Find the total time spent in spaceflight by astronaut James A. Lovell.

88. Find the total time James A. Lovell spent in spaceflight on all Apollo missions.

89. You bought two nontaxable food items at the grocery store: a loaf of bread for $1.89 and a gallon of milk for $2.19. How much change should you receive if you pay with a $10 bill? If you received a five-dollar bill, three quarters, three nickels, and two pennies in change, did you receive the correct change?

90. You bought a new outfit for $28.62 including sales tax. If you pay for your purchase with two $20 bills, how much change should you receive? If you received a $10 bill, a one-dollar bill, a dime, and three pennies in change, did you receive the correct change?

Use a calculator to combine like terms and simplify.

91. $-8.689 + 4.286x - 14.295 - 12.966x + 30.861x$

92. $14.271 - 8.968x + 1.333 - 201.815x + 101.239x$

93. Explain how adding or subtracting decimals is similar to adding or subtracting whole numbers.

94. Can the sum of two negative decimals ever be a positive decimal? Why or why not?

Review Exercises

Multiply. See Sections 1.5 and 4.3.

95. $23 \cdot 2$

96. $46 \cdot 3$

97. $43 \cdot 90$

98. $30 \cdot 32$

99. $\left(\frac{2}{3}\right)^2$

100. $\left(\frac{1}{5}\right)^3$

101. $\frac{12}{7} \cdot \frac{14}{3}$

102. $\frac{25}{36} \cdot \frac{24}{40}$

ANSWERS

91. ______

92. ______

93. ______

94. ______

95. ______

96. ______

97. ______

98. ______

99. ______

100. ______

101. ______

102. ______

5.3

Multiplying Decimals and Circumference of a Circle

TAPE PA 5.3

OBJECTIVES

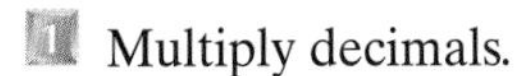

1. Multiply decimals.
2. Multiply by powers of 10.
3. Find the circumference of a circle.

1 Multiplying decimals is similar to multiplying whole numbers. The only difference is that we place a decimal point in the product. To discover where a decimal point is placed in the product, multiply 0.6×0.03. We first write each decimal as an equivalent fraction and then multiply.

$$\underset{\substack{\uparrow \\ \textbf{1}\text{ decimal place}}}{0.6} \times \underset{\substack{\uparrow \\ \textbf{2}\text{ decimal places}}}{0.03} = \frac{6}{10} \times \frac{3}{100}$$

$$= \frac{18}{1000}$$

$$= 0.018 \leftarrow \textbf{3}\text{ decimal places}$$

Next, multiply 0.03×0.002

$$\underset{\substack{\uparrow \\ \textbf{2}\text{ decimal places}}}{0.03} \times \underset{\substack{\uparrow \\ \textbf{3}\text{ decimal places}}}{0.002} = \frac{3}{100} \times \frac{2}{1000}$$

$$= \frac{6}{100{,}000}$$

$$= 0.00006 \leftarrow \textbf{5}\text{ decimal places}$$

Instead of writing decimals as fractions each time we want to multiply, we notice a pattern from these examples and state a rule that we can use.

TO MULTIPLY DECIMALS

Step 1. Multiply the decimals as though they are whole numbers.

Step 2. The decimal point in the product is placed so the number of decimal places in the product is equal to the **sum** of the number of decimal places in the factors.

PRACTICE PROBLEM 1
Multiply: 45.9×0.42.

EXAMPLE 1
Multiply: 23.6×0.78.

Solution:

$$\begin{array}{r l} 23.6 & \text{1 decimal place.} \\ \times\ 0.78 & \text{2 decimal places.} \\ \hline 1888 & \\ 1652 & \\ \hline 18.408 & \text{3 decimal places.} \end{array}$$

PRACTICE PROBLEM 2
Multiply: 0.512×0.6.

EXAMPLE 2
Multiply: 0.283×0.3.

Solution:

$$\begin{array}{r l} 0.283 & \text{3 decimal places.} \\ \times\ \ \ 0.3 & \text{1 decimal place.} \\ \hline 0.0849 & \text{4 decimal places.} \end{array}$$

↑ A 0 is inserted since the product must have 4 decimal places.

PRACTICE PROBLEM 3
Find 0.0721×48.

EXAMPLE 3
Find 0.0531×16.

Solution:

$$\begin{array}{r l} 0.0531 & \text{4 decimal places.} \\ \times\ \ \ \ 16 & \text{0 decimal places.} \\ \hline 3186 & \\ 0531 & \\ \hline 0.8496 & \text{4 decimal places.} \end{array}$$

PRACTICE PROBLEM 4
Multiply: $(5.4)(-1.3)$.

EXAMPLE 4
Multiply: $(-2.6)(0.8)$.

Solution:

Recall that the product of a negative number and a positive number is a negative number.

$$(-2.6)(0.8) = -2.08$$

2 There are some patterns that occur when we multiply a number by a power of ten, such as 10, 100, 1000, 10,000, and so on.

$23.6951 \times 10 = 236.951$ Decimal point moved **1 place** to the **right.**
↑ **1** zero

$23.6951 \times 100 = 2369.51$ Decimal point moved **2 places** to the **right.**
↑ **2** zeros

$23.6951 \times 100{,}000 = 2{,}369{,}510.$ Decimal point moved **5 places** to the **right** (a 0 is inserted).
↑ **5** zeros

Notice that the decimal point is moved the same number of places as there are zeros in the power of 10.

Answers:
1. 19.278
2. 0.3072
3. 3.4608
4. −7.02

5.5

ESTIMATING AND REVIEW OF OPERATIONS

OBJECTIVES

1 Estimate operations on decimals.

2 Simplify expressions involving combinations of operations on decimals.

TAPE PA 5.5

1 Estimating sums, differences, products, and quotients of decimal numbers is an important skill whether you use a calculator or perform decimal operations by hand. When you can estimate results as well as calculate them, you can judge whether the calculations are reasonable. If they aren't, you know you've made an error somewhere along the way.

EXAMPLE 1

Subtract: \$78.62 − \$16.85. Then estimate the difference to see if the proposed result is reasonable by rounding each decimal to the nearest dollar and then subtracting.

Solution:

GIVEN		ESTIMATE
\$78.62	rounds to	\$79
−\$16.85	rounds to	−\$17
\$61.77		\$62

The estimated difference is \$62, so \$61.77 is reasonable.

PRACTICE PROBLEM 1

Subtract: \$65.34 − \$14.68. Then estimate the difference to see if the proposed result is reasonable.

EXAMPLE 2

Multiply: 28.06 × 1.95. Then estimate the product to see if the proposed result is reasonable.

Solution:

GIVEN	ESTIMATE 1	ESTIMATE 2
28.06	28	30
× 1.95	× 2	× 2
14030	56	60
25254		
2806		
54.7170		

The answer 54.7170 is reasonable.

PRACTICE PROBLEM 2

Multiply: 30.26 × 2.98. Then estimate this product to see if the proposed solution is reasonable.

As shown in Example 2, estimated results will vary depending on what estimates are used. Notice that estimating results is a good way to see whether the decimal point has been correctly placed.

Answers:
1. \$50.66
2. 90.1748

PRACTICE PROBLEM 3

Divide: 713.7 ÷ 91.5. Then estimate the quotient to see if the proposed answer is reasonable.

EXAMPLE 3

Divide: 272.356 ÷ 28.4. Then estimate the quotient to see if the proposed result is reasonable.

Solution:

GIVEN

```
        9.59
284.)2723.56
     2556
      1675
      1420
       2556
       2556
          0
```

ESTIMATE

```
    9              9
30)270   or   300)2700
```

The estimate is 9, so 9.59 is reasonable.

PRACTICE PROBLEM 4

Using the figure for Example 4, estimate to the nearest mile how much farther it is between Dodge City and Pratt than it is between Garden City and Dodge City.

EXAMPLE 4

Estimate the distance in miles between Garden City, Kansas, and Wichita, Kansas, by rounding each given distance to the nearest ten.

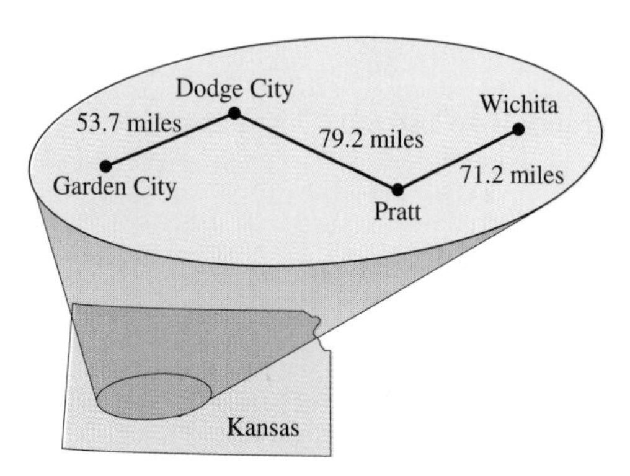

Solution:

CALCULATED DISTANCE		ESTIMATE
53.7	rounds to	50
79.2	rounds to	80
+ 71.2	rounds to	+ 70
		200

The distance between Garden City and Wichita is approximately 200 miles. (The calculated distance is 204.1 miles.)

Answers:
3. 7.8
4. 26 miles

2 The remaining examples review operations on decimals by simplifying expressions that contain more than one operation.

EXAMPLE 5

Simplify $-0.5(8.6 - 1.2)$.

Solution:

Recall the order of operations and simplify inside parentheses first.

$$-0.5(8.6 - 1.2) = -0.5(7.4)$$

Next multiply. Remember that the product of a negative number and a positive number is a negative number.

$$-0.5(7.4) = -3.7$$

PRACTICE PROBLEM 5

Simplify $8.69(80 - 180)$.

EXAMPLE 6

Simplify $(-1.3)^2$.

Solution:

Recall the meaning of an exponent.

$$(-1.3)^2 = (-1.3)(-1.3)$$

Next multiply. Remember that the product of two negative numbers is a positive number.

$$(-1.3)(-1.3) = 1.69$$

PRACTICE PROBLEM 6

Simplify $(-0.7)^2$.

EXAMPLE 7

Simplify $\frac{0.7 + 1.84}{0.4}$.

Solution:

First, simplify the numerator of the fraction; then divide.

$$\frac{0.7 + 1.84}{0.4} = \frac{2.54}{0.4} \quad \text{Simplify numerator.}$$

$$= 6.35 \quad \text{Divide.}$$

PRACTICE PROBLEM 7

Simplify $\frac{8.78 - 2.8}{-20}$.

EXAMPLE 8

Evaluate $-2x + 5$ when $x = 3.8$.

Solution:

Replace x with 3.8 in the expression $-2x + 5$ and simplify.

$$-2x + 5 = -2(3.8) + 5 \quad \text{Replace } x \text{ with 3.8.}$$

$$= -7.6 + 5 \quad \text{Multiply.}$$

$$= -2.6 \quad \text{Add.}$$

PRACTICE PROBLEM 8

Evaluate $1.7y - 2$ when $y = 2.3$.

Answers:

5. -869
6. 0.49
7. -0.299
8. 1.91

PRACTICE PROBLEM 9

Is -2.1 a solution of $3x + 7.5 = 1.2$?

EXAMPLE 9

Determine whether -3.3 is a solution of $2x - 1.2 = -7.8$.

Solution:

Replace x with -3.3 in the equation $2x - 1.2 = -7.8$ and see whether a true statement results.

$$2x - 1.2 = -7.8$$

$$2(-3.3) - 1.2 = -7.8 \quad \text{Replace } x \text{ with } -3.3.$$

$$-6.6 - 1.2 = -7.8 \quad \text{Multiply.}$$

$$-7.8 = -7.8 \quad \text{True.}$$

Since $-7.8 = -7.8$ is a true statement, -3.3 is a solution of $2x - 1.2 = -7.8$.

Answer:

9. yes

5.6

FRACTIONS AND DECIMALS

TAPE PA 5.6

OBJECTIVES

1. Write fractions as decimals.
2. Write decimals as fractions.
3. Compare fractions and decimals.

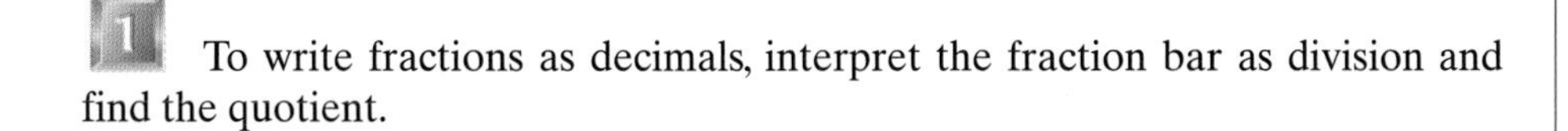

To write fractions as decimals, interpret the fraction bar as division and find the quotient.

> **To write fractions as decimals, divide the numerator by the denominator.**

EXAMPLE 1

Write $\frac{1}{4}$ as a decimal.

Solution:

$$\begin{array}{r} 0.25 \\ 4\overline{)1.00} \\ \underline{8} \\ 20 \\ \underline{20} \\ 0 \end{array}$$

$$\frac{1}{4} = 0.25$$

PRACTICE PROBLEM 1

Write $\frac{2}{5}$ as a decimal.

EXAMPLE 2

Write $\frac{2}{3}$ as a decimal.

Solution:

$$\begin{array}{r} 0.666 \\ 3\overline{)2.000} \\ \underline{1\,8} \\ 20 \\ \underline{18} \\ 20 \\ \underline{18} \\ 2 \end{array}$$

This pattern will continue so that $\frac{2}{3} = 0.6666\ldots$.

A bar can be placed over the digit 6 to indicate that it repeats.

$$\frac{2}{3} = 0.666\ldots = 0.\overline{6}$$

PRACTICE PROBLEM 2

Write $\frac{5}{6}$ as a decimal.

Answers:
1. 0.4
2. $0.8\overline{3} \approx 0.83$

We can also write a decimal approximation for $\frac{2}{3}$. For example, $\frac{2}{3}$ rounded to the nearest hundredth is 0.67. This can be written as $\frac{2}{3} \approx 0.67$.

PRACTICE PROBLEM 3

Write $\frac{1}{9}$ as a decimal. Round to the nearest thousandth.

EXAMPLE 3

Write $\frac{22}{7}$ as a decimal. Round to the nearest hundredth. (The fraction $\frac{22}{7}$ is an approximation for π.)

Solution:

$$\begin{array}{r} 3.142 \approx 3.14 \\ 7\overline{)22.000} \\ \underline{21} \\ 1\,0 \\ \underline{7} \\ 30 \\ \underline{28} \\ 20 \\ \underline{14} \\ 6 \end{array}$$

Carry division out until the thousandths place.

The fraction $\frac{22}{7}$ in decimal form is approximately 3.14.

2 In an earlier section we learned how to write decimals as fractions by using place values.

PRACTICE PROBLEM 4

Write 0.059 as a fraction.

EXAMPLE 4

Write 0.08 as a fraction.

Solution:

0.08 is 8 hundredths, so

$$\underset{\substack{\uparrow \\ \textbf{2}\text{ decimal places}}}{0.08} = \underset{\textbf{2}\text{ zeros}}{\frac{8}{100}} = \frac{2}{25}$$

PRACTICE PROBLEM 5

Write 43.89 as a mixed number.

EXAMPLE 5

Write 39.25 as a mixed number.

Solution:

Write the decimal part, 0.25, as a fraction.

$$\mathbf{0.25} = \frac{\mathbf{25}}{100} = \frac{1}{4}$$

Then $39.25 = 39\frac{1}{4}$.

Answers:

3. 0.111

4. $\frac{59}{1000}$

5. $43\frac{89}{100}$

3 Now we can compare decimals and fractions by writing fractions as equivalent decimals.

EXAMPLE 6

Insert <, >, or = to form a true statement.

$\frac{1}{8}$ 0.12

Solution:

Write $\frac{1}{8}$ as an equivalent decimal and compare decimal places.

Original numbers	$\frac{1}{8}$	0.12
Decimals	0.125	0.12
Compare	0.125 > 0.12	

$\frac{1}{8} > 0.12$

PRACTICE PROBLEM 6

Insert <, >, or = to form a true statement.

$\frac{1}{5}$ 0.25

EXAMPLE 7

Write the numbers in order from smallest to largest.

$\frac{9}{20}$, $\frac{4}{9}$, 0.456

Solution:

Original numbers	$\frac{9}{20}$	$\frac{4}{9}$	0.456
Decimals	0.450	0.444 . . .	0.456
Compare in order	2nd	1st	3rd

PRACTICE PROBLEM 7

Write the numbers in order from smallest to largest.

$\frac{1}{3}$, 0.302, $\frac{3}{8}$

Answers:

6. $\frac{1}{5} < 0.25$

7. 0.302, $\frac{1}{3}$, $\frac{3}{8}$

Then

$$\frac{4}{9}, \frac{9}{20}, 0.456$$

Next, we solve a problem that involves fractions and decimals.

PRACTICE PROBLEM 8

Find the area of the triangle.

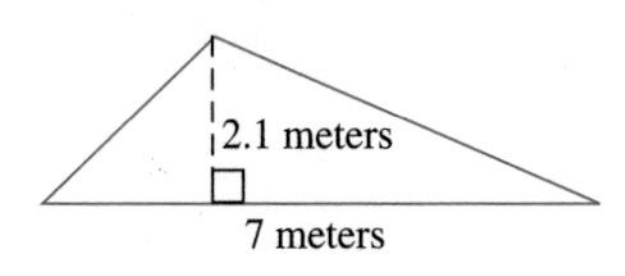

EXAMPLE 8

The formula for finding the area of a triangle is Area $= \frac{1}{2} \cdot$ base $\cdot$ height. Find the area of the triangle shown.

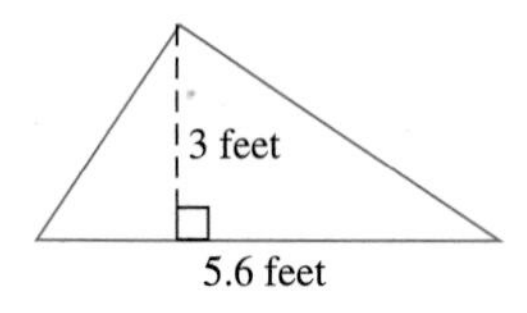

Solution:

$$\begin{aligned} \text{Area} &= \frac{1}{2} \cdot \text{base} \cdot \text{height} \\ &= \frac{1}{2} \cdot 5.6 \cdot 3 \\ &= 0.5 \cdot 5.6 \cdot 3 \quad \text{Write } \frac{1}{2} \text{ as the decimal 0.5.} \\ &= 8.4 \end{aligned}$$

The area of the triangle is 8.4 square feet.

Answer:
8. 7.35 square meters

5.7

EQUATIONS CONTAINING DECIMALS

TAPE PA 5.7

OBJECTIVE

1 Solve equations with decimals.

1 In this section, we continue our work with decimals and algebra by solving equations containing decimals. First, we review the steps given earlier for solving an equation.

STEPS FOR SOLVING AN EQUATION IN *X*

Step 1. If fractions are present, multiply both sides of the equation by the LCD of the fractions.

Step 2. If parentheses are present, use the distributive property.

Step 3. Combine any like terms on each side of the equation.

Step 4. Use the addition property of equality to rewrite the equation so that variable terms are on one side of the equation and constant terms are on the other side.

Step 5. Divide both sides by the numerical coefficient of x to solve.

Step 6. Check the answer in the **original equation.**

EXAMPLE 1

Solve for x: $x - 1.5 = 8$.

Solution:

To isolate x (get x by itself on one side of the equation), add 1.5 to both sides.

$x - 1.5 = 8$ Original equation.

$x - 1.5 + 1.5 = 8 + 1.5$ Add 1.5 to both sides.

$x = 9.5$ Simplify.

To check, replace x with 9.5 in the **original equation.**

$x - 1.5 = 8$ Original equation.

$9.5 - 1.5 = 8$ Replace x with 9.5.

$8 = 8$ True.

Since $8 = 8$ is a true statement, 9.5 is a solution of the equation.

PRACTICE PROBLEM 1

Solve for z: $z + 0.9 = 1.3$.

Answer:
1. 0.4

PRACTICE PROBLEM 2
Solve for x: $0.17x = -0.34$.

EXAMPLE 2
Solve for y: $-2y = 6.7$.

Solution:
To solve for y, divide both sides by the coefficient of y, which is -2.

$$-2y = 6.7 \quad \text{Original equation.}$$
$$\frac{-2y}{-2} = \frac{6.7}{-2} \quad \text{Divide both sides by } -2.$$
$$y = -3.35 \quad \text{Simplify.}$$

To check, replace y with -3.35 in the original equation.

$$-2y = 6.7 \quad \text{Original equation.}$$
$$-2(-3.35) = 6.7 \quad \text{Replace } y \text{ with } -3.35.$$
$$6.7 = 6.7 \quad \text{True.}$$

Thus -3.35 is a solution of the equation $-2y = 6.7$.

PRACTICE PROBLEM 3
Solve for x:
$6.3 - 5x = 3(x + 2.9)$.

EXAMPLE 3
Solve for x: $5(x - 0.36) = -x + 2.4$.

Solution:
First, use the distributive property to distribute the factor 5.

$$5(x - 0.36) = -x + 2.4 \quad \text{Original equation.}$$
$$5x - 1.8 = -x + 2.4 \quad \text{Apply the distributive property.}$$

Next, isolate x on the left side of the equation by adding 1.8 to both sides of the equation and then adding x to both sides of the equation.

$$5x - 1.8 + 1.8 = -x + 2.4 + 1.8 \quad \text{Add 1.8 to both sides.}$$
$$5x = -x + 4.2 \quad \text{Simplify.}$$
$$5x + x = -x + 4.2 + x \quad \text{Add } x \text{ to both sides.}$$
$$6x = 4.2 \quad \text{Simplify.}$$
$$\frac{6x}{6} = \frac{4.2}{6} \quad \text{Divide both sides by 6.}$$
$$x = 0.7 \quad \text{Simplify.}$$

To verify that 0.7 is the solution, replace x with 0.7 in the original equation.

Instead of solving equations with decimals, sometimes it may be easier to first rewrite the equation so it contains integers only. Recall that multiplying a decimal by a power of 10, such as 10, 100, or 1000, has the effect of moving the decimal point to the right. We can use the multiplication property of equality to multiply both sides of the equation through by an appropriate power of 10. The resulting equivalent equation contains integers only.

Answers:
2. -2
3. -0.3

EXAMPLE 4

Solve for y: $0.5y + 2.3 = 1.65$.

Solution:

Multiply the equation through by 100. This will move the decimal point in each term two places to the right.

$0.5y + 2.3 = 1.65$	Original equation.
$100\,(0.5y + 2.3) = 100\,(1.65)$	Multiply both sides by 100.
$100(0.5y) + 100(2.3) = 100(1.65)$	Apply the distributive property.
$50y + 230 = 165$	Simplify.

Now our equation contains integers only. Finish solving by adding -230 to both sides.

$50y + 230 = 165$	
$50y + 230 + (-230) = 165 + (-230)$	Add -230 to both sides.
$50y = -65$	Simplify.
$\frac{50y}{50} = \frac{-65}{50}$	Divide both sides by 50.
$y = -1.3$	Simplify.

Check to see that -1.3 is the solution.

PRACTICE PROBLEM 4

Solve for y: $0.2y + 2.6 = 4$.

Answer:

4. 7

> REMINDER When finding the area of figures, be sure all measurements are changed to the same unit before calculations are made.

EXAMPLE 3

Find the area of the figure.

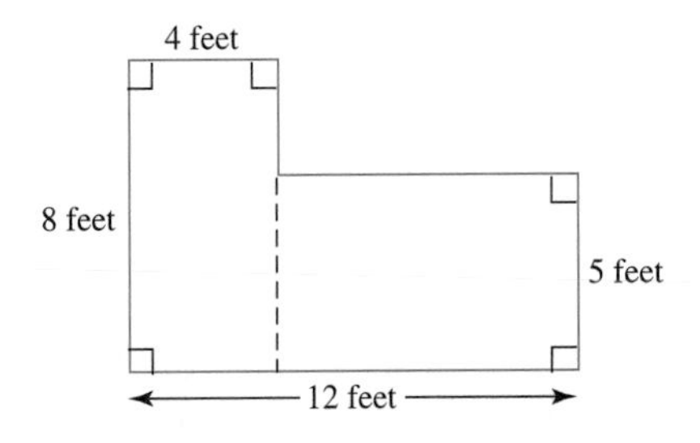

Solution:

The figure is composed of two rectangles, so to find its area, we find the sum of the areas of the two rectangles.

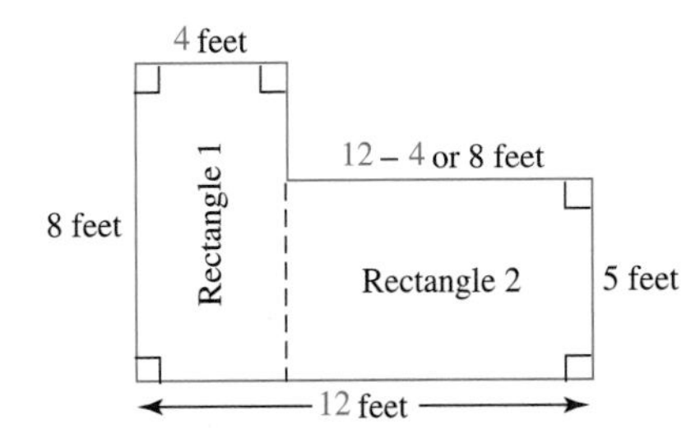

$$\begin{aligned}\text{The area of rectangle 1} &= l \cdot w \\ &= 8 \cdot 4 \\ &= 32 \text{ square feet}\end{aligned}$$

Notice that the length of rectangle 2 is 12 feet − 4 feet or 8 feet.

$$\begin{aligned}\text{The area of rectangle 2} &= l \cdot w \\ &= 8 \cdot 5 \\ &= 40 \text{ square feet}\end{aligned}$$

$$\begin{aligned}\text{The area of the figure} &= \text{area of rectangle 1} + \text{area of rectangle 2} \\ &= 32 \text{ square feet} + 40 \text{ square feet} \\ &= 72 \text{ square feet}\end{aligned}$$

EXAMPLE 4

Find the area of a circle with a radius of 3 feet. Find the exact area and an approximation. Use 3.14 as an approximation for π.

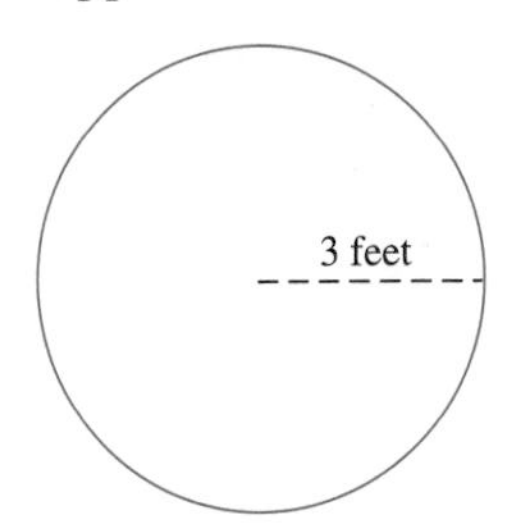

PRACTICE PROBLEM 3

Find the area of the figure.

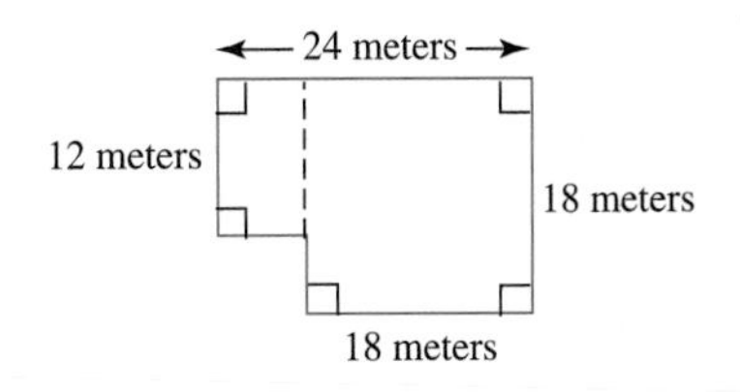

PRACTICE PROBLEM 4

Find the area of the given circle. Find the exact area and an approximation. Use 3.14 as an approximation for π.

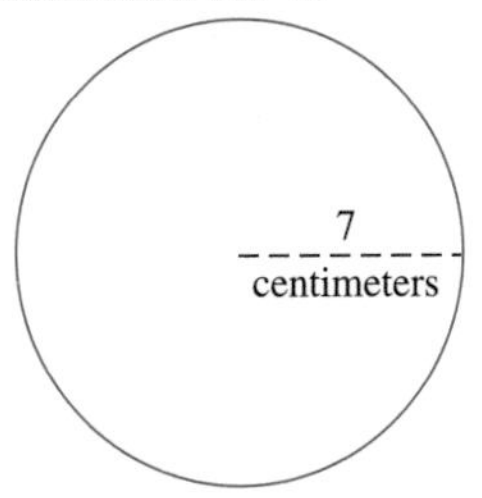

Answers:

3. 396 sq. in.
4. 49π sq. cm $\approx$ 153.86 sq. cm

Solution:

Let $r = 3$ feet and use the formula

$$\begin{aligned} \text{Area} &= \pi r^2 \\ &= \pi(3 \text{ feet})^2 \\ &= 9\pi \text{ square feet} \end{aligned}$$

To approximate this area, we substitute 3.14 for π.

$$\begin{aligned} 9\pi \text{ square feet} &\approx 9(3.14) \text{ square feet} \\ &= 28.26 \text{ square feet} \end{aligned}$$

The **exact** area of the circle is 9π square feet, which is **approximately** 28.26 square feet.

2 Volume is a measure of the space of a region. The volume of a box or can, for example, is the amount of space inside. Volume can be used to describe the amount of juice in a pitcher or the amount of concrete needed to pour a foundation for a house.

The volume of a solid is the number of **cubic units** in the solid. A cubic centimeter and a cubic inch are illustrated.

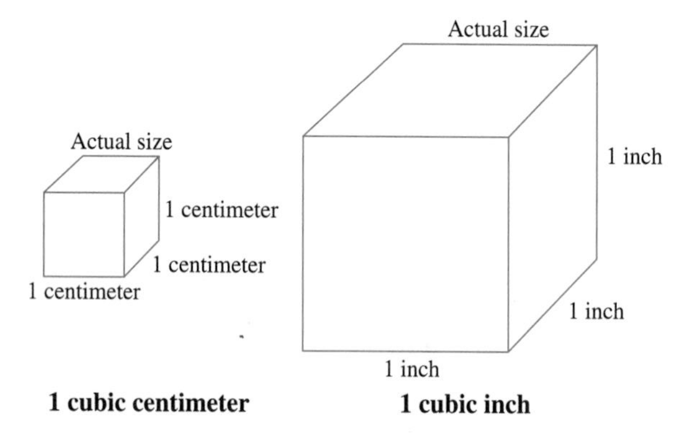

Before the formulas for volume are listed, we will review some basic solids. See Appendix E for more review.

A **rectangular solid** or box has length, width, and height. A brick is an example of a rectangular solid.

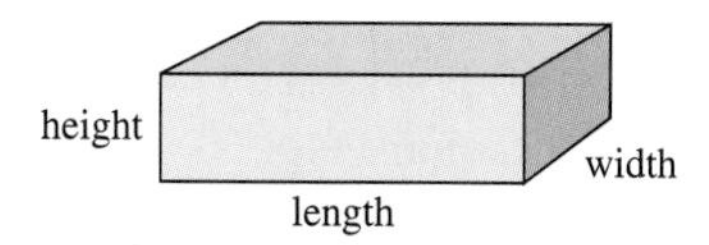

A **cube** is a rectangular solid or box with length = width = height. A die is an example of a cube.

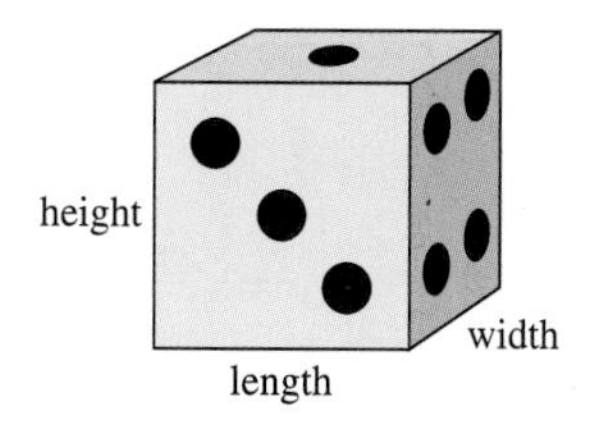

A **sphere** has radius. A beach ball is an example of a sphere.

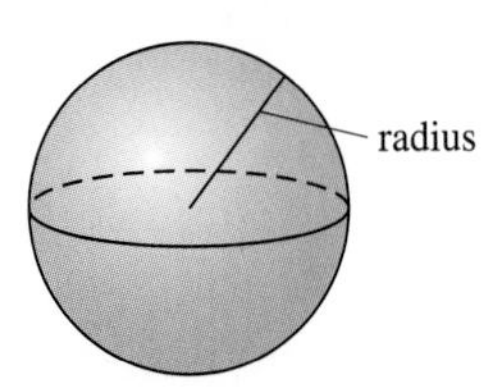

A **right circular cylinder** has height and radius. An example of a circular cylinder is a can.

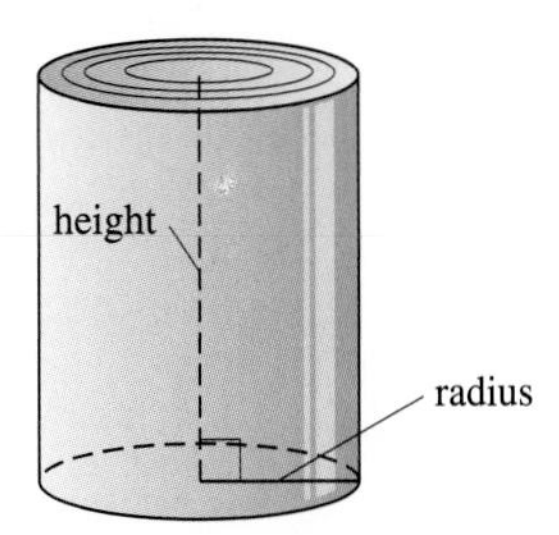

A **right circular cone** has height and radius. An example of a cone is an ice cream cone.

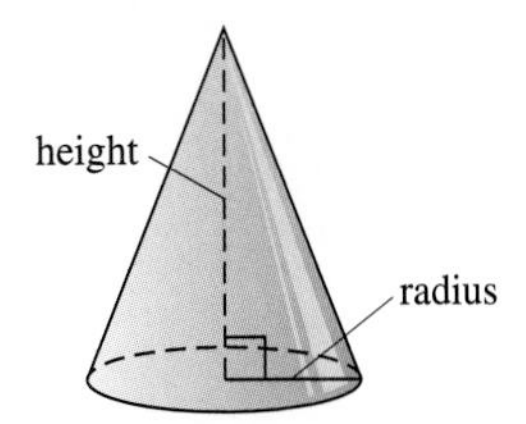

A **square-based pyramid** has height and a square for a base. An example of a pyramid is the Great Pyramid of Cheops.

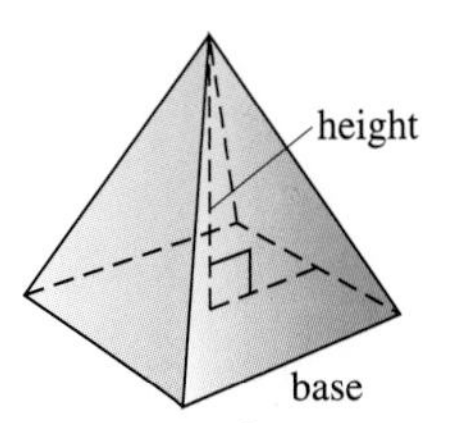

3 Formulas for finding the volumes of the solids we have just reviewed are given next.

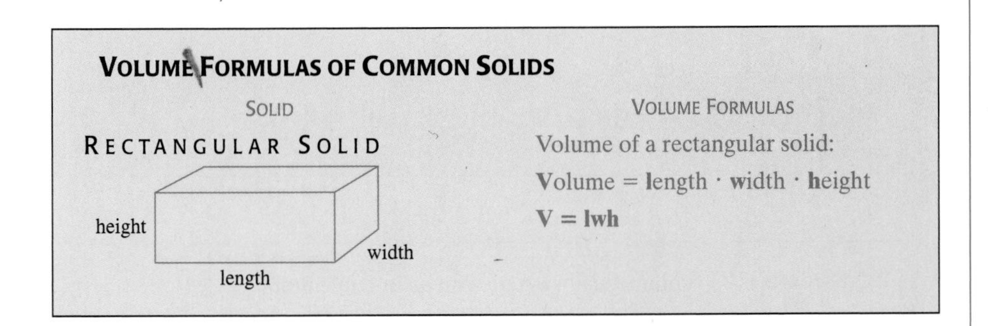

VOLUME FORMULAS OF COMMON SOLIDS

SOLID	VOLUME FORMULAS
RECTANGULAR SOLID	Volume of a rectangular solid: **V**olume = **l**ength · **w**idth · **h**eight **V = lwh**

VOLUME FORMULAS OF COMMON SOLIDS

SOLID	VOLUME FORMULAS
CUBE 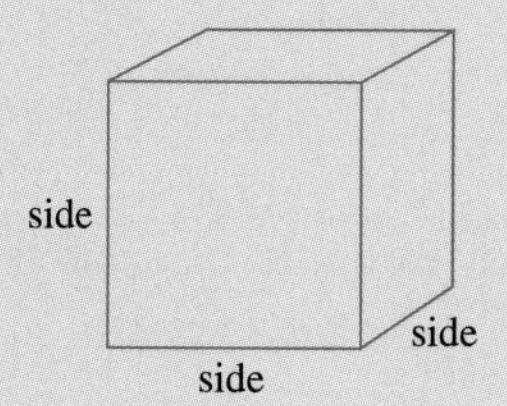	Volume of a cube: Volume = side · side · side $\mathbf{V = s^3}$
SPHERE 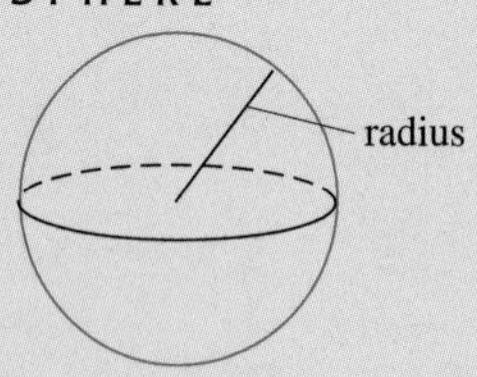	Volume of a sphere: Volume $= \frac{4}{3} \cdot \pi \cdot (\text{radius})^3$ $\mathbf{V = \frac{4}{3}\pi r^3}$
CIRCULAR CYLINDER 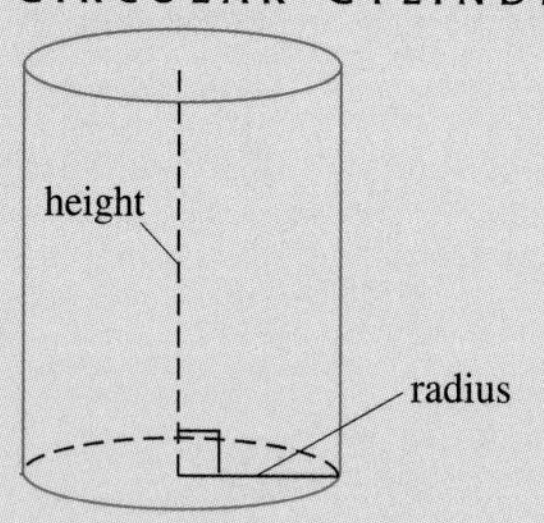	Volume of a circular cylinder: Volume $= \pi \cdot (\text{radius})^2 \cdot \text{height}$ $\mathbf{V = \pi r^2 h}$
CONE 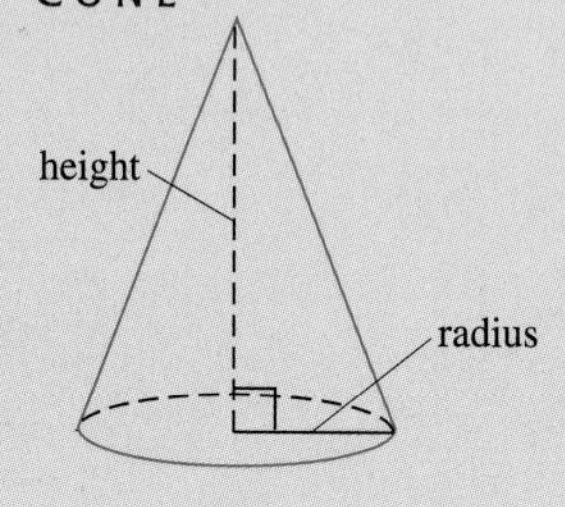	Volume of a cone: Volume $= \frac{1}{3} \cdot \pi \cdot (\text{radius})^2 \cdot \text{height}$ $\mathbf{V = \frac{1}{3}\pi r^2 h}$
SQUARE-BASED PYRAMID 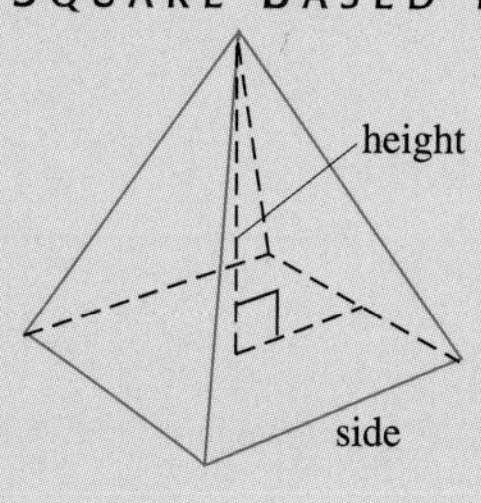	Volume of a square-based pyramid: Volume $= \frac{1}{3} \cdot (\text{side})^2 \cdot \text{height}$ $\mathbf{V = \frac{1}{3}s^2 h}$

REMINDER Volume is always measured in cubic units.

EXAMPLE 5

Find the volume of a rectangular box that is 12 inches long, 6 inches wide, and 3 inches high.

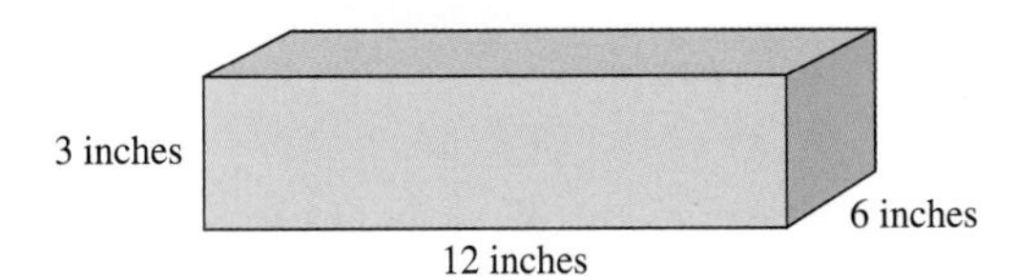

Solution:

Volume = length · width · height or

$$\text{Volume} = (12 \text{ inches}) \cdot (6 \text{ inches}) \cdot (3 \text{ inches}) = 216 \text{ cubic inches.}$$

The volume of the rectangular box is 216 cubic inches.

EXAMPLE 6

Approximate the volume of a ball of radius 3 inches. Use the approximation $\frac{22}{7}$ for π.

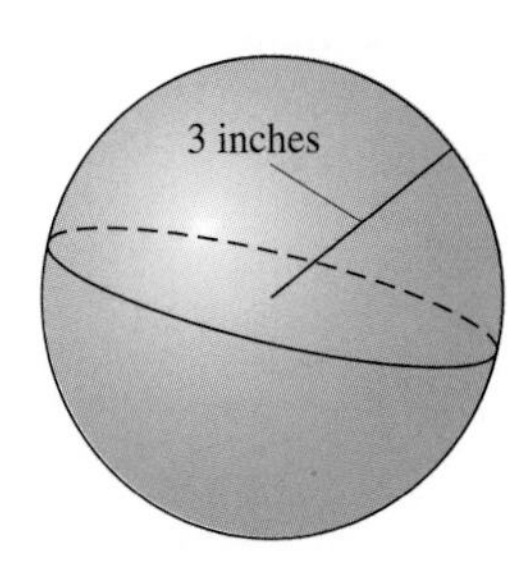

Solution:

Since the radius is 3 inches,

$$\text{Volume} = \frac{4}{3} \cdot \pi \cdot (\text{radius})^3$$

or approximately

$$\approx \frac{4}{3} \cdot \frac{22}{7} (3 \text{ inches})^3$$

$$= \frac{4}{3} \cdot \frac{22}{7} \cdot 27 \text{ cubic inches}$$

$$= \frac{4 \cdot 22 \cdot 3 \cdot 9}{3 \cdot 7} \text{ cubic inches}$$

$$= \frac{792}{7} \quad \text{or} \quad 113\frac{1}{7} \text{ cubic inches}$$

The volume is **approximately** $113\frac{1}{7}$ cubic inches.

PRACTICE PROBLEM 5

Draw a diagram and find the volume of a rectangular box that is 5 feet long, 2 feet wide, and 4 feet deep.

PRACTICE PROBLEM 6

Draw a diagram and approximate the volume of a ball of radius $\frac{1}{2}$ centimeter. Use the approximation $\frac{22}{7}$ for π.

Answers:

5. 40 cu. ft

6. $\frac{11}{21}$ cu. cm

PRACTICE PROBLEM 7

Approximate the volume of a cylinder of radius 5 inches and height 7 inches. Use the approximation 3.14 for π.

EXAMPLE 7

Approximate the volume of a can that has a $3\frac{1}{2}$-inch radius and a height of 6 inches. Use the approximation $\frac{22}{7}$ for π.

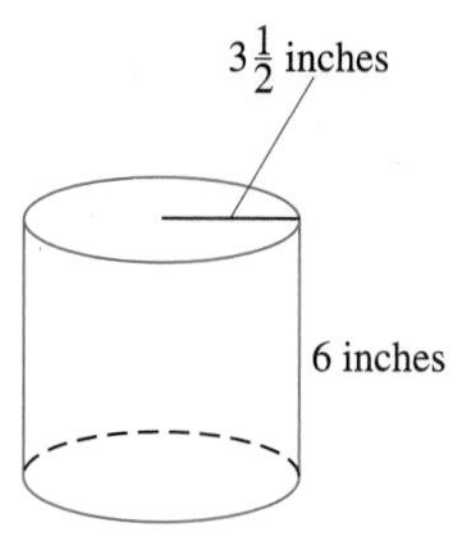

Solution:

Using the formula for a circular cylinder, we have

$$\text{Volume} = \pi \cdot (\text{radius})^2 \cdot \text{height}$$

$$3\frac{1}{2} = \frac{7}{2}$$

$$= \pi \cdot \left(\frac{7}{2}\text{ inches}\right)^2 \cdot 6\text{ inches}$$

or approximately

$$\approx \frac{22}{7} \cdot \frac{49}{4} \cdot 6\text{ cubic inches}$$

$$= 231\text{ cubic inches}$$

The volume is approximately 231 cubic inches.

Answer:

7. 549.5 cu. in.

EXERCISE SET 5.9

*Find the area of the geometric figure. If the figure is a circle, give an exact area and then use the given **approximation** for π to approximate the area. See Examples 1–4.*

1. 2 meters; Rectangle; 3.5 meters

2. 2.75 feet; Rectangle; 7 feet

3.

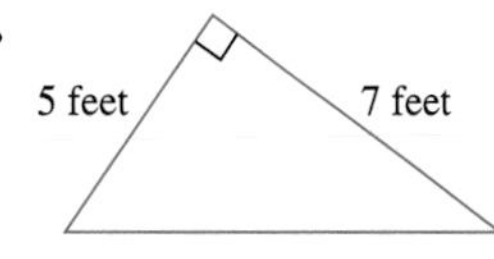

4.

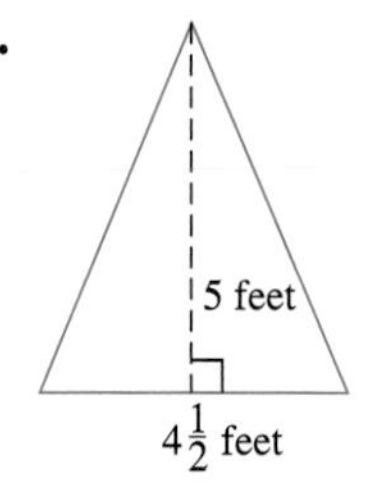

5.

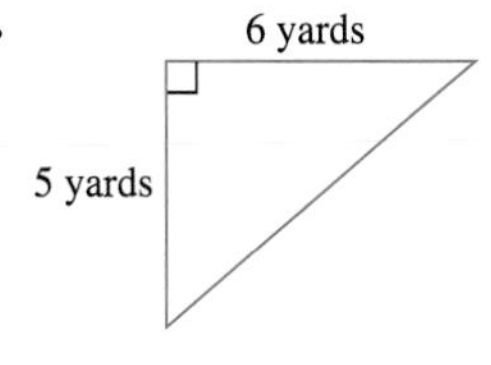

6.

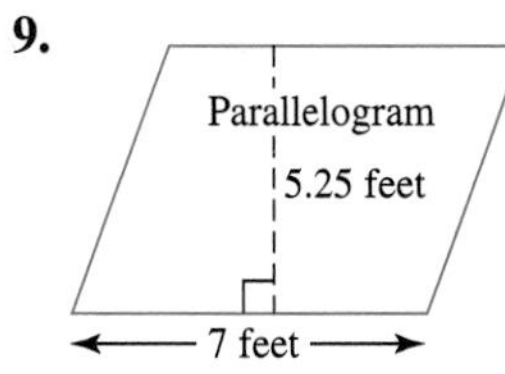

7. Use 3.14 for π.

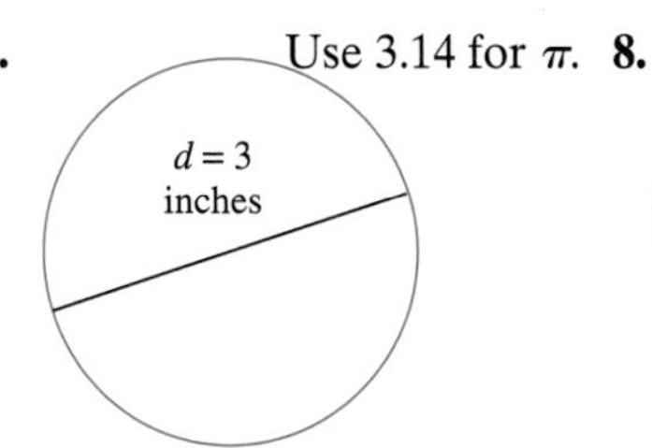

8. Use $\frac{22}{7}$ for π.

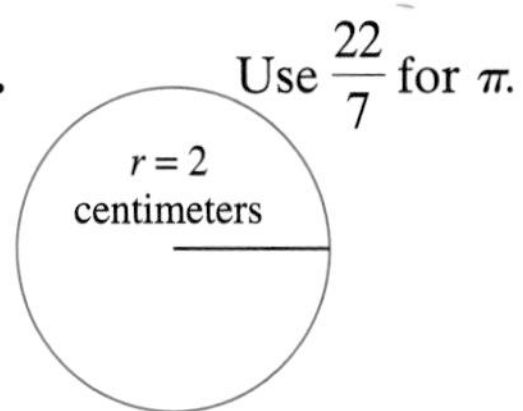

9. Parallelogram; 5.25 feet; 7 feet

10.

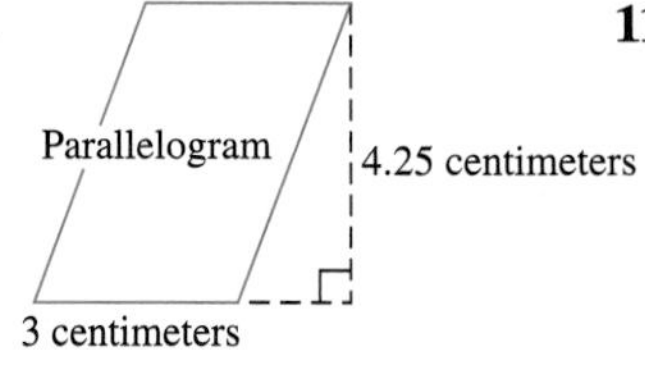

11.

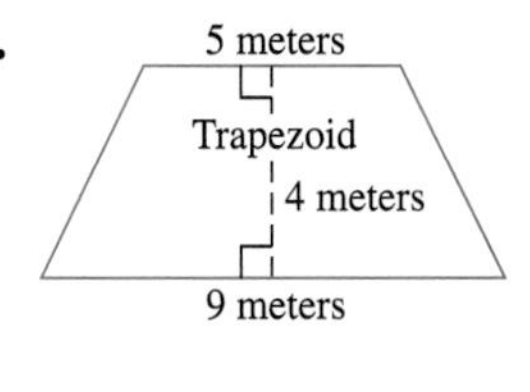

12.

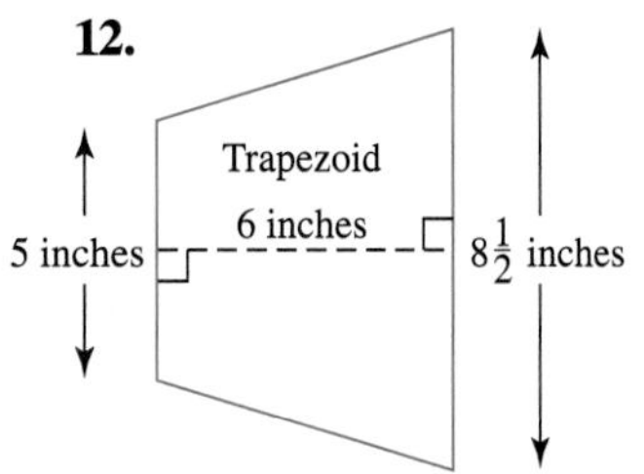

13.

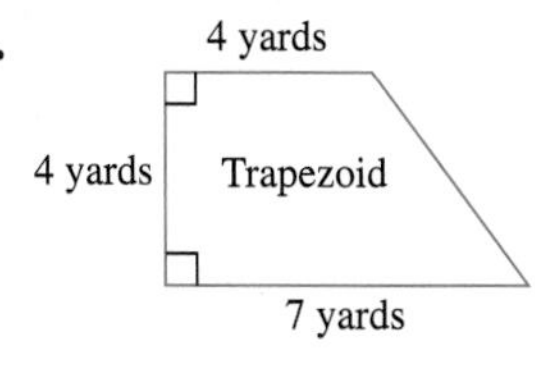

14.

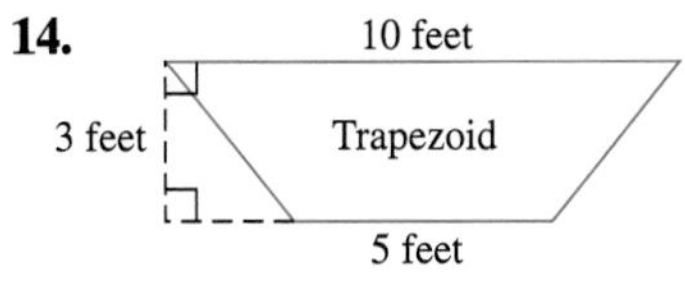

15.

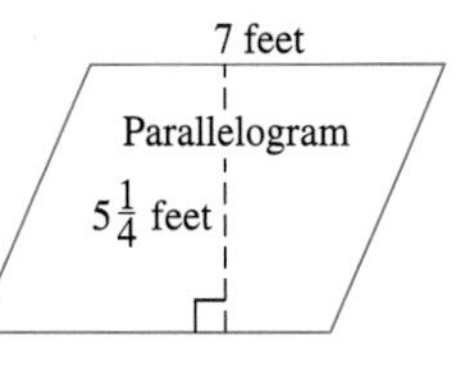

16.

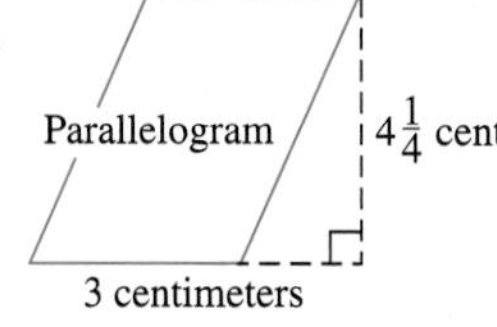

17.

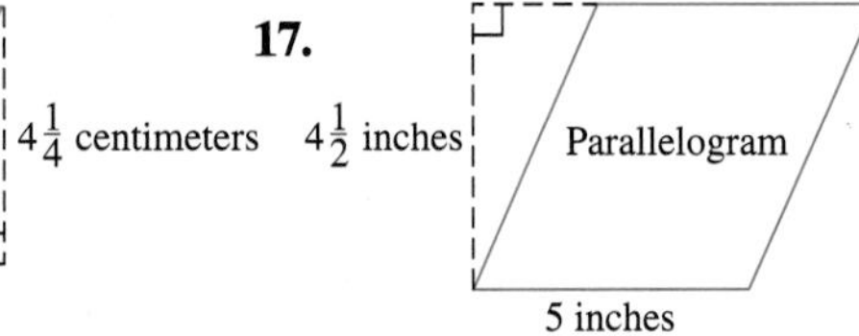

18.

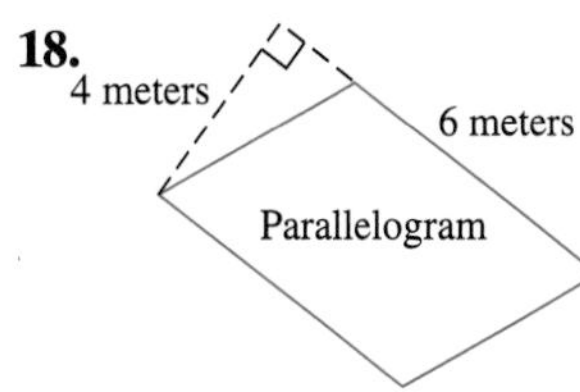

ANSWERS

1. ______
2. ______
3. ______
4. ______
5. ______
6. ______
7. ______
8. ______
9. ______
10. ______
11. ______
12. ______
13. ______
14. ______
15. ______
16. ______
17. ______
18. ______

ANSWERS

19. ____________

20. ____________

21. ____________

22. ____________

23. ____________

24. ____________

25. ____________

26. ____________

27. ____________

28. ____________

19.

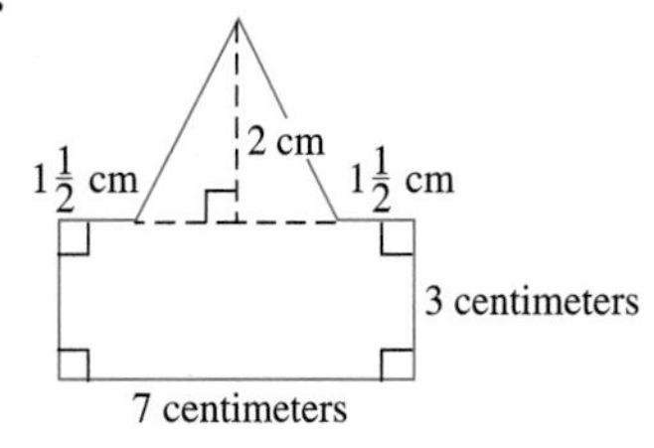

20.

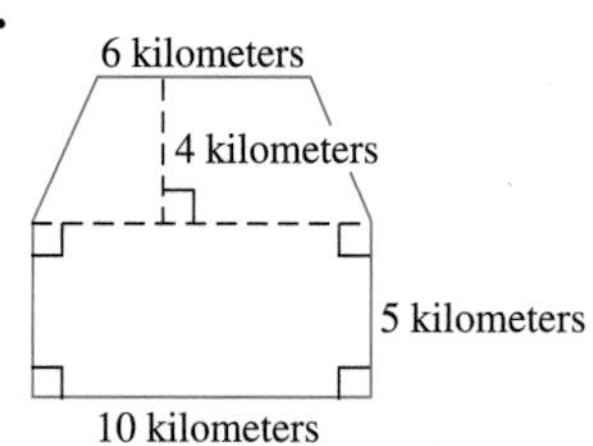

21.

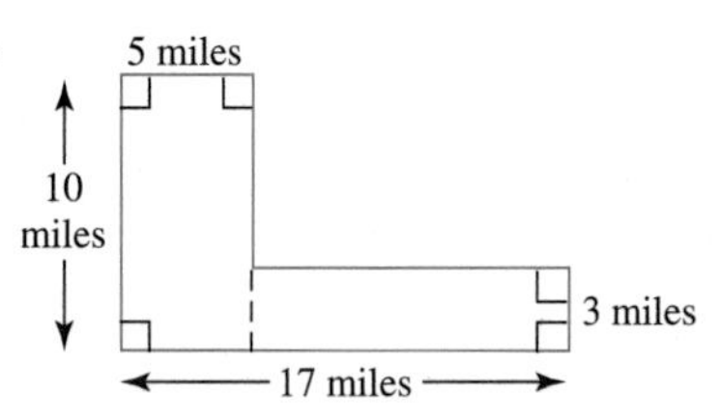

22.

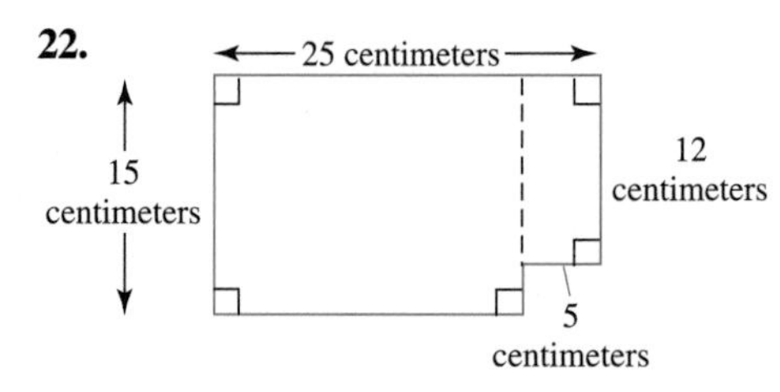

23.

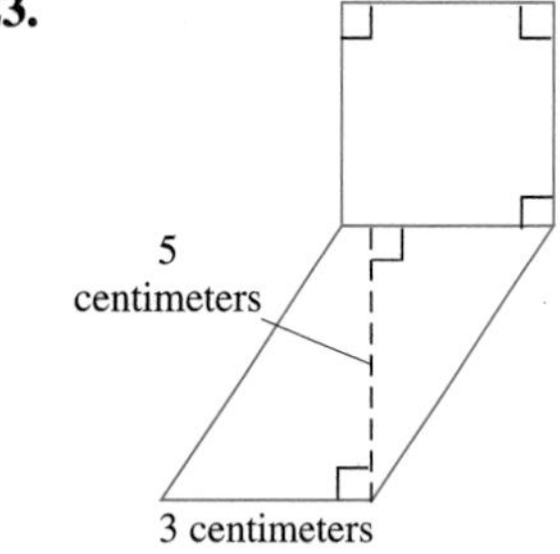

24.

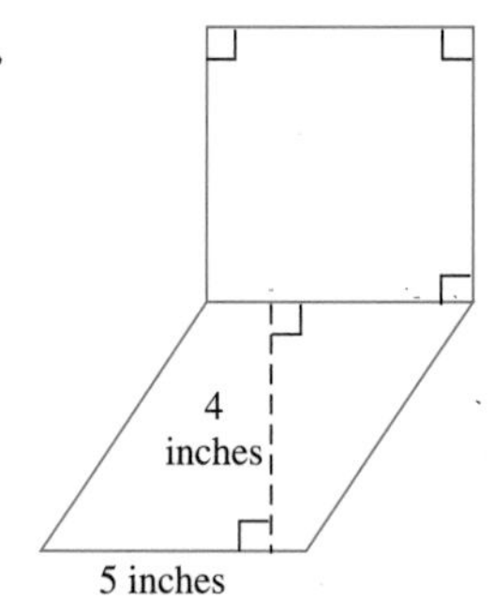

Find the volume of the solid. See Examples 5–7. Use $\frac{22}{7}$ as an approximation for π.

25.

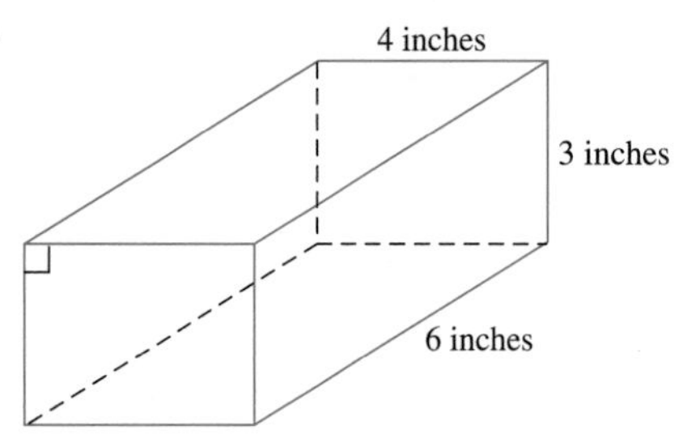

26.

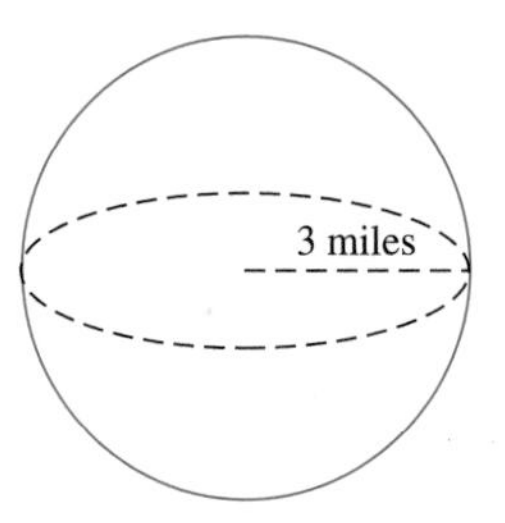

27.

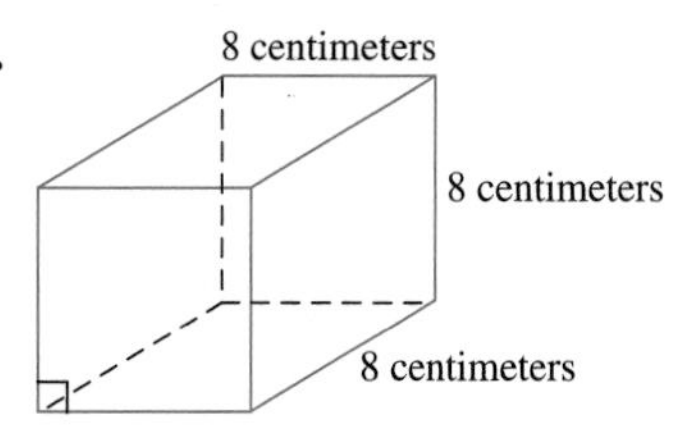

28.

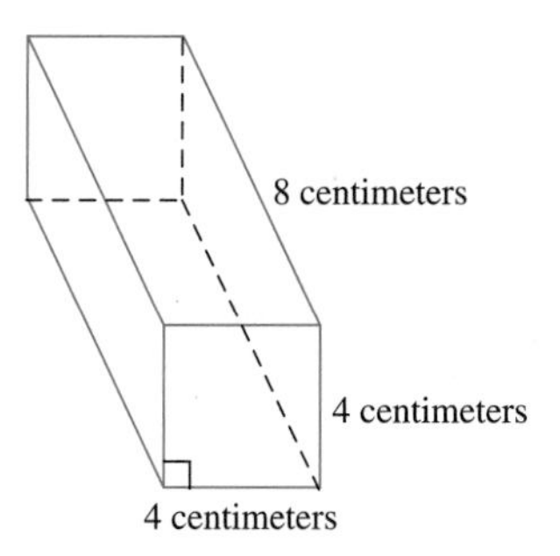

(5.8) *Simplify.*

96. $\sqrt{64}$ **97.** $\sqrt{144}$ **98.** $-2\sqrt{36}$

99. $3\sqrt{25} + \sqrt{4}$ **100.** $\sqrt{\frac{4}{25}}$ **101.** $\sqrt{\frac{1}{100}}$

Find the unknown length of each given right triangle. If necessary round to the nearest tenth.

102. leg = 12, leg = 5 **103.** leg = 20, leg = 21 **104.** leg = 9, hypotenuse = 14

105. leg = 124, hypotenuse = 155 **106.** leg = 66, leg = 56

107. Find the length to the nearest hundredth of the diagonal of a square that has a side of length 20 centimeters.

108. Find the height of the building rounded to the nearest tenth.

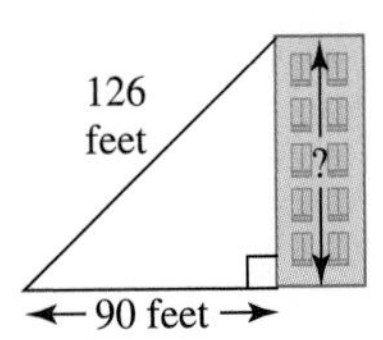

(5.9) *Find the area of the figure.*

109.

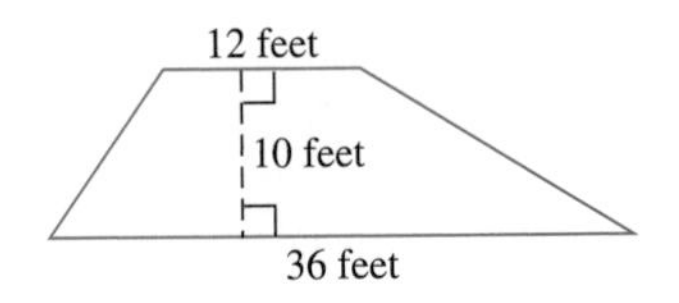

110.

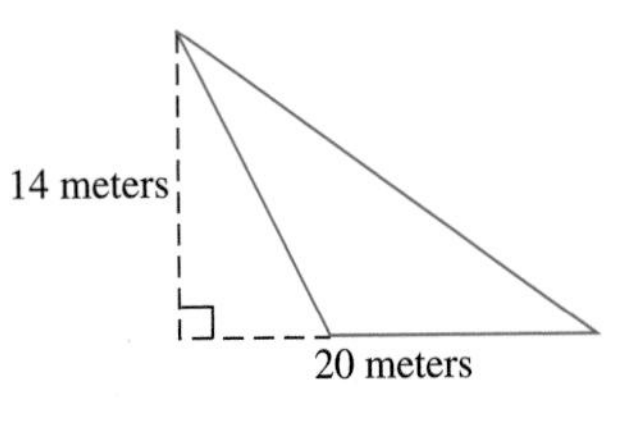

111.

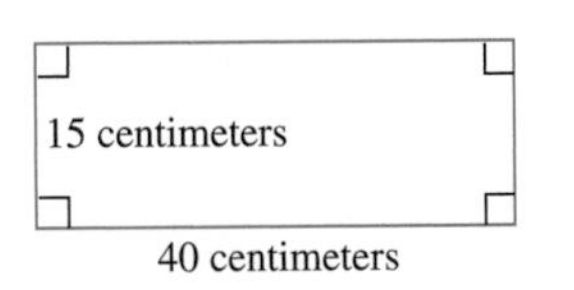

112.

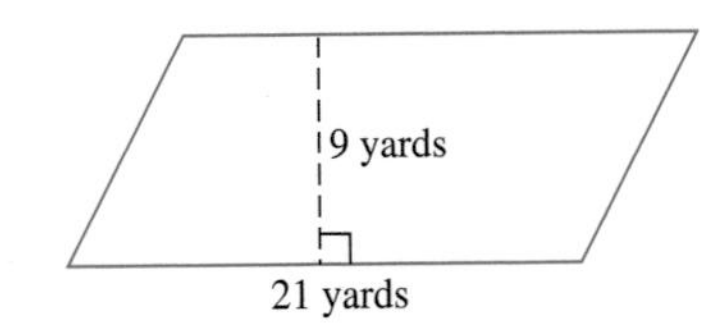

ANSWERS

96. ______
97. ______
98. ______
99. ______
100. ______
101. ______
102. ______
103. ______
104. ______
105. ______
106. ______
107. ______
108. ______
109. ______
110. ______
111. ______
112. ______

ANSWERS

113. ____________

114. ____________

115. ____________

116. ____________

117. ____________

118. ____________

119. ____________

120. ____________

121. ____________

113.

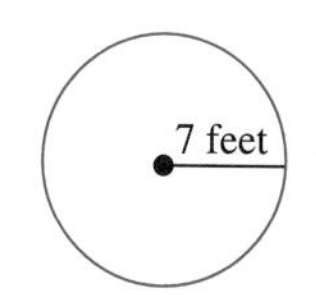

114.

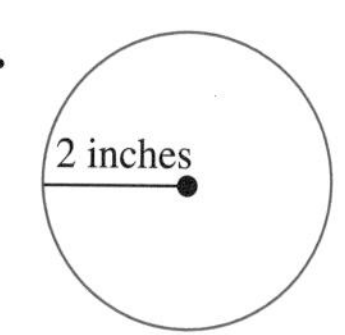

115.

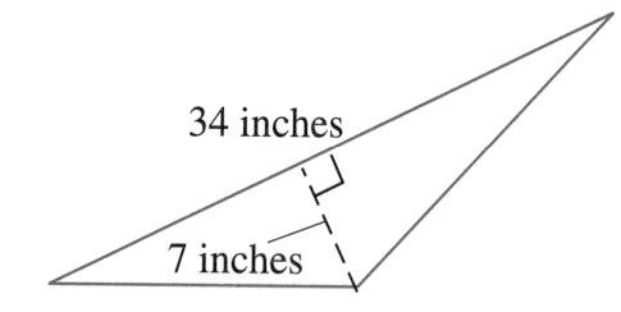

116.

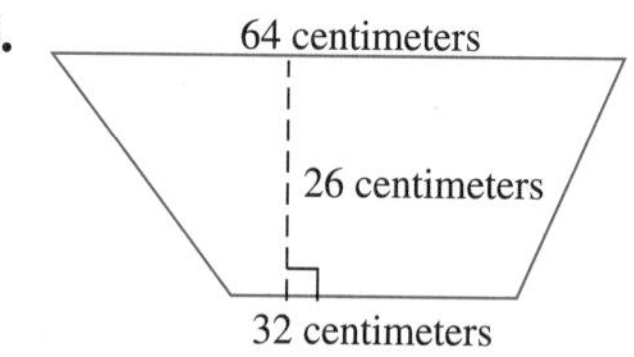

117.

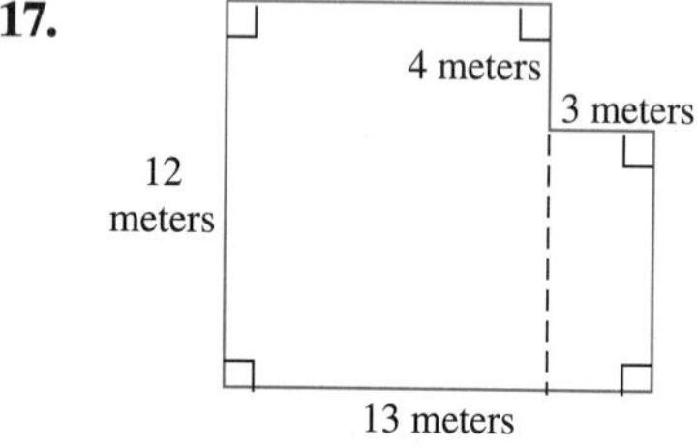

118. The amount of sealer necessary to seal a driveway depends on the area. Find the area of a rectangular driveway 36 feet by 12 feet.

119. Find how much carpet is necessary to cover the floor of the room shown.

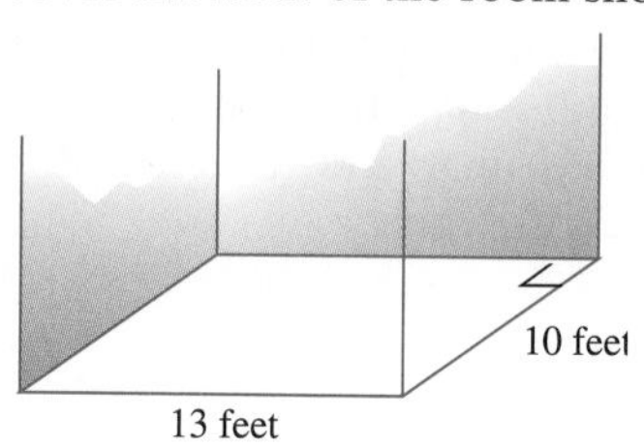

Find the volume of the solid.

120.

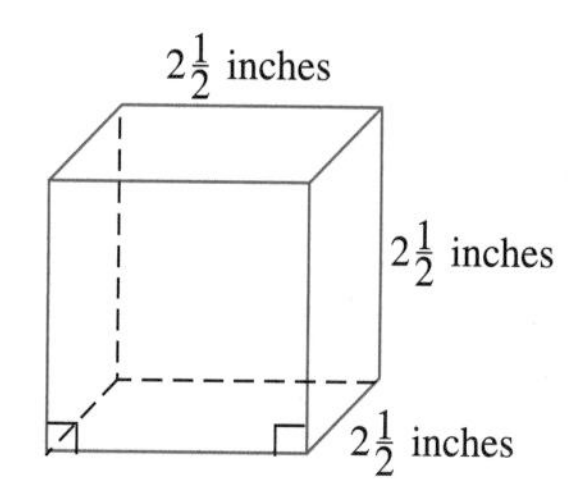

121.

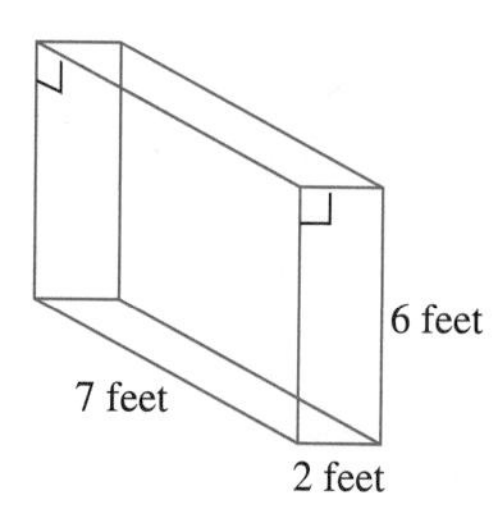

122. Use $\pi \approx \frac{22}{7}$.

20 centimeters

50 centimeters

123. Use $\pi \approx \frac{22}{7}$.

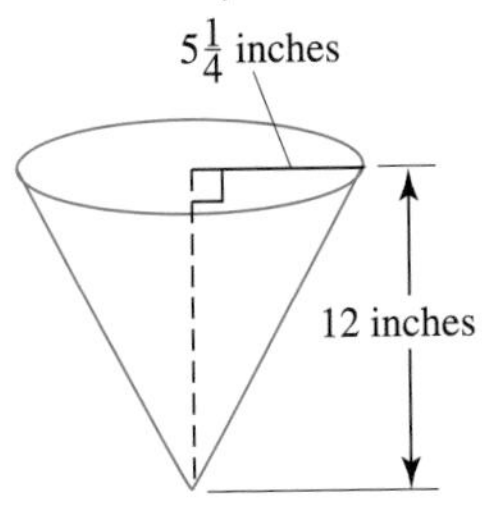

124. Find the volume of a pyramid with a square base 2 feet on a side and a height of 2 feet.

125. Approximate the volume of a tin can 8 inches high and 3.5 inches in radius. Use 3.14 for π.

126. A chest has 3 drawers. If each drawer has inside measurements of $2\frac{1}{2}$ feet by $1\frac{1}{2}$ feet by $\frac{2}{3}$ feet, find the total capacity of the chest of drawers.

127. A cylindrical canister for a shop vacuum is 2 feet tall and 1 foot in **diameter.** Find its exact capacity.

128. Find the volume of air in a rectangular room 15 feet by 12 feet with a 7-foot ceiling.

129. A mover has two boxes left for packing. Both are cubical, one 3 feet on a side and the other 1.2 feet on a side. Find their combined capacity.

ANSWERS

122. ______

123. ______

124. ______

125. ______

126. ______

127. ______

128. ______

129. ______

6.2

RECTANGULAR COORDINATE SYSTEM

TAPE PA 6.2

OBJECTIVES

1. Plot points on a rectangular coordinate system.
2. Know the meaning of a solution of a two-variable equation.
3. Determine whether ordered pairs are solutions of equations.
4. Complete ordered-pair solutions of equations.

1 In the last section, we saw how bar and line graphs can be used to show relationships between items listed on the horizontal and vertical axes. We can use this same horizontal and vertical axis idea to describe the location of points in a plane.

The system that we use to describe the location of points in a plane is called the **rectangular coordinate system.** It consists of two number lines, one horizontal and one vertical, intersecting at the point 0 on each number line. This point of intersection is called the **origin.** We call the horizontal number line the ***x*-axis** and the vertical number line the ***y*-axis**. Notice that the axes divide the plane into four regions, called **quadrants.** They are numbered as shown.

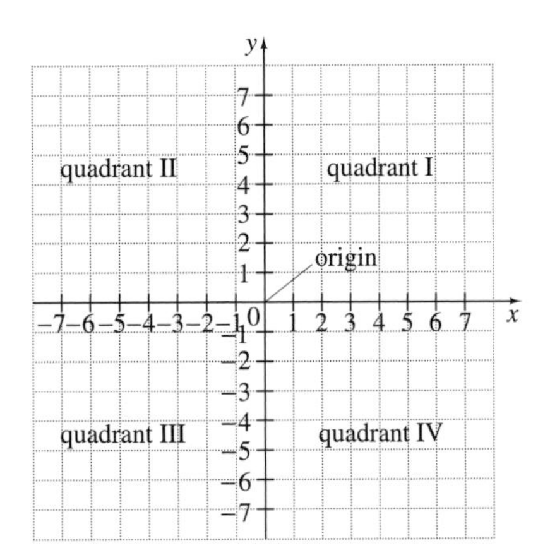

Every point in the rectangular coordinate system corresponds to an **ordered pair of numbers,** such as (3, 4). The first number, 3, of an ordered pair is associated with the x-axis and is called the ***x*-coordinate** or ***x*-value.** The second number, 4, is associated with the y-axis and is called the ***y*-coordinate** or ***y*-value.** To find the **single point** on the rectangular coordinate system corresponding to the ordered pair (3, 4), start at the origin. Move 3 three units in the positive direction along the x-axis. From there, move 4 units in the positive direction parallel to the y-axis. This process of locating a point on the rectangular coordinate system is called **plotting the point.** Since the origin is located at 0 on the x-axis and 0 on the y-axis, the origin corresponds to the ordered pair $(0, 0)$.

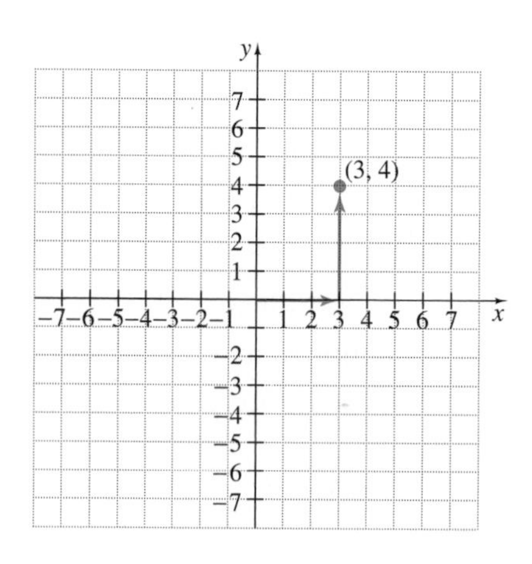

In general, to plot the ordered pair (a, b), start at the origin. Next,

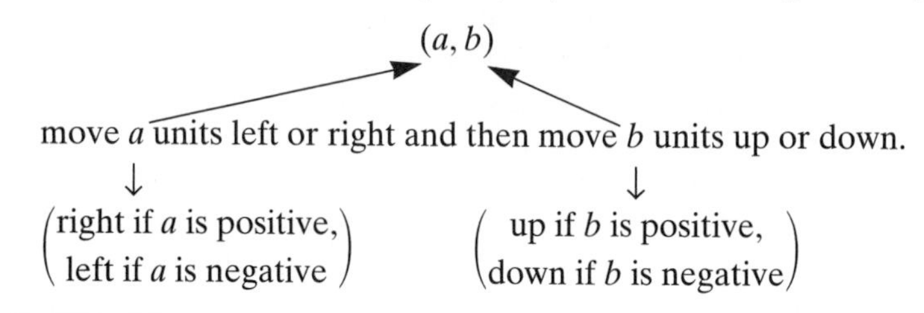

REMINDER Since the first number or x-coordinate of an ordered pair is associated with the x-axis, it tells how many units to move left or right. Similarly, the second number or y-coordinate tells how many units to move up or down. For example, to plot $(-1, 5)$, start at the origin. Move 1 unit left (because the x-coordinate is negative) and then 5 units up and draw a dot at that point. This dot is the graph of the point that corresponds to the ordered pair $(-1, 5)$.

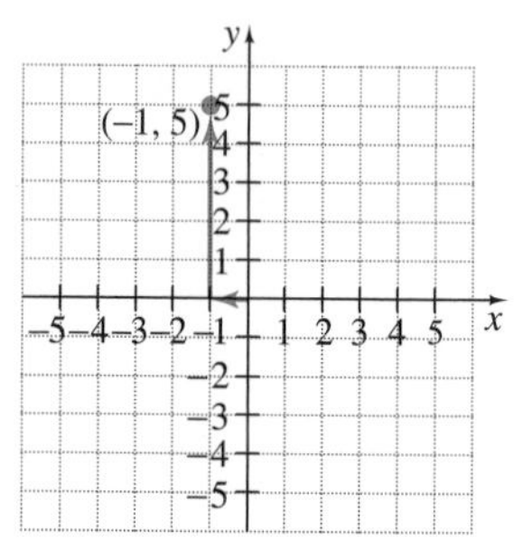

Here are some more plotted points with their corresponding ordered pairs.

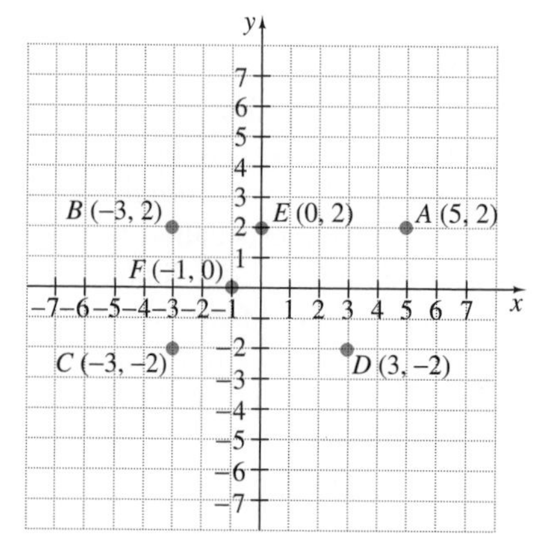

Solution:

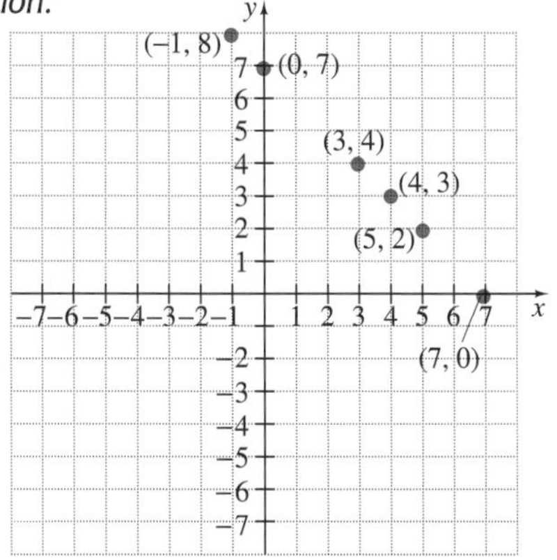

Notice that the points in Example 4 all seem to lie on the same line. We will discuss this more in the next section.

An equation such as $x + y = 7$ is called a **linear equation in two variables.** It is called a **linear equation** because the exponent on x and y is an understood 1. The equation is called an equation **in two variables** because it contains two different variables, x and y.

4 Before we can graph linear equations in two variables, we need to practice finding ordered-pair solutions of these equations. If one coordinate of an ordered-pair solution is known, the other coordinate can be determined. To find the unknown coordinate, replace the appropriate variable with the known coordinate in the equation. Doing so results in an equation with one variable that we can solve.

EXAMPLE 5

Complete the ordered-pair solutions of the equation $2x = y$.

a. (3,) **b.** (, 0) **c.** (−2,)

Solution:

a. In the ordered pair (3,), the x-value is 3. To find the corresponding y-value, let $x = 3$ in the equation $2x = y$ and solve for y.

$2x = y$ Original equation.
$2(3) = y$ Replace x with 3.
$6 = y$ Solve for y.

To check, replace x with 3 and y with 6 in the original equation to see that a true statement results.

$2x = y$ Original equation.
$2(3) = 6$ Let $x = 3$ and $y = 6$.
$6 = 6$ True.

The ordered pair solution is (3, 6).

b. Replace y with 0 in the equation and solve for x.

$2x = y$ Original equation.
$2x = 0$ Replace y with 0.
$\frac{2x}{2} = \frac{0}{2}$ Solve for x.
$x = 0$ Simplify.

The ordered-pair solution is (0, 0), the origin.

PRACTICE PROBLEM 5

Complete the ordered-pair solutions of the equation $x + y = 10$.

a. (5,) **b.** (0,) **c.** (, −2)

Answers:
5a. (5, 5) **b.** (0, 10) **c.** (12, −2)

c. Replace x with -2 in the equation and solve for y.

$2x = y$	Original equation.
$2(-2) = y$	Replace x with -2.
$-4 = y$	Solve for y.

The ordered-pair solution is $(-2, -4)$.

EXERCISE SET 6.2

Plot points corresponding to the ordered pairs on the same set of axes. See Example 1.

1. $(1, 3)$, $(-2, 4)$, $(0, 7)$, $(-5, 0)$, $(-6, -3)$, $(5, -5)$

2. $(5, 2)$, $(3, -4)$, $(-1, -1)$, $(0, -6)$, $(4, 0)$, $(-2, 4)$

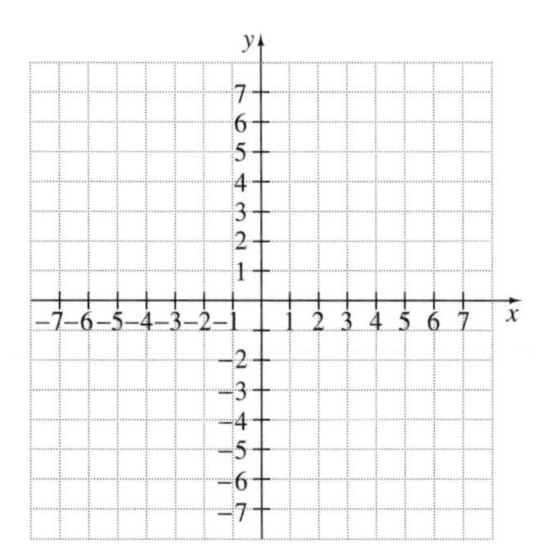

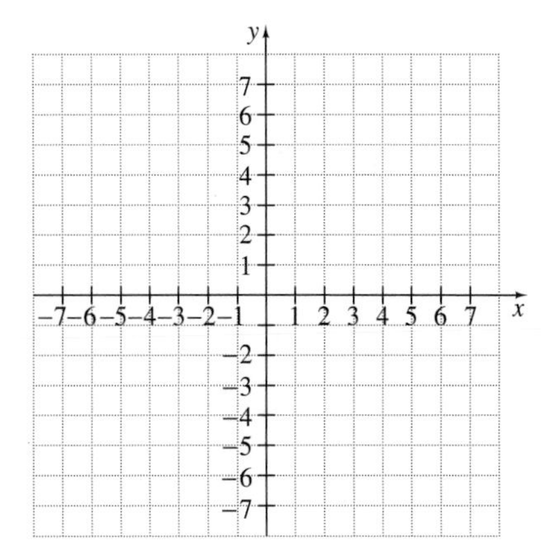

3. $\left(2\frac{1}{2}, 3\right)$, $(0, -3)$, $(-4, -6)$, $\left(-1, 5\frac{1}{2}\right)$, $(1, 0)$, $(3, -5)$

4. $\left(5, \frac{1}{2}\right)$, $\left(-3\frac{1}{2}, 0\right)$, $(-1, 4)$, $(4, -1)$, $(0, 2)$, $(-5, -5)$

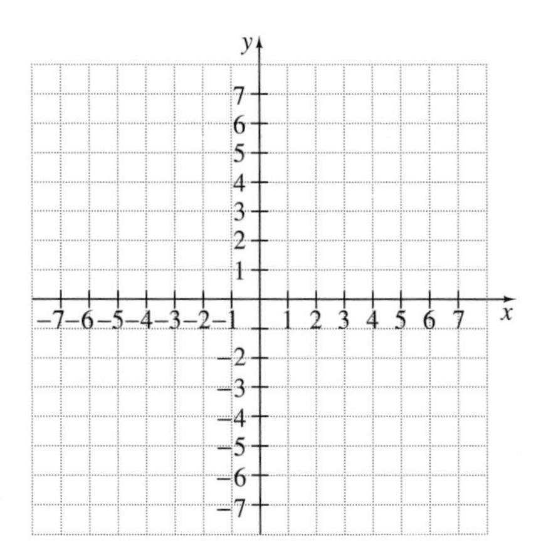

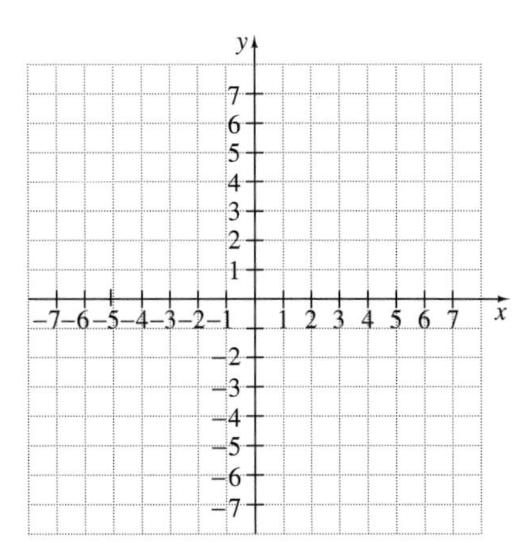

Find the x- and y-coordinates of each labeled point. See Example 2.

5.

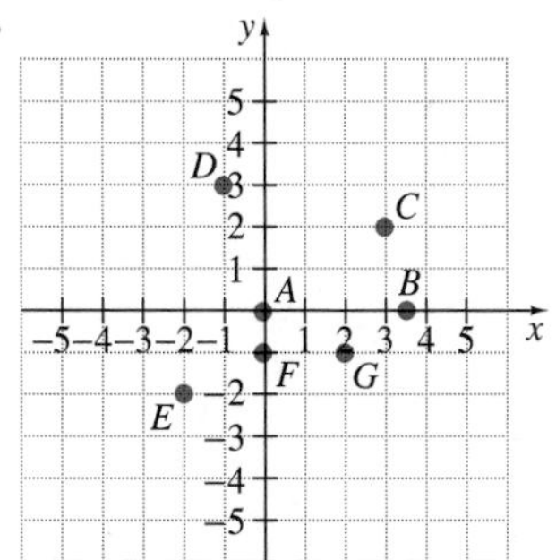

6.

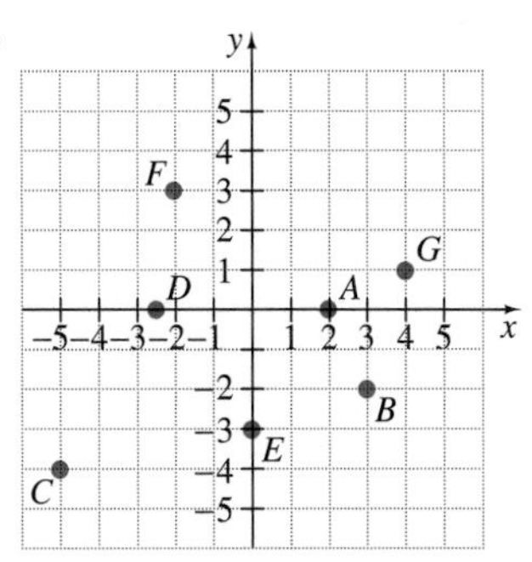

Determine whether each ordered pair is a solution of the given linear equation. See Example 3.

7. $(0, 0); y = -10x$

8. $(1, 7); x = 7y$

9. $(1, 2); x - y = 3$

10. $(-1, 9); x + y = 8$

11. $(-2, -3); y = 2x + 1$

12. $(1, 1); y = -x$

ANSWERS

5. ______

6. ______

7. ______

8. ______

9. ______

10. ______

11. ______

12. ______

ANSWERS

19. ______
20. ______
21. ______
22. ______
23. ______
24. ______
25. ______
26. ______
27. ______
28. ______
29. ______
30. ______

Plot the three ordered-pair solutions of the given equation. See Example 4.

13. $2x + y = 5; (1, 3), (0, 5), (3, -1)$

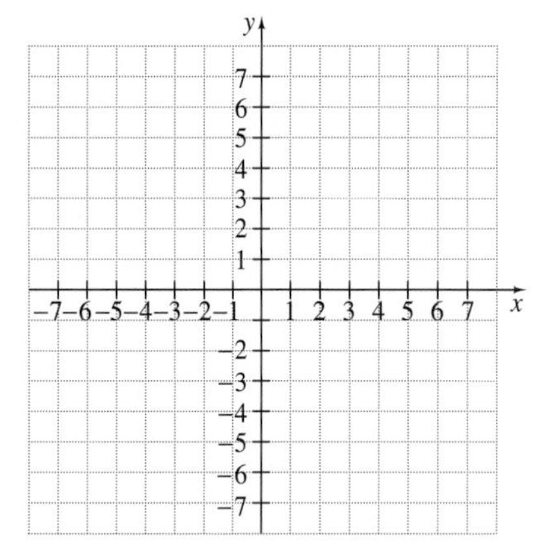

14. $x + y = 5; (-1, 6), (0, 5), (4, 1)$

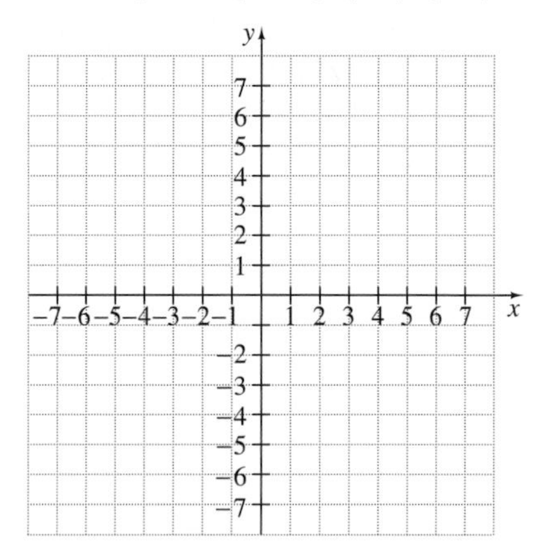

15. $x = 5y; (5, 1), (0, 0), (-5, -1)$

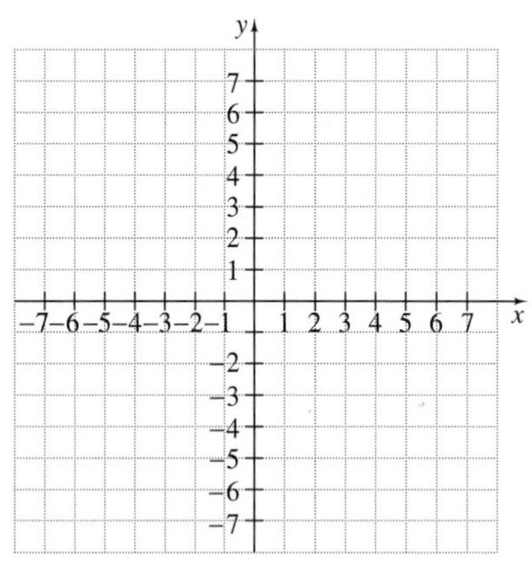

16. $y = -3x; (1, -3), (2, -6), (-1, 3)$

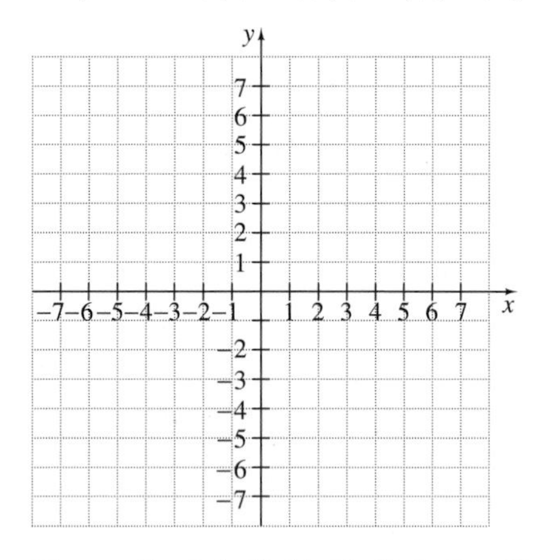

17. $x - y = 7; (8, 1), (2, -5), (0, -7)$

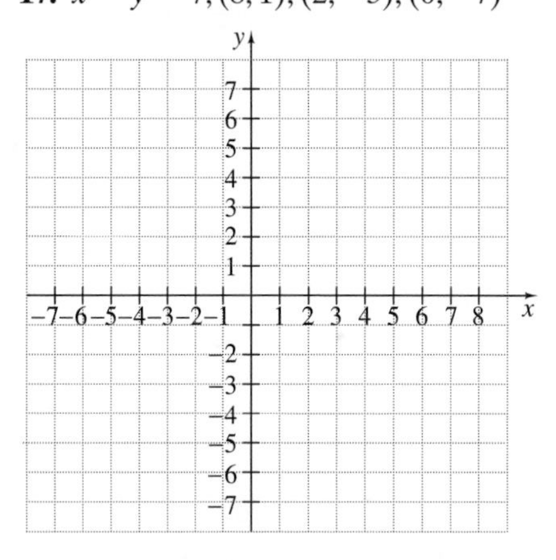

18. $y = 3x + 1; (0, 1), (1, 4), (-1, -2)$

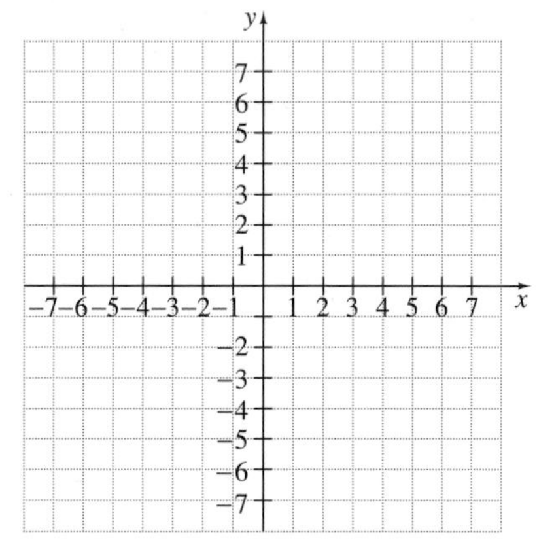

Complete the ordered-pair solutions of the given equations. See Example 5.

19. $y = 8x; (1, \ \ ), (0, \ \ ), (\ \ , -16)$

20. $x = -7y; (\ \ , 2), (14, \ \ ), (\ \ , -1)$

21. $x + y = 14; (2, \ \ ), (\ \ , -8), (0, \ \ )$

22. $x - y = 8; (0, \ \ ), (\ \ , 0), (5, \ \ )$

23. $y = x + 5; (1, \ \ ), (\ \ , 7), (3, \ \ )$

24. $x = y; (-8, \ \ ), (\ \ , 3), (100, \ \ )$

Determine whether the ordered pair is a solution of the given equation.

25. $(2, -8); y = -4x$

26. $(9, 1); x = 9y$

27. $(5, 0); 2x + 3y = 10$

28. $(1, 1); 4x - 5y = -1$

29. $(3, 1); x - 5y = -1$

30. $(0, 2); x - 7y = -15$

Complete the ordered-pair solutions of the given equations.

31. $y = 3x - 5; (1, \ ), (2, \ ), (3, \ )$

32. $x = -12y; (\ , -1), (\ , 1), (36, \ )$

33. $y = -x; (\ , 0), (2, \ ), (\ , 2)$

34. $y = 5x + 1; (0, \ ), (-1, \ ), (2, \ )$

35. $x + y = -2; (-2, \ ), (1, \ ), (\ , 5)$

36. $x - y = -3; (\ , 0), (0, \ ), (4, \ )$

Recall that the axes divide the plane into 4 quadrants as shown. If a and b are both positive numbers, answer the following questions true or false.

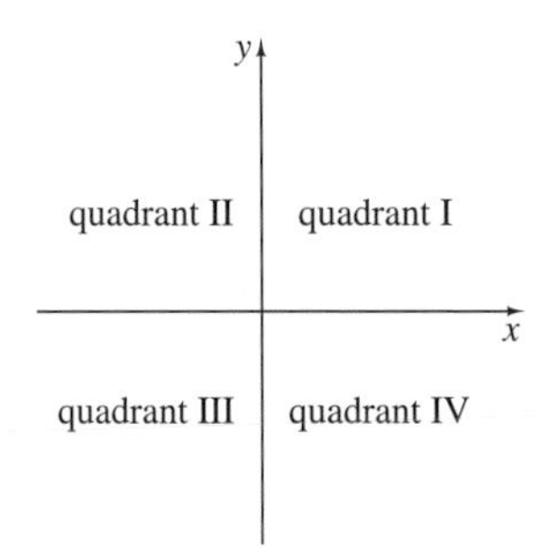

37. $(a, \ b)$ lies in quadrant I.

38. $(-a, \ -b)$ lies in quadrant IV.

39. $(0, \ b)$ lies on the y-axis.

40. $(a, \ 0)$ lies on the x-axis.

41. $(0, \ -b)$ lies on the x-axis.

42. $(-a, \ 0)$ lies on the y-axis.

43. $(-a, \ b)$ lies in quadrant III.

44. $(a, \ -b)$ lies in quadrant IV.

Review Exercises

Perform the indicated operations on decimals. See Sections 5.2 to 5.4.

45. $5.6 - 3.9$

46. $5 + 2.54 + 8.7$

47. 5.6×3.9

48. $0.56 \div 0.8$

49. $(0.236)(-100)$

50. $44.72 \div 100$

ANSWERS

31. ______

32. ______

33. ______

34. ______

35. ______

36. ______

37. ______

38. ______

39. ______

40. ______

41. ______

42. ______

43. ______

44. ______

45. ______

46. ______

47. ______

48. ______

49. ______

50. ______

TAPE PA 6.3

6.3

GRAPHING LINEAR EQUATIONS

OBJECTIVES

1. Define linear equations in two variables.
2. Graph linear equations by plotting points.
3. Graph linear equations by finding three ordered-pair solutions.

1 Now that we know how to plot points in a rectangular coordinate system and how to find ordered-pair solutions, we are ready to graph linear equations in two variables. First, we give a formal definition.

LINEAR EQUATION IN TWO VARIABLES

A linear equation in two variables is an equation that can be written in the form

$$ax + by = c$$

where a, b, and c are numbers, and a and b are not both 0.

In the last section, we recognized that a linear equation in two variables has many solutions. For the linear equation $x + y = 7$, we listed, for example, the solutions $(3, 4)$, $(0, 7)$, $(-1, 8)$, $(7, 0)$, $(5, 2)$, and $(4, 3)$. Are these all the solutions? No. There are infinitely many solutions of the equation $x + y = 7$ since there are infinitely many pairs of numbers whose sum is 7. Every linear equation in two variables has infinitely many ordered-pair solutions. Since it is impossible to list every solution, we graph the solutions instead.

2 Fortunately, the pattern described by the solutions of a linear equation makes "seeing" the solutions possible by graphing. This is so because **all the solutions of a linear equation in two variables correspond to points on a single straight line.** If we plot just a few of these points and draw the straight line connecting them, we have a complete graph of all the solutions.

To graph the equation $x + y = 7$, then, we plot a few ordered-pair solutions, say $(3, 4)$, $(0, 7)$, and $(-1, 8)$.

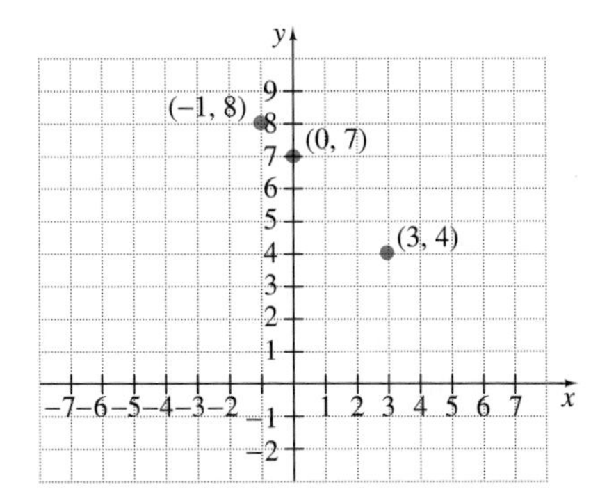

x	y
0	0
1	5
2	10

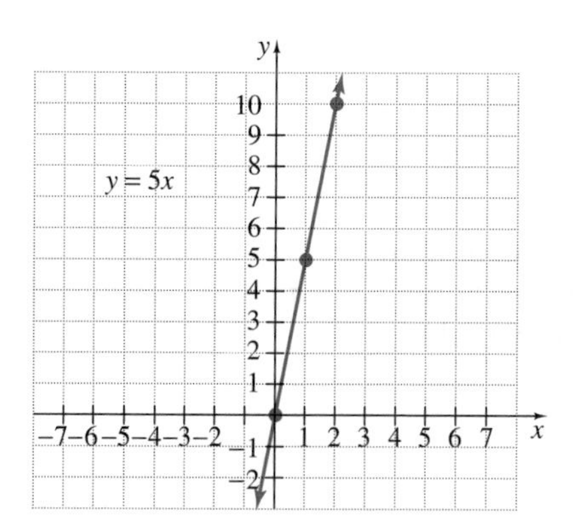

Next, we graph a few special linear equations.

PRACTICE PROBLEM 4

Graph $y = -1$.

EXAMPLE 4

Graph $y = 4$.

Solution:

The equation $y = 4$ can be written as $0x + y = 4$. When the equation is written in this form, notice that no matter what value we choose for x, y is always 4.

$$0 \cdot x + y = 4$$
$$0 \cdot (\text{any number}) + y = 4$$
$$0 + y = 4$$
$$y = 4$$

Fill in a table listing ordered-pair solutions of $y = 4$. Choose any three x-values. The y-values must be 4. Plot the ordered-pair solutions and graph $y = 4$.

x	y
0	4
1	4
−2	4

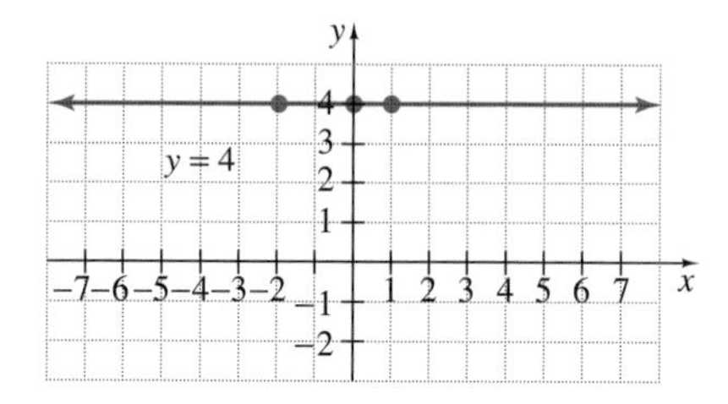

The graph is a horizontal line that crosses the y-axis at 4.

HORIZONTAL LINES

If a is a number, then the graph of $\boldsymbol{y = a}$ is a horizontal line that crosses the y-axis at a.

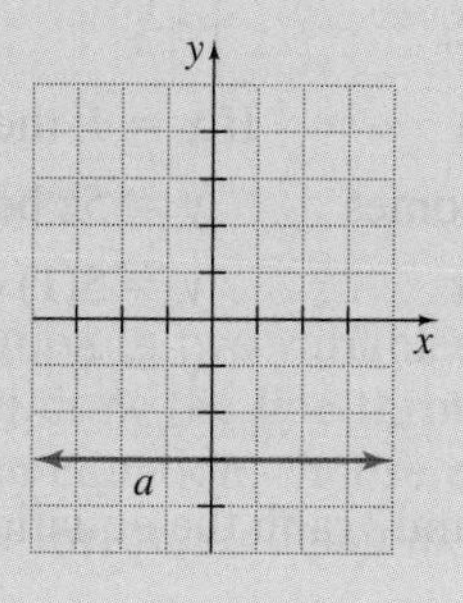

Answer:

4.

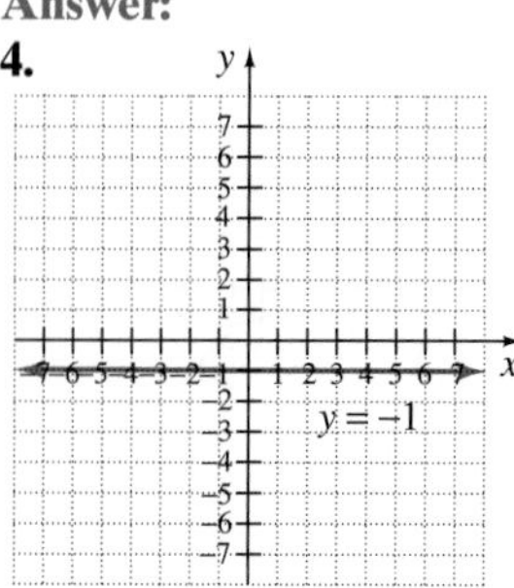

EXAMPLE 5

Graph $x = -2$.

Solution:

The equation $x = -2$ can be written as $x + 0y = -2$. No matter what y-value we choose, x is always -2. Fill in a table listing ordered-pair solutions of $x = -2$. Choose any three y-values. The x-values must be -2. Plot the ordered-pair solutions and graph $x = -2$.

x	y
−2	3
−2	0
−2	−2

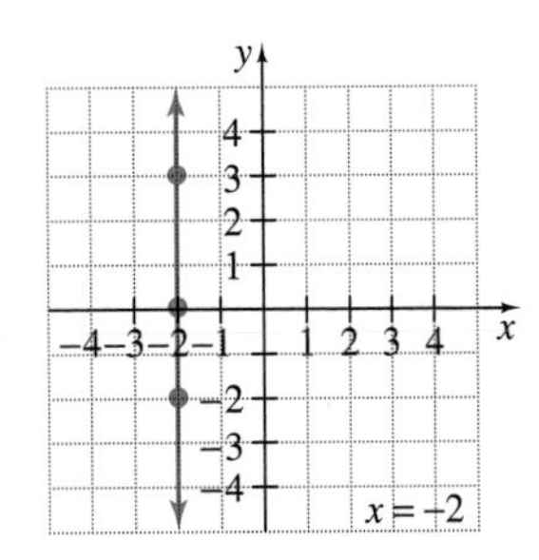

The graph is a vertical line that crosses the x-axis at -2.

VERTICAL LINES

If a is a number, then the graph of $x = a$ is a vertical line that crosses the x-axis at a.

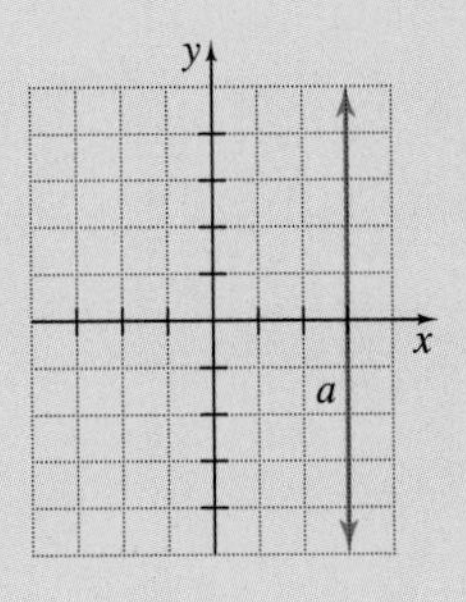

PRACTICE PROBLEM 5

Graph $x = 5$.

Answer:

5.

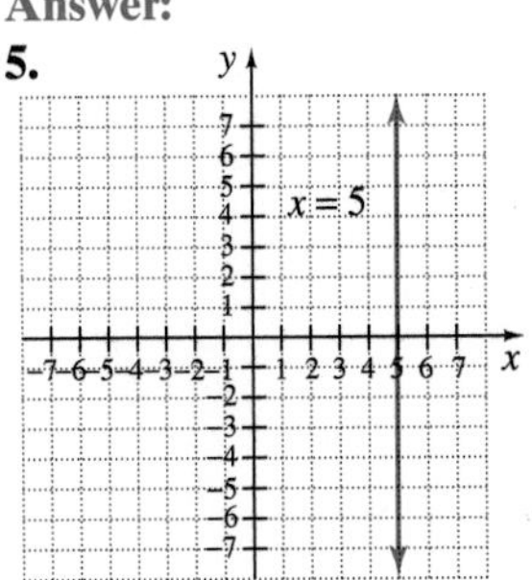

EXERCISE SET 6.3

Graph each equation. See Examples 1 to 3.

1. $x + y = 6$

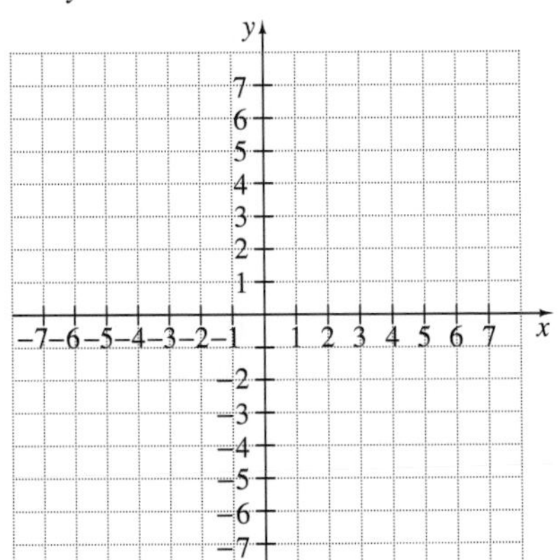

2. $x + y = 7$

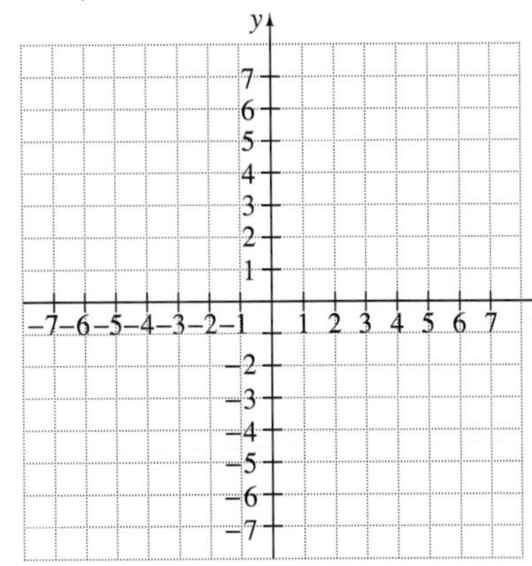

3. $x - y = -2$

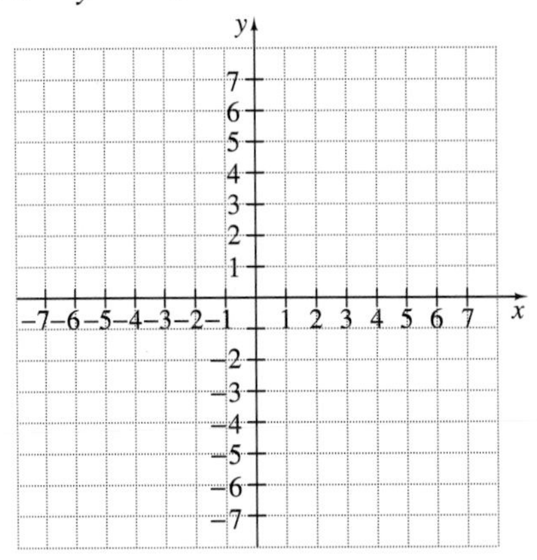

4. $y - x = 6$

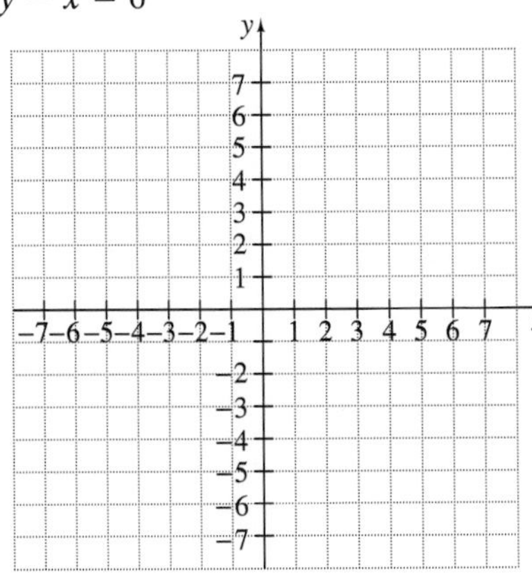

5. $y = 4x$

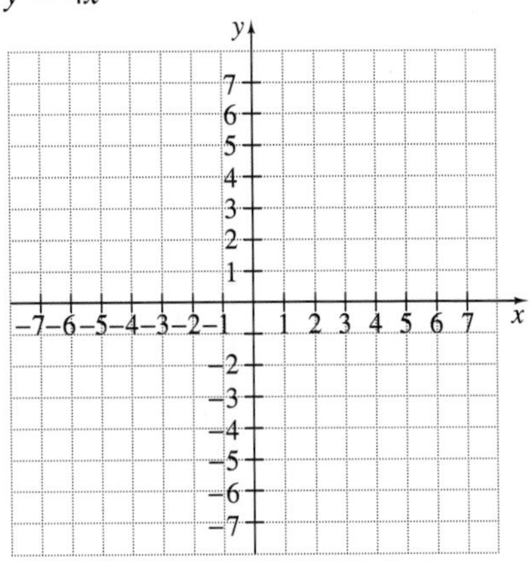

6. $x = 2y$

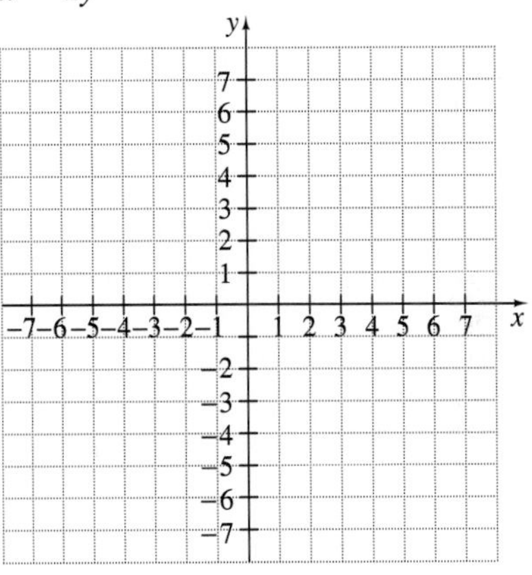

7. $y = 2x - 1$

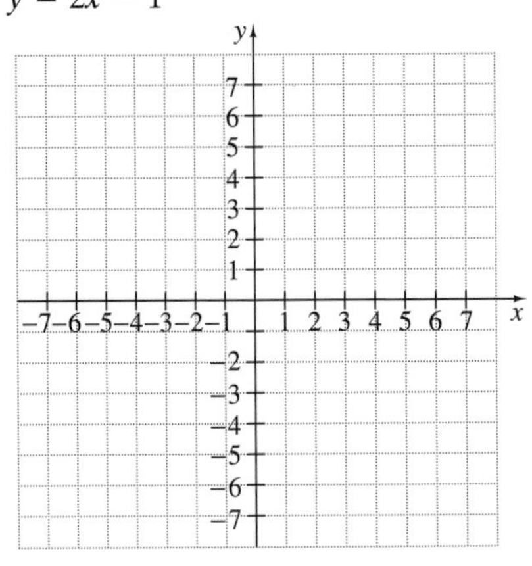

8. $y = x + 5$

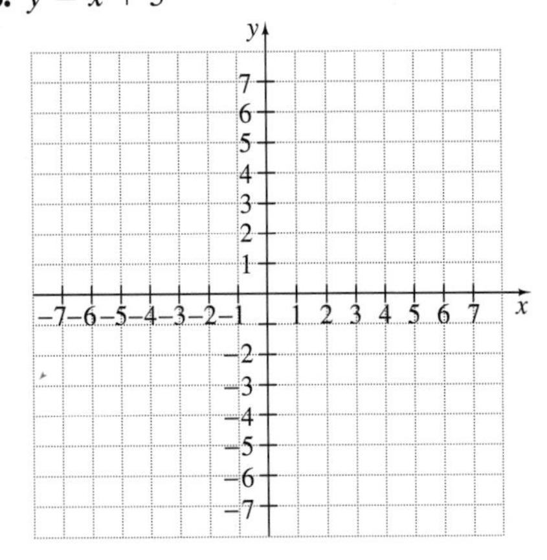

Graph each equation. See Examples 4 and 5.

9. $x = 5$

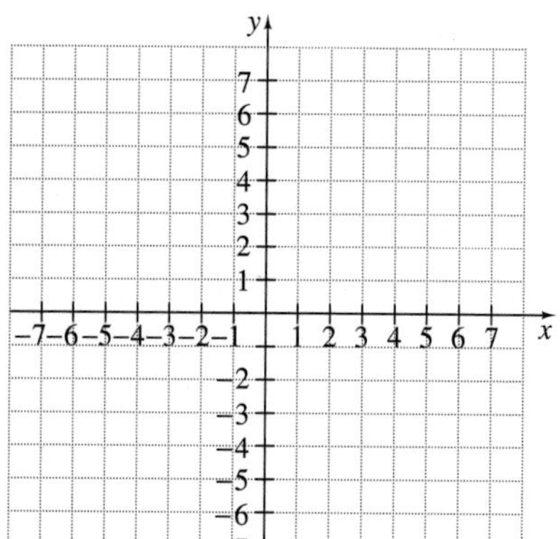

10. $y = 1$

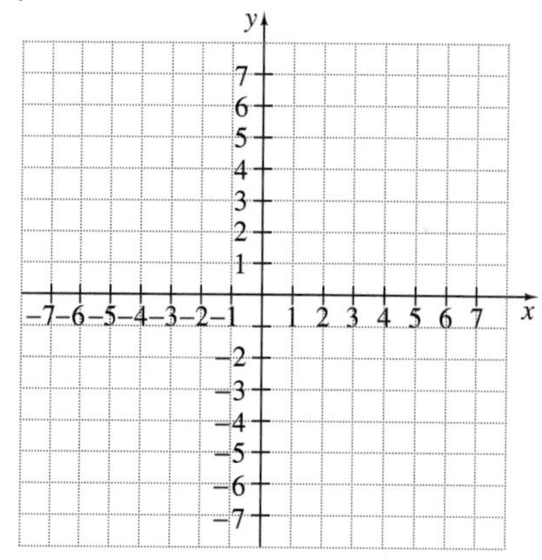

11. $y = -3$

12. $x = -7$

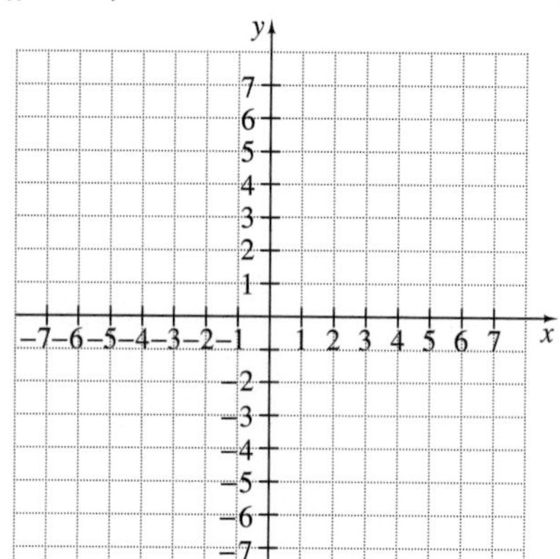

13. $x = 0$

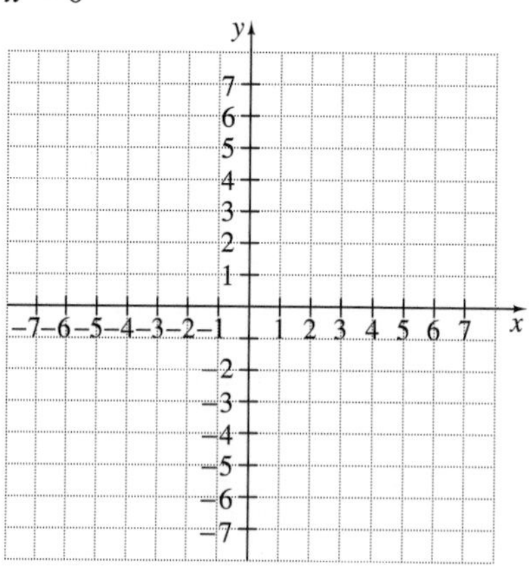

14. $y = 0$

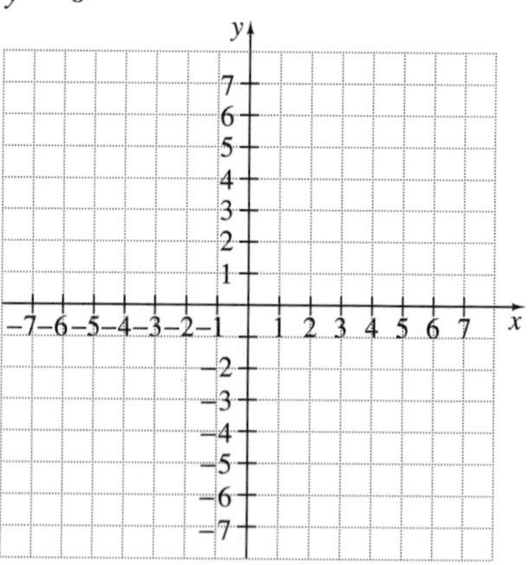

Graph each equation.

15. $y = -2x$

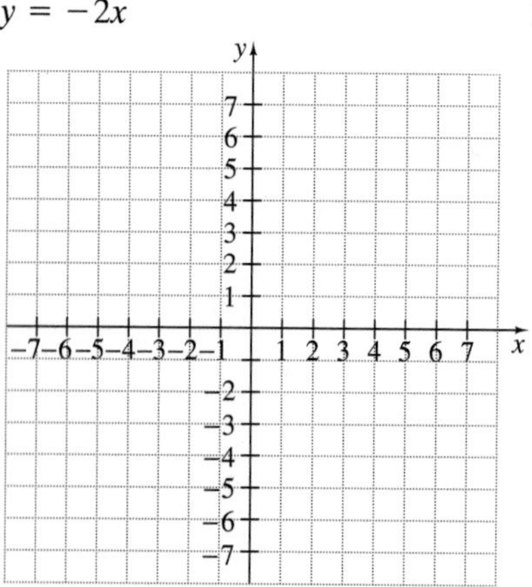

16. $x = y$

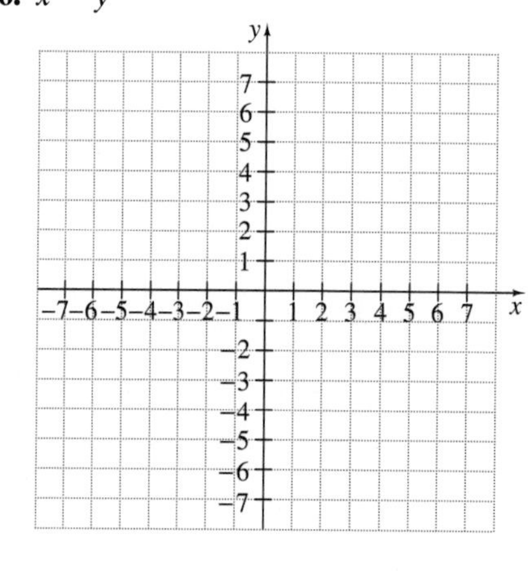

17. $y = -2$

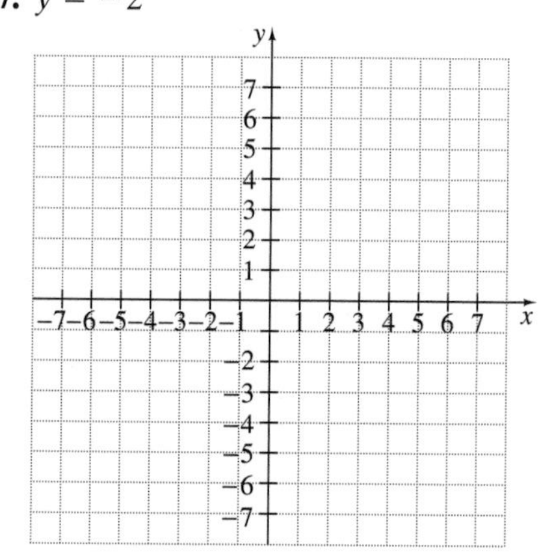

6.4

AVERAGES, MEDIANS, AND MODES

TAPE PA 6.4

OBJECTIVES

1. Find the average of a list of numbers.
2. Find the median of a list of numbers.
3. Find the mode of a list of numbers.

1 Sometimes we want to summarize data by displaying them in a graph, but sometimes it is also desirable to be able to describe a set of data, or a set of numbers, by a single "middle" number. Three such **measures of central tendency** are the average, the median, and the mode.

The most common measure of central tendency is the average (sometimes called the arithmetic mean or the mean). Recall that we first introduced finding the average of a list of numbers in Section 1.6.

> The **average** of a set of number items is the sum of the items divided by the number of items.

EXAMPLE 1

Seven students in a psychology class conducted an experiment on mazes. Each student was given a pencil and asked to successfully complete the same maze. The timed results are below.

STUDENT	ANN	THANH	CARLOS	JESSE	MELINDA	RAMZI	DAYNI
Time (seconds)	13.2	11.8	10.7	16.2	15.9	13.8	18.5

a. Who completed the maze in the shortest time? Who completed the maze in the longest time?

b. Find the average time.

c. How many students took longer than the average time? How many students took shorter than the average time?

Solution:

a. Carlos completed the maze in 10.7 seconds, the shortest time. Dayni completed the maze in 18.5 seconds, the longest time.

b. To find the average, find the sum of the number items and divide by 7, the number of items.

$$\textit{average} = \frac{13.2 + 11.8 + 10.7 + 16.2 + 15.9 + 13.8 + 18.5}{7} = \frac{100.1}{7} = 14.3$$

c. Three students, Jesse, Melinda, and Dayni, had times longer than the average time. Four students, Ann, Thanh, Carlos, and Ramzi had times shorter than the average time.

PRACTICE PROBLEM 1

Find the average of the following test scores: 77, 85, 86, 91, and 88.

Answer:

1. 85.4

Often in college, the calculation of a *grade point average* GPA is a *weighted mean* and is calculated as follows.

PRACTICE PROBLEM 2

Find the grade point average if the following grades were earned in one semester.

GRADE	CREDIT HOURS
A	2
C	4
B	5
D	2
A	2

EXAMPLE 2

The following grades were earned by a student during one semester. Find the student's grade point average.

COURSE	GRADE	CREDIT HOURS
Prealgebra	A	3
Biology	B	3
English	A	3
PE	C	1
Social Studies	D	2

Solution:

To calculate the grade point average, we need to know the point values for the different possible grades. The point values of grades commonly used in colleges and universities are given below.

A: 4, B: 3, C: 2, D: 1, F: 0

To find the grade point average, multiply the number of credit hours for each course by the point value of each grade. The grade point average is the sum of these products divided by the sum of the credit hours.

COURSE	GRADE	POINT VALUE OF GRADE	CREDIT HOURS	(POINT VALUE) • (CREDIT HOURS)
Prealgebra	A	4	3	12
Biology	B	3	3	9
English	A	4	3	12
PE	C	2	1	2
Social Studies	D	1	2	2
		Totals:	12	37

$$\textit{grade point average} = \frac{37}{12} \approx 3.08 \textit{ rounded} \text{ to two decimal places}$$

The student earned a grade point average of 3.08.

2 A second measure of central tendency is called the median.

> The **median** of an ordered set of numbers is the middle number. If the number of items is even, the median is the mean of the two middle numbers.

Answer:
2. 2.73

EXAMPLE 3

Find the median of the following list of numbers: 25, 54, 56, 57, 60, 71, 98.

Solution:

The median is the middle number, 57.

PRACTICE PROBLEM 3

Find the median of the following list of numbers: 7, 9, 13, 23, 24, 35, 38, 41, 43.

EXAMPLE 4

Find the median of the following list of scores: 67, 91, 75, 86, 55, 91.

Solution:

First, list the scores in numerical order and then find the middle number.

55, 67, 75, 86, 91, 91

Since there is an even number of scores, there are two middle numbers. The median is the average of the two middle numbers.

$$median = \frac{75 + 86}{2} = 80.5$$

The median is 80.5.

PRACTICE PROBLEM 4

Find the median of the following list of scores: 43, 89, 78, 65, 95, 95, 88, 71.

The last common measure of central tendency is called the mode.

The **mode** of a set of numbers is the number that occurs most often. (It is possible for a set of numbers to have more than one mode or to have no mode.)

EXAMPLE 5

Find the mode of the following list of numbers: 11, 14, 14, 16, 31, 56, 65, 77, 77, 78, 79.

Solution:

There are two numbers that occur the most often. They are 14 and 77. This list of numbers has two modes, 14 and 77.

PRACTICE PROBLEM 5

Find the mode of the following list of numbers: 9, 10, 10, 13, 15, 15, 15, 17, 18, 18, 20.

EXAMPLE 6

Find the median and the mode of the following set of numbers. These numbers were high temperatures for fourteen consecutive days in a city in Montana.

76, 80, 85, 86, 89, 87, 82, 77, 76, 79, 82, 89, 89, 92

Solution:

First, write the numbers in order.

76, 76, 77, 79, 80, 82, 82, 85, 86, 87, 89, 89, 89, 92

Since there is an even number of items, the median is the mean of the two middle numbers.

$$\text{median} = \frac{82 + 85}{2} = 83.5$$

The mode is 89, since 89 occurs most often.

PRACTICE PROBLEM 6

Find the median and the mode of the following list of numbers: 26, 31, 15, 15, 26, 30, 16, 18, 15, 35.

Answers:
3. 24
4. 83
5. 15
6. median: 22; mode: 15

GROUP ACTIVITY

SURVEYS

1. Conduct a survey of 30 students in one of your classes. Ask each student to report his or her age.
2. Find the difference between the ages of the youngest and oldest survey respondent (this difference between the largest and smallest value is called the *range*). Divide the range into five or six equal age categories. Tally the number of your respondents that fall into each category. Make a bar graph of your results. What does this graph tell you about the ages of your survey respondents?
3. Find the average age of your survey respondents.
4. Find the median age of your survey respondents.
5. Find the mode of the ages of your survey respondents.
6. Compare the mean, median, and mode of your age data. Are these measures similar? Which is largest? Which is smallest? If there is a noticeable difference between any of these measures, can you explain why?

CHAPTER 6 HIGHLIGHTS

DEFINITIONS AND CONCEPTS	EXAMPLES
SECTION 6.2 RECTANGULAR COORDINATE SYSTEM	
The **rectangular coordinate system** consists of two number lines intersecting at the point 0 on each number line. The horizontal number line is called the ***x*-axis** and the vertical number line is called the ***y*-axis.** Every point in the rectangular coordinate system corresponds to an **ordered pair of numbers** such as $(2, \; -2)$ ↑ *x*-coordinate or *x*-value; ↑ *y*-coordinate or *y*-value An ordered pair is a **solution** of an equation if the equation is a true statement when the variables are replaced by the coordinates of the ordered pair.	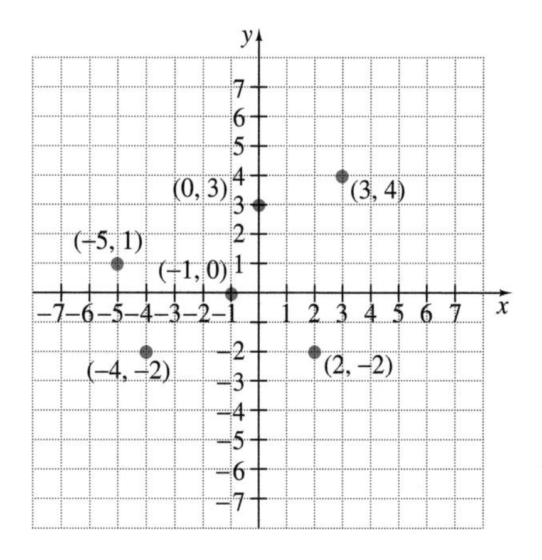 Is $(2, -1)$ a solution of $5x - y = 11$? $5x - y = 11$ $5(2) - (-1) = 11$ Replace x with 2 and y with -1. $10 + 1 = 11$ Multiply. $11 = 11$ True. Yes, $(2, -1)$ is a solution of $5x - y = 11$.

DEFINITIONS AND CONCEPTS	EXAMPLES

SECTION 6.3 GRAPHING LINEAR EQUATIONS

Linear Equation in Two Variables

A linear equation in two variables is an equation that can be written in the form

$$ax + by = c$$

where a, b, and c are numbers, and a and b are not both 0.

To graph a linear equation in two variables, find three ordered pair solutions and draw the line through the plotted points.

Graph $y = 4x$.

Let $x = 1$: $y = 4(1)$, $y = 4$

Let $x = -1$: $y = 4(-1)$, $y = -4$

Let $x = 0$: $y = 4(0)$, $y = 0$

x	y
1	4
−1	−4
0	0

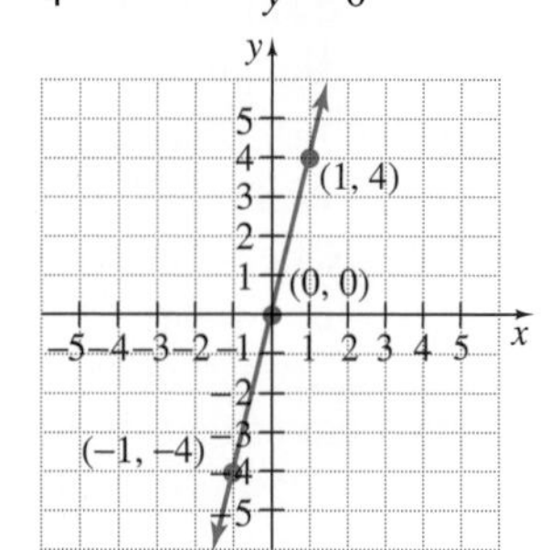

The graph of $y = a$ is a horizontal line that crosses the y-axis at a.

The graph of $x = a$ is a vertical line that crosses the x-axis at a.

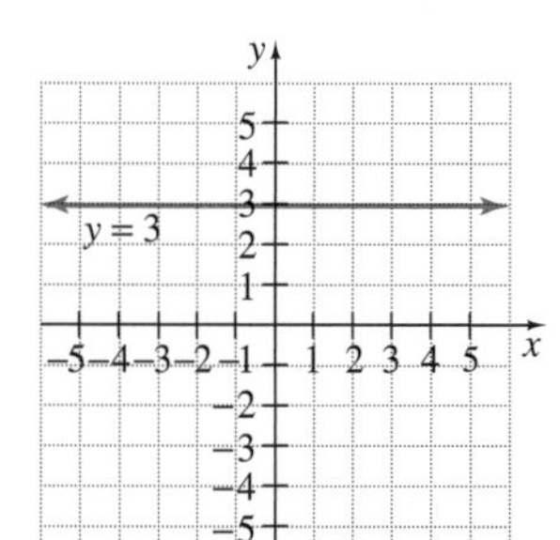

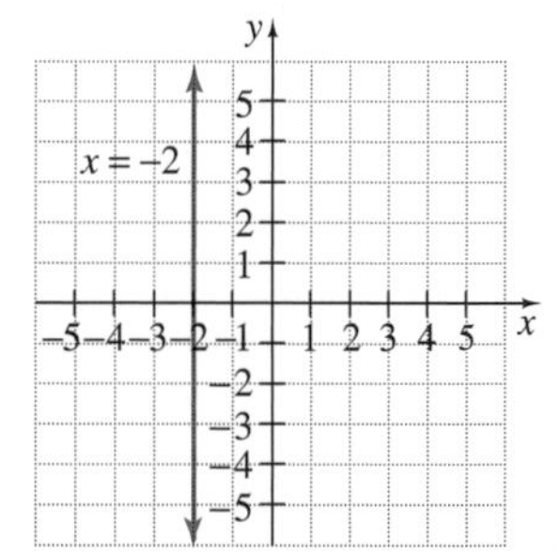

SECTION 6.4 AVERAGES, MEDIANS, AND MODES

The **average** of a set of number items is

$$\text{average} = \frac{\textit{sum of items}}{\textit{number of items}}$$

The **median** of an ordered set of numbers is the middle number. If the number of items is even, the median is the mean of the two middle numbers.

The **mode** of a set of numbers is the number that occurs most often. (A set of numbers may have no mode or more than one mode.)

Find the average, median, and mode of the set of numbers: 33, 35, 35, 43, 68, 68

$$\text{average} = \frac{33 + 35 + 35 + 43 + 68 + 68}{6} = 47$$

The median is the average of the two middle numbers:

$$\text{median} = \frac{35 + 43}{2} = 39$$

There are two modes because there are two numbers that both occur twice:

modes: 35 and 68

CHAPTER 6 REVIEW

(6.1) *The pictograph shows the number of new homes constructed, by state.*

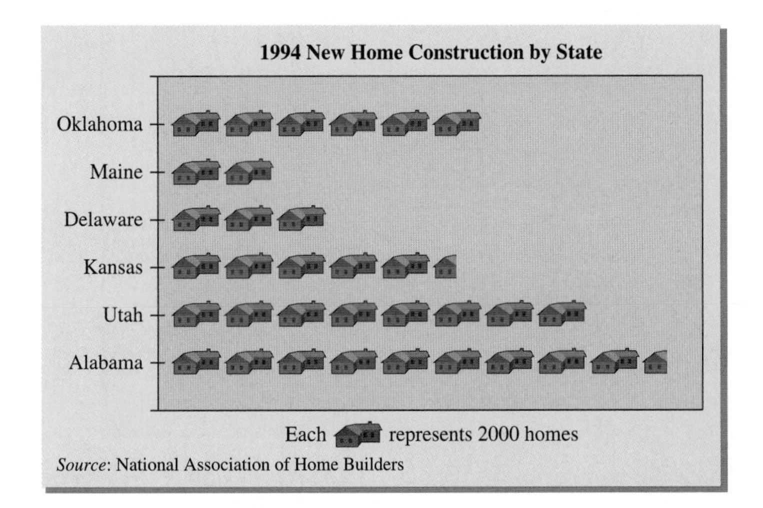

Source: National Association of Home Builders

1. How many new homes were constructed in Oklahoma in 1994?
2. How many new homes were constructed in Kansas in 1994?
3. Which state shown had the most new homes constructed?
4. Which state shown had the fewest new homes constructed?
5. Which state(s) shown had more than 13,000 new homes constructed?
6. Which state(s) shown had fewer than 8000 new homes constructed?

This bar graph shows percent of persons age 25 or more who completed four or more years of college.

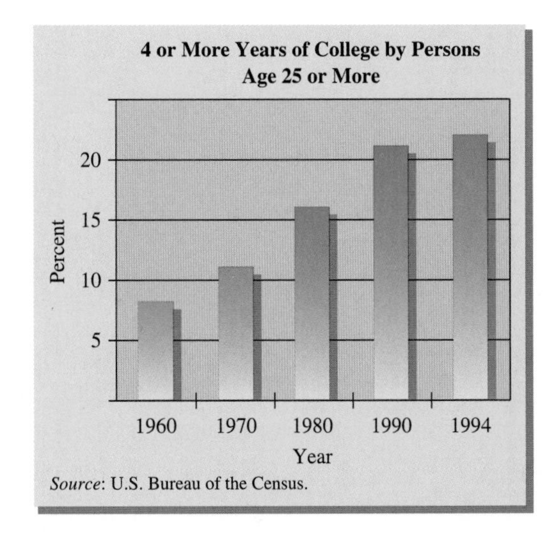

Source: U.S. Bureau of the Census.

7. Approximate the percent of persons who completed four or more years of college in 1960.
8. What year shown had the greatest percent of persons completing four or more years of college?
9. What years shown had 15% or more of persons completing four or more years of college?
10. Describe any patterns you notice in this graph.

ANSWERS

1. ______
2. ______
3. ______
4. ______
5. ______
6. ______
7. ______
8. ______
9. ______
10. ______

ANSWERS

11. ____________

12. ____________

13. ____________

14. ____________

15. ____________

16. ____________

17. ____________

18. ____________

19. ____________

This double line graph shows the number of deaths per 100,000 persons from motor vehicle accidents or firearms.

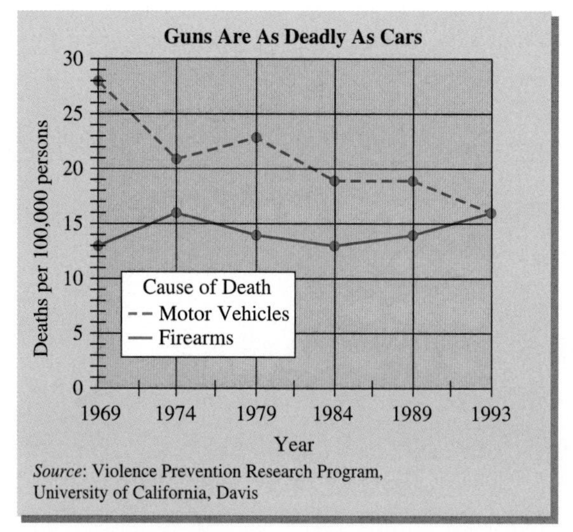

11. Approximate the number of deaths per 100,000 persons caused by firearms in 1969.

12. Approximate the number of deaths per 100,000 persons caused by motor vehicle accidents in 1969.

13. In what year(s) is the number of deaths by firearms decreasing?

14. In what year(s) is the number of deaths by motor vehicle accidents increasing?

15. Describe the meaning of the single data point for 1993.

(6.2) *Complete and graph the ordered-pair solutions of the given equation.*

16. $x = -7y; (0,\quad), (\quad, -1), (-7,\quad)$

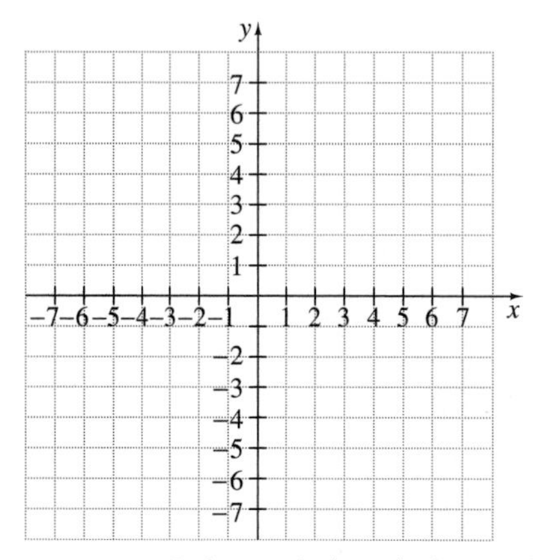

17. $y = 3x - 2; (0,\quad), (1,\quad), (-2,\quad)$

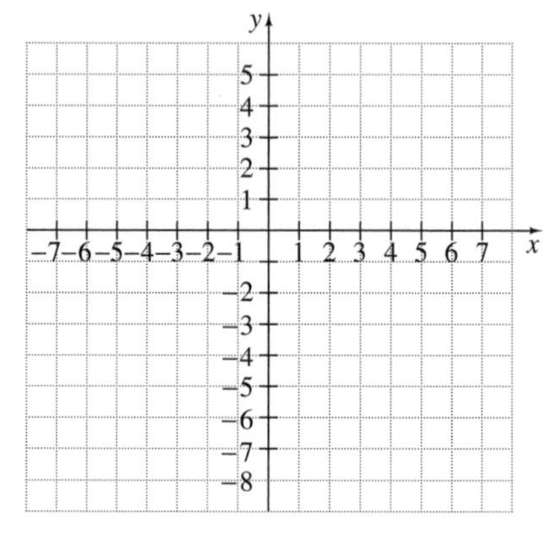

18. $x + y = -9; (-1,\quad), (\quad, 0), (-5,\quad)$

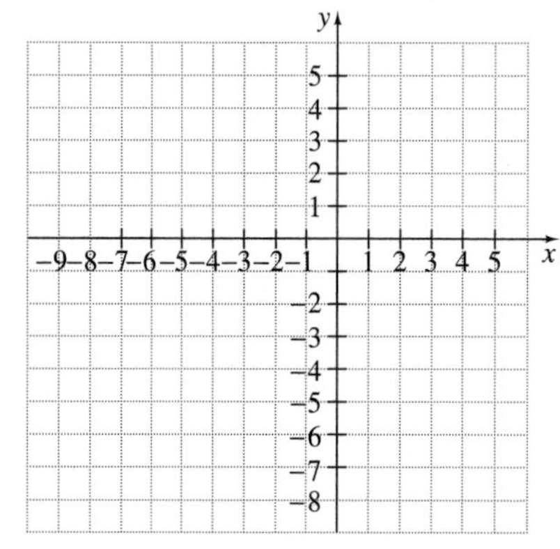

19. $x - y = 3; (4,\quad), (0,\quad), (\quad, 3)$

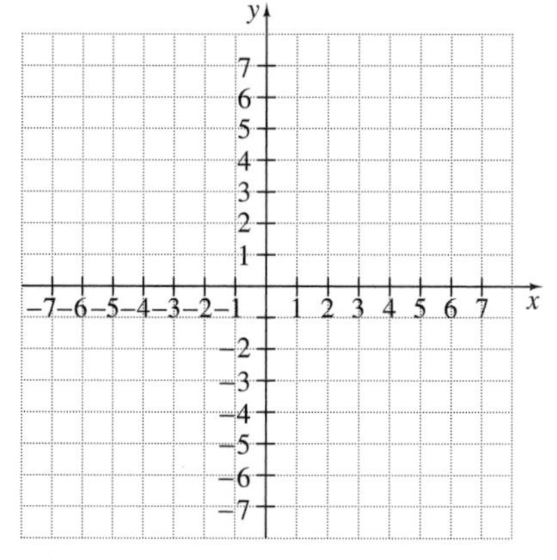

20. $y = 3x; (1, \quad), (-2, \quad), (\quad, 0)$

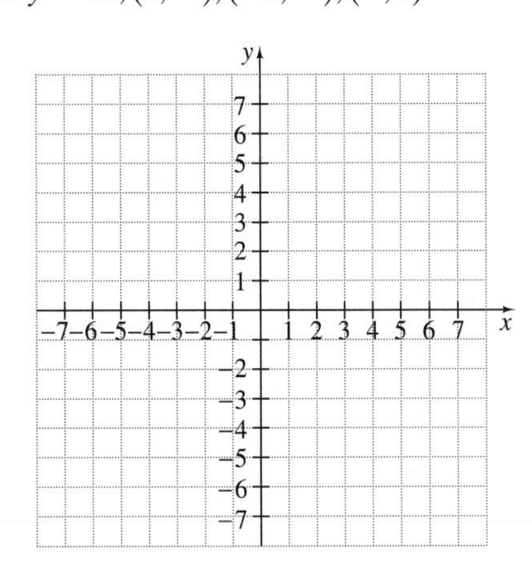

21. $x = y + 6; (1, \quad), (6, \quad), (-1, \quad)$

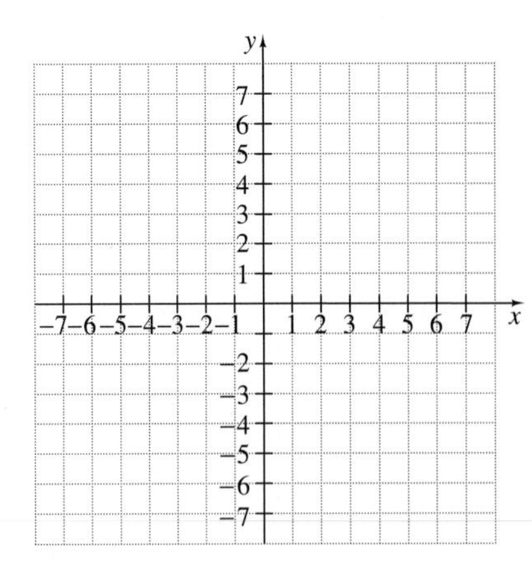

(6.3) *Graph each linear equation.*

22. $x = -6$

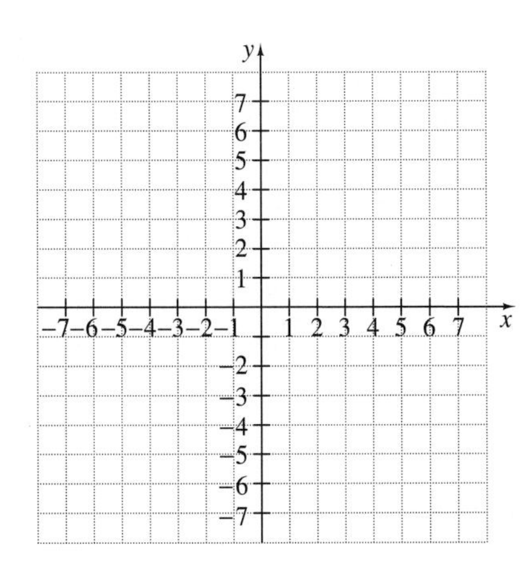

23. $y = 0$

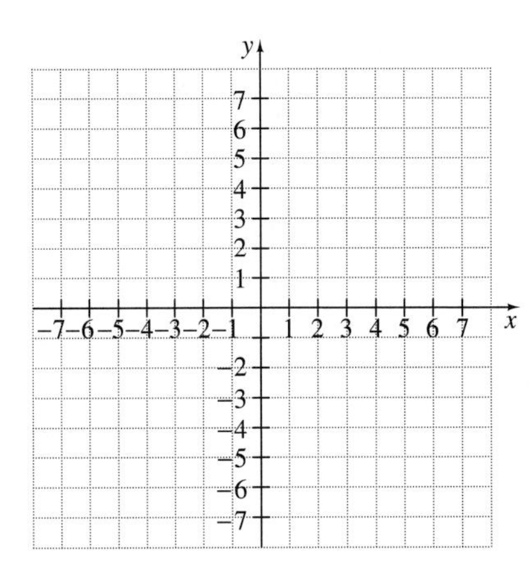

24. $x + y = 11$

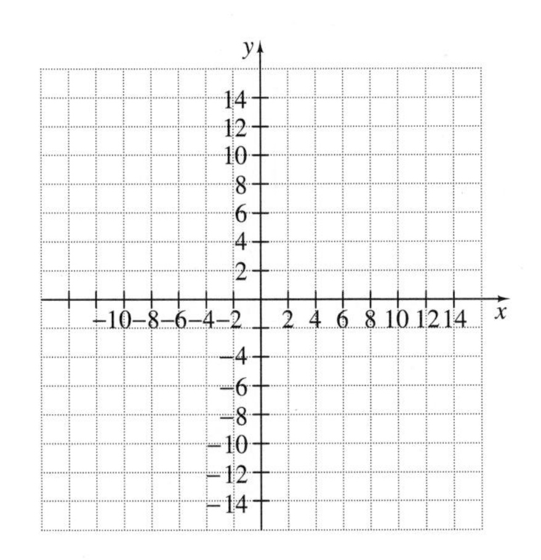

25. $x - y = 11$

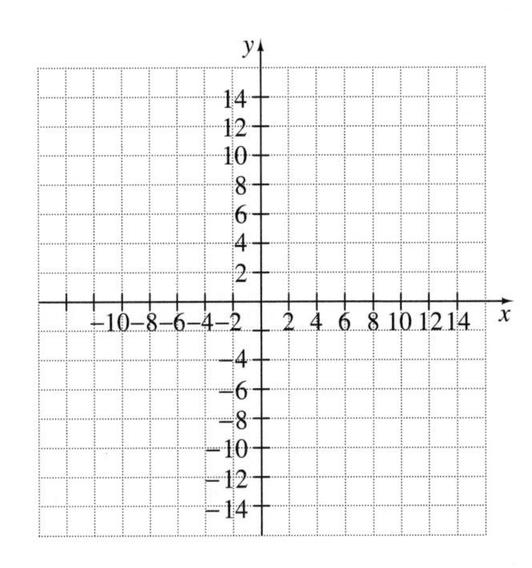

ANSWERS

20. ____________________

21. ____________________

ANSWERS

32. ____________

33. ____________

34. ____________

35. ____________

26. $y = 4x - 2$

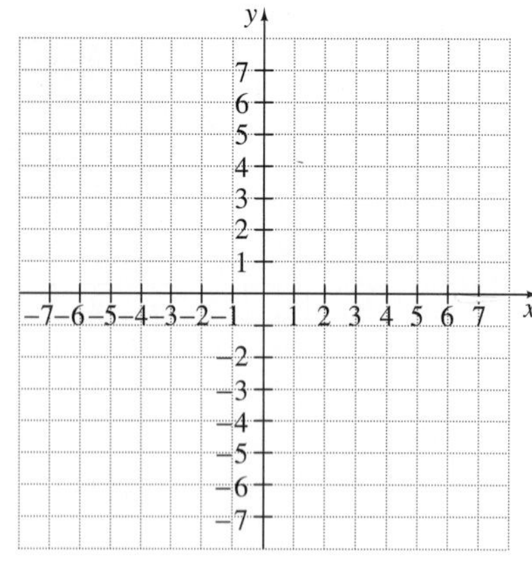

27. $y = 5x$

28. $x = -2y$

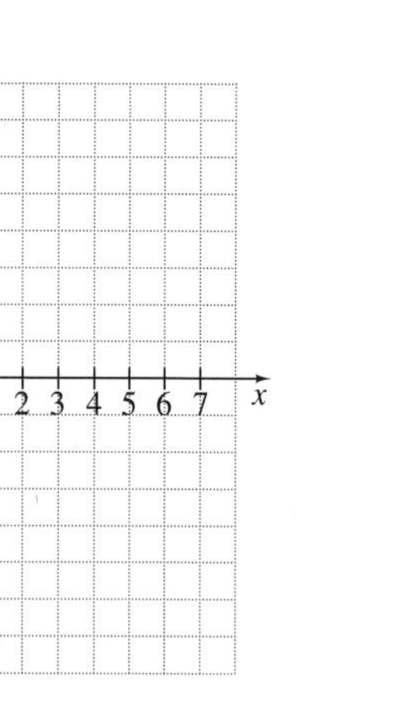

29. $x + y = -1$

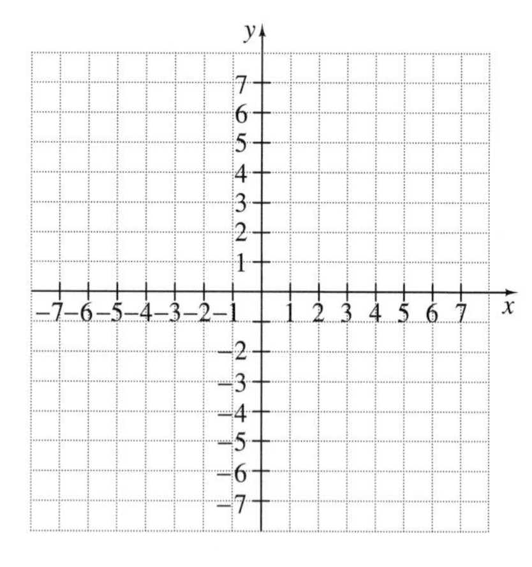

30. $2x - 3y = 12$

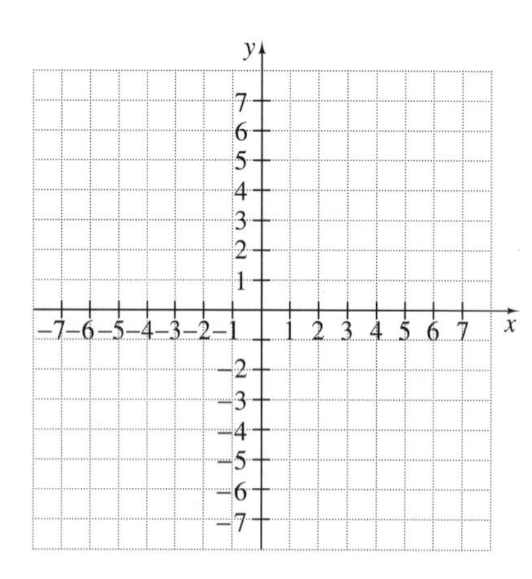

31. $x = \frac{1}{2}y$

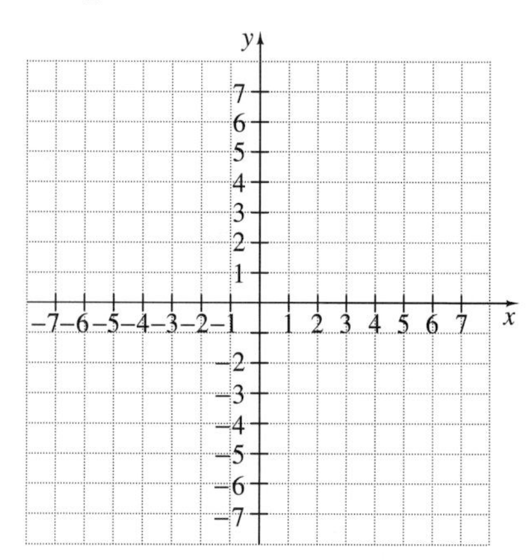

(6.4) *Find the average, the median, and any mode(s) for each list of numbers.*

32. 13, 23, 33, 14, 6

33. 45, 21, 60, 86, 64

34. \$14,000, \$20,000, \$12,000, \$20,000, \$36,000, \$45,000

35. 560, 620, 123, 400, 410, 300, 400, 780, 430, 450

For Exercises 36 and 37 the grades are given for a student for a particular semester. Find the grade point average. If necessary round the grade point average to the nearest hundredth.

36.

GRADE	CREDIT HOURS
A	3
A	3
C	2
B	3
C	1

37.

GRADE	CREDIT HOURS
B	3
B	4
C	2
D	2
B	3

ANSWERS

36. ____________

37. ____________

CHAPTER 6 TEST

The pictograph below shows the money collected each week from a wrapping paper fundraiser.

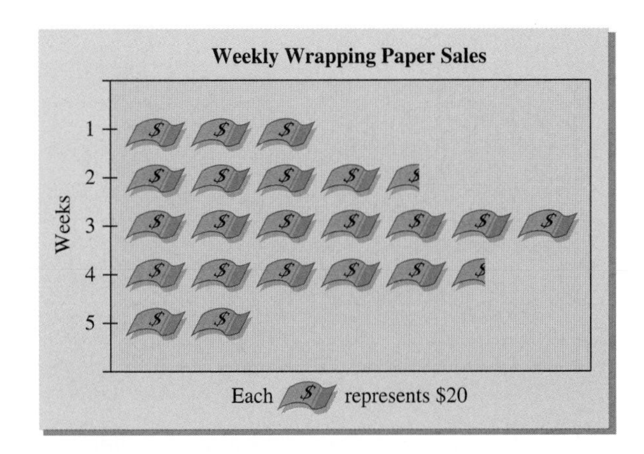

1. How much money was collected during the second week?

2. During which week was the most money collected? How much money was collected during that week?

3. What was the total money collected for the fundraiser?

The bar graph below shows the normal monthly precipitation in centimeters for Chicago.

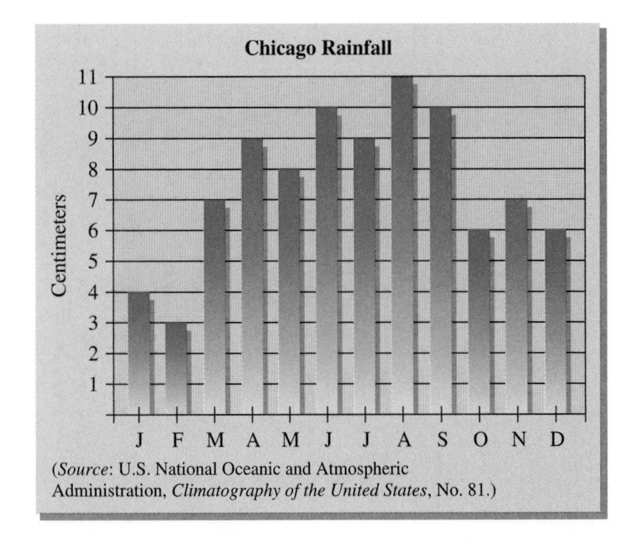

(*Source*: U.S. National Oceanic and Atmospheric Administration, *Climatography of the United States*, No. 81.)

4. During which months did Chicago have greater than 9 centimeters of rainfall?

5. During which month did Chicago have the least amount of rainfall? How much rain fell during that month?

6. During which month(s) did 7 centimeters of rain fall?

ANSWERS

1. ______

2. ______

3. ______

4. ______

5. ______

6. ______

ANSWERS

7. ______

8. ______

9. ______

10. ______

11. ______

12. ______

Find the coordinates of each point.

7. A **8.** B

9. C **10.** D

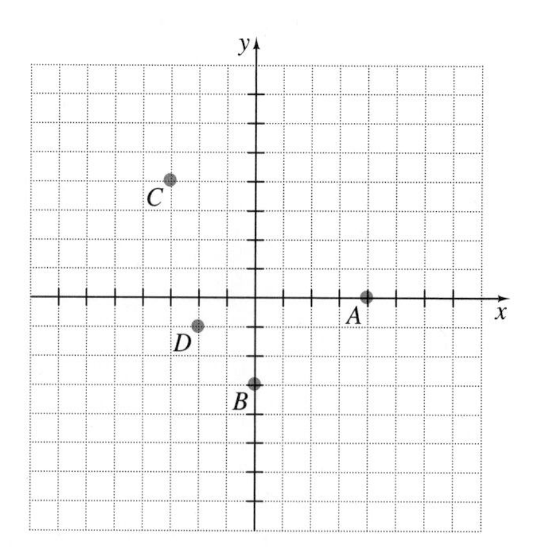

Complete and graph the ordered-pair solutions of the given equation.

11. $x = -6y$; $(0,\quad)$, $(\quad, 1)$, $(12,\quad)$

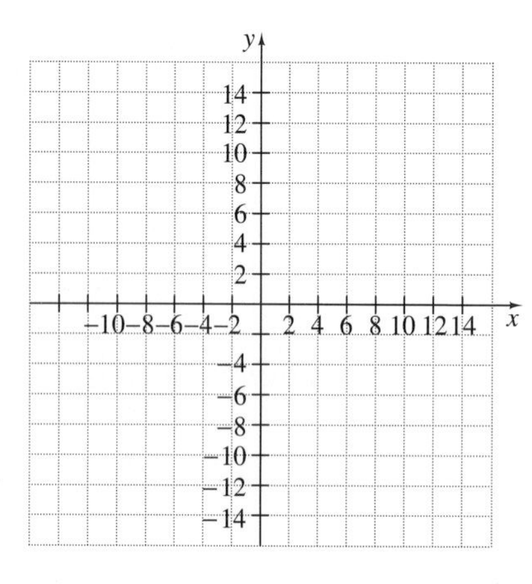

12. $y = 7x - 4$; $(2,\quad)$, $(-1,\quad)$, $(0,\quad)$

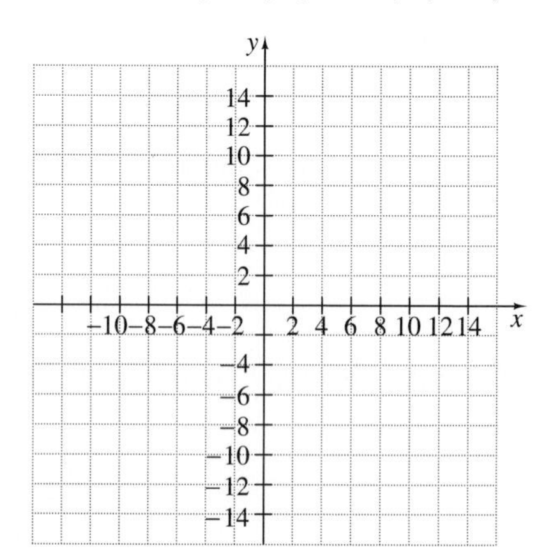

Graph each linear equation.

13. $y + x = -4$

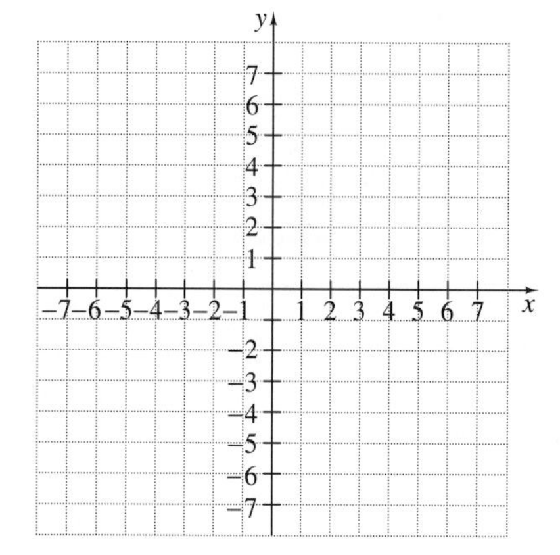

14. $y = -4$

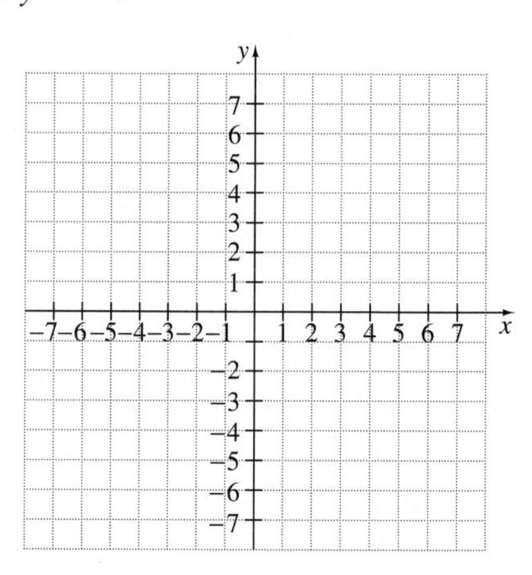

ANSWERS

19. ______

20. ______

21. ______

15. $y = 3x - 5$

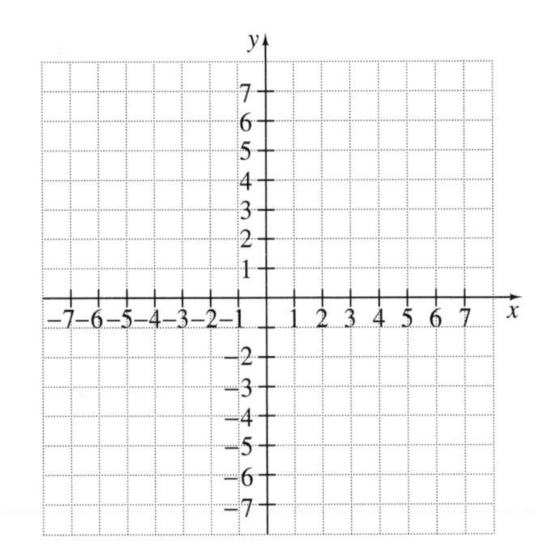

16. $x = 5$

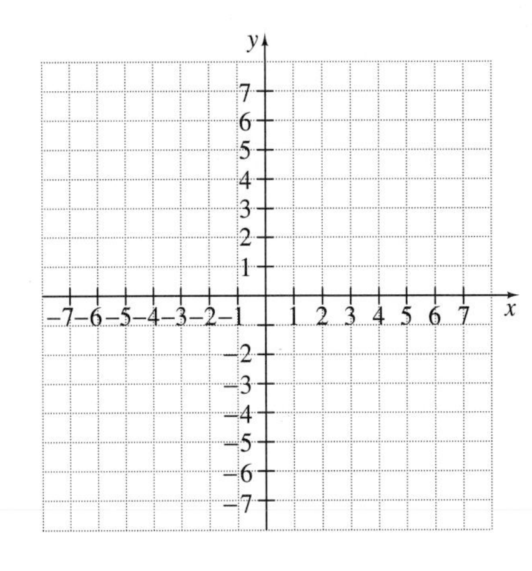

17. $y = -\frac{1}{2}x$

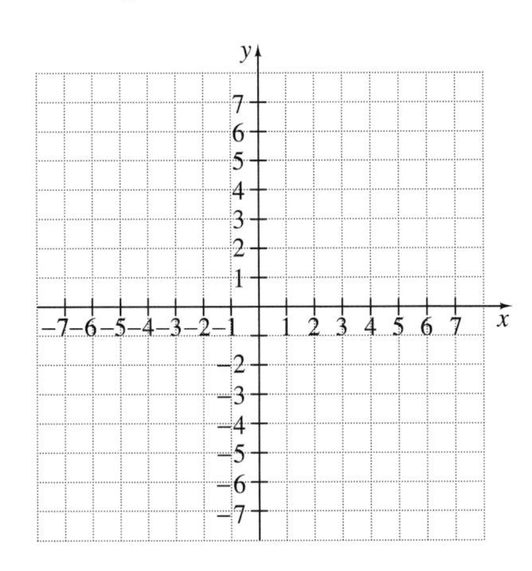

18. $3x - 2y = 12$

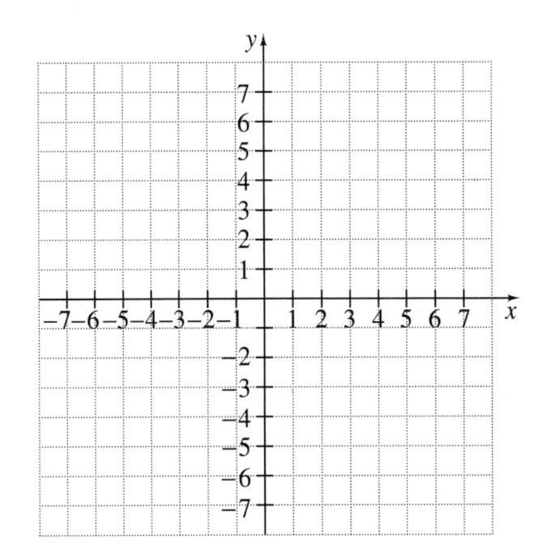

Find the average, median, and mode of each list of numbers.

19. 26, 32, 42, 43, 49

20. 8, 10, 16, 16, 14, 12, 12, 13

Find the grade point average. If necessary, round to the nearest hundredth.

21.

GRADE	CREDIT HOURS
A	3
B	3
C	3
B	4
A	1

EXAMPLE 6

Solve for y: $\frac{14}{y} = \frac{12}{16}$.

Solution:

$$\frac{14}{y} = \frac{12}{16}$$

$12y = 224$ Cross multiply.

$\frac{12y}{12} = \frac{224}{12}$ Divide both sides by 12.

$y = \frac{56}{3}$ Simplify the fraction.

Check to see that the solution is $\frac{56}{3}$.

PRACTICE PROBLEM 6

Solve for z: $\frac{15}{z} = \frac{8}{10}$.

EXAMPLE 7

Solve for x: $\frac{1.6}{1.1} = \frac{x}{0.3}$. Round the solution to the nearest hundredth.

Solution:

$$\frac{1.6}{1.1} = \frac{x}{0.3}$$

$1.1x = 0.48$ Cross multiply.

$\frac{1.1x}{1.1} = \frac{0.48}{1.1}$ Divide both sides by 1.1.

$x \approx 0.44$ Round to the nearest hundredth.

PRACTICE PROBLEM 7

Solve for y: $\frac{3.4}{1.8} = \frac{y}{3}$. Round the solution to the nearest tenth.

Answer:

6. $\frac{75}{4}$ or 18.75

7. 5.7

EXERCISE SET 7.2

Write the sentence as a proportion. See Example 1.

1. 10 diamonds is to 6 opals as 5 diamonds is to 3 opals.

2. 8 books is to 6 courses as 4 books is to 3 courses.

3. 3 printers is to 12 computers as 1 printer is to 4 computers.

4. 4 hit songs is to 16 releases as 1 hit song is to 4 releases.

5. 1 raisin is to 5 cornflakes as 8 raisins is to 40 cornflakes.

6. 1 conviction is to 3 arrests as 12 convictions is to 36 arrests.

7. 6 eagles is to 58 sparrows as 3 eagles is to 29 sparrows.

8. 12 errors is to 8 pages as 1.5 errors is to 1 page.

9. $2\frac{1}{4}$ cups of flour is to 24 cookies as $6\frac{3}{4}$ cups flour is to 72 cookies.

10. $1\frac{1}{2}$ cups milk is to 10 bagels as $\frac{3}{4}$ cup milk is to 5 bagels.

Determine whether the proportion is a true proportion. See Example 2.

11. $\frac{15}{9} = \frac{5}{3}$

12. $\frac{8}{6} = \frac{20}{15}$

13. $\frac{8}{6} = \frac{9}{7}$

14. $\frac{7}{12} = \frac{4}{7}$

15. $\frac{9}{36} = \frac{2}{8}$

16. $\frac{8}{24} = \frac{3}{9}$

17. $\frac{5}{8} = \frac{625}{1000}$

18. $\frac{30}{50} = \frac{600}{1000}$

19. $\frac{0.8}{0.3} = \frac{0.2}{0.6}$

20. $\frac{0.7}{0.4} = \frac{0.3}{0.1}$

ANSWERS

1. ______
2. ______
3. ______
4. ______
5. ______
6. ______
7. ______
8. ______
9. ______
10. ______
11. ______
12. ______
13. ______
14. ______
15. ______
16. ______
17. ______
18. ______
19. ______
20. ______

ANSWERS

21. ______
22. ______
23. ______
24. ______
25. ______
26. ______
27. ______
28. ______
29. ______
30. ______
31. ______
32. ______
33. ______
34. ______
35. ______
36. ______
37. ______
38. ______
39. ______
40. ______
41. ______
42. ______
43. ______
44. ______
45. ______
46. ______
47. ______

21. $\frac{4.2}{8.4} = \frac{5}{10}$ **22.** $\frac{8}{10} = \frac{5.6}{0.7}$ **23.** $\frac{3.6}{4.2} = \frac{7.3}{8.6}$ **24.** $\frac{5.1}{3.1} = \frac{10.2}{6.2}$

25. $\frac{\frac{4}{9}}{\frac{5}{8}} = \frac{\frac{8}{15}}{\frac{3}{4}}$ **26.** $\frac{\frac{5}{2}}{\frac{3}{4}} = \frac{\frac{3}{2}}{\frac{2}{3}}$ **27.** $\frac{3\frac{1}{2}}{5} = \frac{7}{10}$ **28.** $\frac{4\frac{3}{4}}{3} = \frac{2}{1\frac{1}{4}}$

29. $\frac{2\frac{2}{5}}{\frac{2}{3}} = \frac{\frac{10}{9}}{\frac{1}{4}}$ **30.** $\frac{5\frac{5}{8}}{\frac{5}{3}} = \frac{4\frac{1}{2}}{1\frac{1}{5}}$

Solve the proportion for the variable. Approximate the solution when indicated. See Examples 3 to 7.

31. $\frac{x}{5} = \frac{6}{10}$ **32.** $\frac{x}{3} = \frac{12}{9}$ **33.** $\frac{30}{10} = \frac{15}{y}$ **34.** $\frac{25}{100} = \frac{7}{y}$

35. $\frac{z}{8} = \frac{50}{100}$ **36.** $\frac{12}{18} = \frac{z}{21}$ **37.** $\frac{n}{6} = \frac{8}{15}$ **38.** $\frac{24}{n} = \frac{60}{96}$

39. $\frac{12}{10} = \frac{x}{16}$ **40.** $\frac{18}{54} = \frac{3}{x}$ **41.** $\frac{n}{\frac{6}{5}} = \frac{4\frac{1}{6}}{6\frac{2}{3}}$ **42.** $\frac{8}{\frac{1}{3}} = \frac{24}{n}$

43. $\frac{\frac{3}{4}}{12} = \frac{y}{48}$ **44.** $\frac{\frac{11}{4}}{\frac{25}{8}} = \frac{7\frac{3}{5}}{y}$ **45.** $\frac{\frac{2}{3}}{\frac{6}{9}} = \frac{12}{z}$ **46.** $\frac{z}{24} = \frac{\frac{5}{8}}{3}$

47. $\frac{n}{0.6} = \frac{0.05}{12}$

48. $\frac{0.2}{0.7} = \frac{8}{n}$

49. $\frac{3.5}{12.5} = \frac{7}{z}$

50. $\frac{7.8}{13} = \frac{z}{2.6}$

51. $\frac{3.2}{0.3} = \frac{x}{1.4}$. Round to the nearest tenth.

52. $\frac{1.8}{n} = \frac{2.5}{8.4}$. Round to the nearest tenth.

53. $\frac{z}{5.2} = \frac{0.08}{6}$. Round to the nearest hundredth.

54. $\frac{4.25}{6.03} = \frac{5}{y}$. Round to the nearest hundredth.

55. $\frac{9}{11} = \frac{x}{4}$. Round to the nearest tenth.

56. $\frac{24}{x} = \frac{7}{3}$. Round to the nearest thousandth.

57. $\frac{43}{17} = \frac{8}{z}$. Round to the nearest thousandth.

58. $\frac{n}{12} = \frac{18}{7}$. Round to the nearest hundredth.

59. $\frac{x}{7} = \frac{0}{8}$

60. $\frac{0}{2} = \frac{z}{3.5}$

61. $\frac{x}{1150} = \frac{588}{483}$

62. $\frac{585}{x} = \frac{117}{474}$

63. $\frac{222}{1515} = \frac{37}{x}$

64. $\frac{1425}{1062} = \frac{x}{177}$

65. Explain the difference between a ratio and a proportion.

66. Explain how to solve a proportion for such as $\frac{x}{18} = \frac{12}{8}$.

Review Exercises

Insert $<$ or $>$ to make a true statement. See Sections 2.1 and 4.1.

67. $-8 \quad 8$

68. $7 \quad -7$

69. $-2 \quad -3$

70. $0 \quad -2$

71. $-5 \quad 0$

72. $-5\frac{1}{3} \quad -6\frac{2}{3}$

73. $-1\frac{1}{2} \quad -2\frac{1}{2}$

74. $-4 \quad -1$

ANSWERS

48. ____
49. ____
50. ____
51. ____
52. ____
53. ____
54. ____
55. ____
56. ____
57. ____
58. ____
59. ____
60. ____
61. ____
62. ____
63. ____
64. ____
65. ____
66. ____
67. ____
68. ____
69. ____
70. ____
71. ____
72. ____
73. ____
74. ____

7.3

PROBLEM SOLVING WITH PROPORTIONS AND SIMILAR TRIANGLES

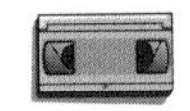

TAPE PA 7.3

OBJECTIVES

1. Solve problems by writing proportions.
2. Solve problems involving similar triangles by writing proportions.

1 Writing proportions is a powerful tool for solving problems in almost every field, including business, chemistry, biology, health sciences, and engineering, and in daily life, too. Given a specified ratio (or rate) of two quantities, a proportion can be used to determine an unknown quantity.

EXAMPLE 1

On a Chamber of Commerce map of Abita Springs, 5 miles corresponds to 2 inches. How many miles correspond to 7 inches?

Solution:

1. UNDERSTAND the problem. To do so, read and reread the problem. You may want to draw a diagram to better understand the problem.

5 miles	5 miles	5 miles	?	= a little over 15 miles
2 inches	2 inches	2 inches	1 inch	= 7 inches

From the diagram, we can see that our solution should be a little over 15 miles.

2. ASSIGN a variable.

Let $x =$ the number of miles represented by 7 inches.

3. TRANSLATE. Since 5 miles corresponds to 2 inches as x miles corresponds to 7 inches, we have the proportion:

$$\begin{matrix}\text{miles} \rightarrow \\ \text{inches} \rightarrow\end{matrix} \frac{5}{2} = \frac{x}{7} \begin{matrix}\leftarrow \text{miles} \\ \leftarrow \text{inches}\end{matrix}$$

4. SOLVE.

$$\frac{5}{2} = \frac{x}{7}$$

$2x = 35$ Cross multiply.

$\frac{2x}{2} = \frac{35}{2}$ Divide both sides by 2.

$$x = \frac{35}{2} \text{ or } 17\frac{1}{2}$$

5. CHECK. Replace x with $\frac{35}{2}$.

$$\frac{2}{5} = \frac{7}{\frac{35}{2}}$$

$$7 \cdot 5 = \frac{35}{2} \cdot 2$$

$35 = 35$ True.

PRACTICE PROBLEM 1

On an architect's blueprint, 1 inch corresponds to 12 feet. How long is a wall represented by a $3\frac{1}{2}$-inch line on the blueprint?

Answer:

1. 42 feet

Thus, 7 inches corresponds to $17\frac{1}{2}$ miles. Is this answer reasonable? Since 2 inches corresponds to 5 miles, $3 \cdot 2$ inches or 6 inches corresponds to $3 \cdot 5$ miles or 15 miles, and 7 inches corresponds to a little over 15 miles, so the answer $17\frac{1}{2}$ miles is reasonable.

6. STATE your conclusions. The solution checks and is reasonable, so 7 inches corresponds to $17\frac{1}{2}$ miles.

> REMINDER We can also solve Example 1 by writing the proportion $\frac{5 \text{ miles}}{2 \text{ inches}} = \frac{x \text{ miles}}{7 \text{ inches}}$. Although other proportions may be used to solve Example 1, we will solve by writing proportions so that the numerators have the same unit measures and the denominators have the same unit measures.

PRACTICE PROBLEM 2

An auto mechanic recommends that 3 ounces of isopropyl alcohol be mixed with a tankful of gas (14 gallons) to increase the octane of the gasoline for better engine performance. At this rate, how many gallons of gas can be treated with a 16-ounce bottle of alcohol?

EXAMPLE 2

The standard dose of an antibiotic is 4 cc (cubic centimeters) for every 25 pounds (lb) of body weight. At this rate, find the standard dose for a 140-lb woman.

Solution:

1. UNDERSTAND. Read and reread the problem. You may want to draw a diagram to better understand the problem and to estimate a reasonable solution.

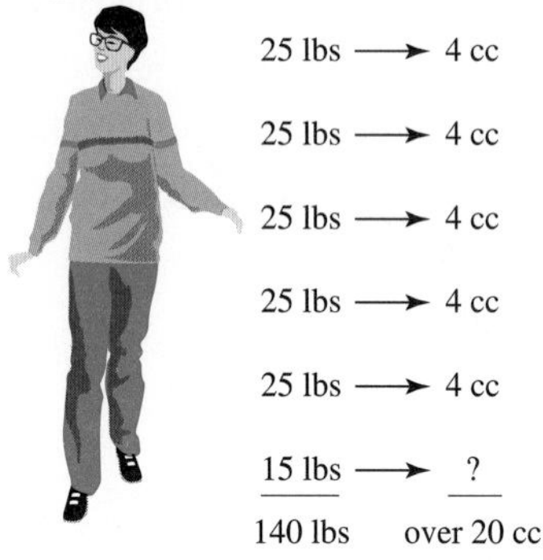

From the diagram, we can see that a reasonable solution is a little over 20 cc.

2. ASSIGN. Let x = number of cc's needed for a 140-lb woman.
3. TRANSLATE. From the problem, we know that 4 cc is to 25 pounds as x cc is to 140 pounds, or

$$\begin{matrix}\text{cubic centimeters} \rightarrow \\ \text{pounds} \rightarrow\end{matrix} \frac{4}{25} = \frac{x}{140} \begin{matrix}\leftarrow \text{cubic centimeters} \\ \leftarrow \text{pounds}\end{matrix}$$

4. SOLVE.

$$\frac{4}{25} = \frac{x}{140}$$

$25 \cdot x = 4 \cdot 140$	Cross multiply.
$25x = 560$	Simplify.
$\frac{25x}{25} = \frac{560}{25}$	Divide both sides by 25.
$x = 22.4$	Simplify.

Answer:

2. $74\frac{2}{3}$ gal

5. CHECK that replacing x with 22.4 makes the proportion true. Recall that this answer is reasonable.
6. STATE. The dosage needed for a 140-lb woman is 22.4 cc.

EXAMPLE 3

A 50-pound bag of fertilizer covers 2400 square feet of lawn. How many bags of fertilizer are needed to cover a town square containing 15,360 square feet of lawn? Round the solution up to the nearest whole bag.

Solution:

1. UNDERSTAND. Read and reread the problem. Draw a picture.

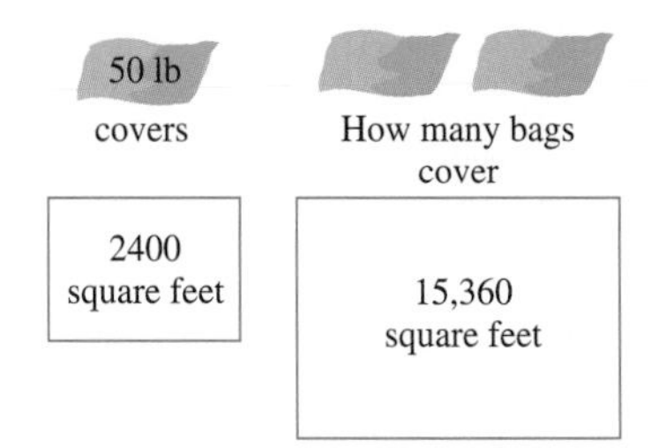

2. ASSIGN. Let $x =$ the number of bags needed to cover 15,360 square feet.
3. TRANSLATE.

$$\begin{array}{r} \text{bags} \rightarrow \\ \text{square feet} \rightarrow \end{array} \frac{1}{2400} = \frac{x}{15{,}360} \begin{array}{l} \leftarrow \text{bags} \\ \leftarrow \text{square feet} \end{array}$$

4. SOLVE $\frac{1}{2400} = \frac{x}{15{,}360}$

$$2400 \cdot x = 1 \cdot 15{,}360 \quad \text{Cross multiply.}$$

$$2400x = 15{,}360 \quad \text{Simplify.}$$

$$\frac{2400x}{2400} = \frac{15{,}360}{2400} \quad \text{Divide both sides by 2400.}$$

$$x = 6.4 \text{ bags} \quad \text{Simplify and attach units.}$$

5. CHECK that replacing x with 6.4 makes the proportion true. Is the answer reasonable?

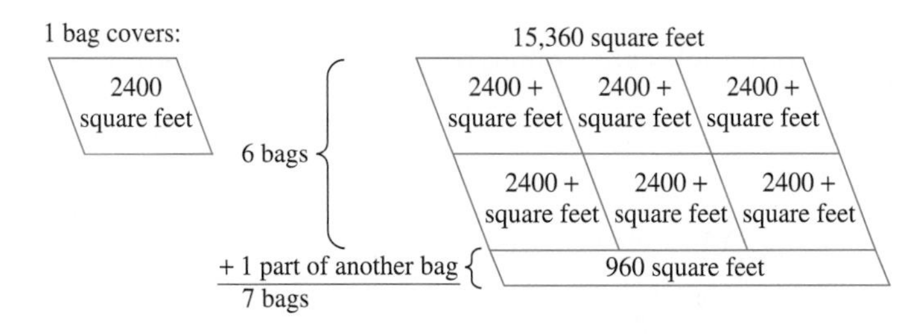

Yes. Since we must buy whole bags of fertilizer, 7 bags are needed.

6. STATE. To cover 15,360 square feet of lawn, 7 bags are needed.

2 Next, we look at a special type of application involving proportions. These special applications involve similar triangles. Similar triangles are found in art, engineering, architecture, biology, and chemistry. Similar triangles have exactly the same shape, although not necessarily the same size. Examine these pairs of similar triangles.

PRACTICE PROBLEM 3

If a gallon of paint covers 400 square feet, how many gallons must be bought to paint a retaining wall 260 feet long and 4 feet high? Round the solution up to the nearest whole gallon.

Answer:
3. 3 gal

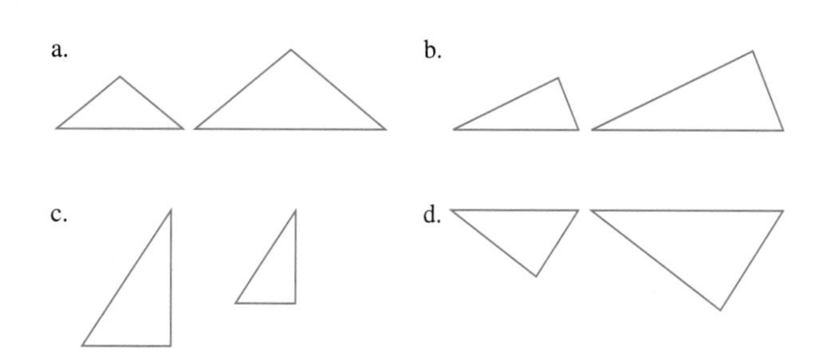

It can be shown that corresponding angles of similar triangles are equal and the ratios of the lengths of corresponding sides are equal. For example, two similar triangles are shown.

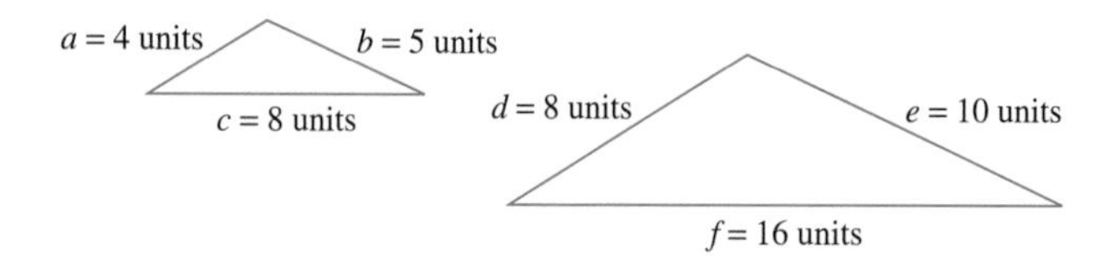

Side a corresponds to side d.

Side b corresponds to side e.

Side c corresponds to side f.

Write the ratios of corresponding sides.

$$\frac{a}{d} = \frac{4}{8} = \frac{1}{2}, \quad \frac{b}{e} = \frac{5}{10} = \frac{1}{2}, \quad \frac{c}{f} = \frac{8}{16} = \frac{1}{2}$$

All these ratios are equal. In this section, we will position similar triangles so they are oriented the same.

We can use the fact that the ratios of the lengths of corresponding sides of similar triangles are equal to find unknown lengths of sides. To do so, we write a proportion that can then be solved.

PRACTICE PROBLEM 4

If the following two triangles are similar, find the unknown length y.

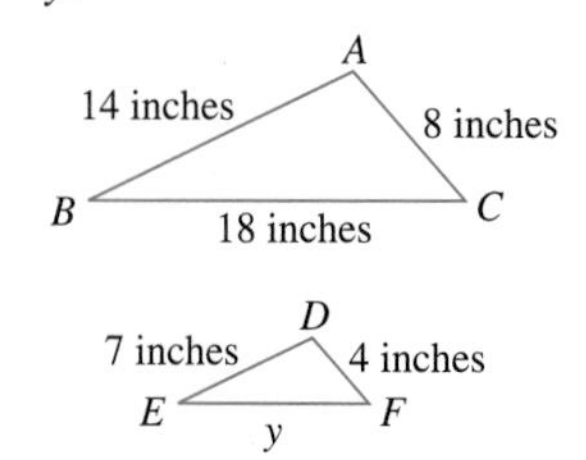

Answer:
4. 9 inches

EXAMPLE 4

If the following two triangles are similar, find the unknown length x.

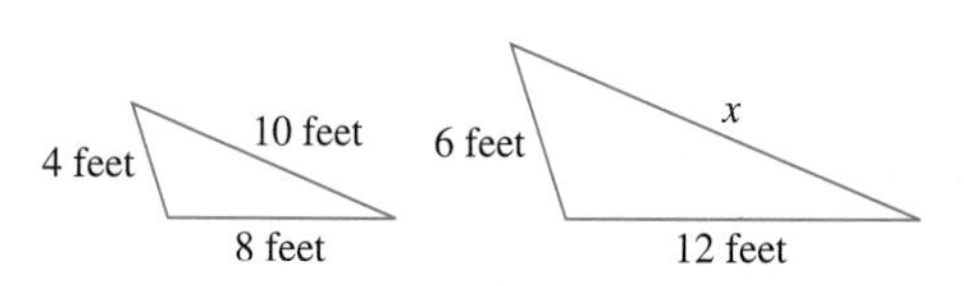

Solution:

Let x represent the length of the unknown side. Since the triangles are similar, the ratios of the corresponding sides are equal.

corresponding sides of triangles $\left[\frac{10}{x} = \frac{4}{6} \right]$ corresponding sides of triangles

Solve this proportion for x.

$$\frac{10}{x} = \frac{4}{6}$$

$$4 \cdot x = 10 \cdot 6$$

$4x = 60$ Simplify.

$\frac{4x}{4} = \frac{60}{4}$ Divide both sides by 4.

$x = 15$ Simplify.

Thus, x is 15 feet.

Many applications involve a diagram containing similar triangles. Surveyors, astronomers, and many other professionals use ratios of similar triangles continually in their work.

EXAMPLE 5

Mel is a 6-foot-tall park ranger who needs to know the height of a particular tree. He notices that when the shadow of the tree is 69 feet long, his own shadow is 9 feet long. Find the height of the tree.

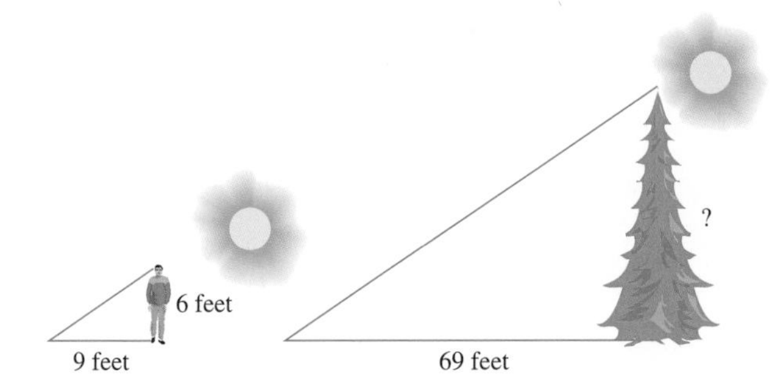

Solution:

1. UNDERSTAND. Read and reread the problem. Notice that the triangle formed by the sun's rays, Mel, and his shadow is similar to the triangle formed by the sun's rays, the tree, and its shadow.
2. ASSIGN. Let $x =$ the height of the tree.
3. TRANSLATE. Write a proportion from the similar triangles formed.

$$\text{height of man} \rightarrow \frac{6}{x} = \frac{9}{69} \leftarrow \text{length of man's shadow}$$
$$\text{height of tree} \rightarrow \qquad \leftarrow \text{length of tree's shadow}$$

$$\text{or } \frac{6}{x} = \frac{3}{23} \text{ (ratio in lowest terms)}$$

4. SOLVE for x.

$$\frac{6}{x} = \frac{3}{23}$$

$$x \cdot 3 = 6 \cdot 23$$

$$3x = 138$$

$$\frac{3x}{3} = \frac{138}{3}$$

$$x = 46$$

PRACTICE PROBLEM 5

Tammy, a firefighter, needs to estimate the height of a burning building. She estimates the length of her shadow to be 8 feet long and the length of the building's shadow to be 60 feet long. Find the height of the building if she is 5 feet tall.

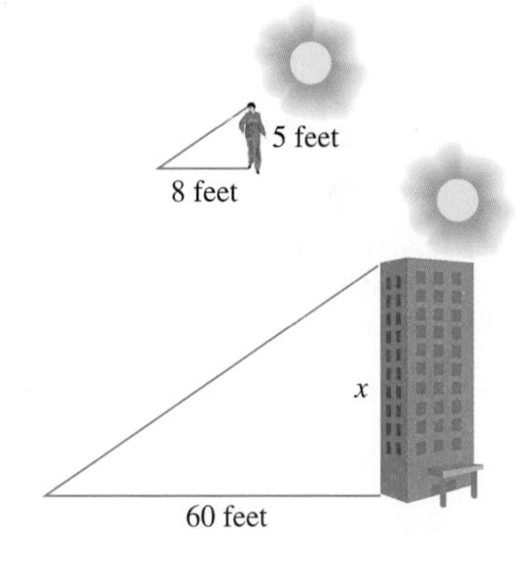

Answer:
5. 37.5 feet

5. CHECK to see that replacing x with 46 in the proportion makes the proportion true.
6. STATE. The height of the tree is 46 feet.

TAPE PA 7.5

7.5

WEIGHT AND MASS

OBJECTIVES

1. Define U.S. units of weight and convert from one unit to another.
2. Used mixed units of weight.
3. Perform arithmetic operations on units of weight.
4. Define metric units of mass and convert from one unit to another.
5. Perform arithmetic operations on units of mass.

WEIGHT

1 Whenever we talk about how heavy an object is, we are concerned with the object's **weight.** We discuss weight when we refer to a 12-ounce box of Rice Krispies, an overweight 19-pound tabby cat, or a barge hauling 24 tons of garbage.

The most common units of weight in the U.S. measurement system are the **ounce,** the **pound,** and the **ton.** The following is a summary of equivalencies between units of weight.

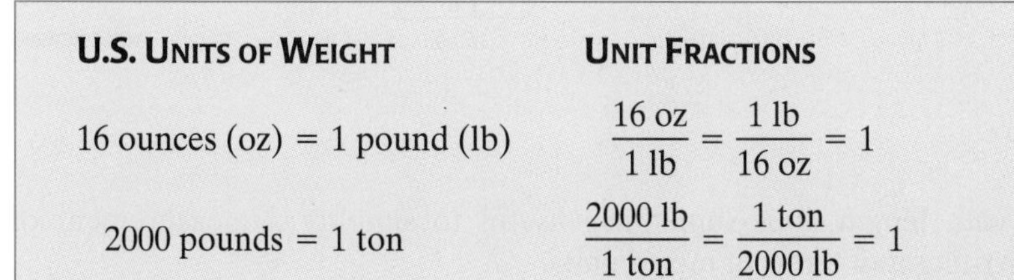

U.S. UNITS OF WEIGHT	UNIT FRACTIONS
16 ounces (oz) = 1 pound (lb)	$\frac{16\text{ oz}}{1\text{ lb}} = \frac{1\text{ lb}}{16\text{ oz}} = 1$
2000 pounds = 1 ton	$\frac{2000\text{ lb}}{1\text{ ton}} = \frac{1\text{ ton}}{2000\text{ lb}} = 1$

Unit fractions that equal 1 will be used to convert between units of weight in the U.S. system. When converting using unit fractions, recall that the numerator of a unit fraction should contain units we are converting to and the denominator should contain original units.

To convert 40 ounces to pounds, multiply by $\frac{1\text{ lb}}{16\text{ oz}}$. ← units converting to / ← original units

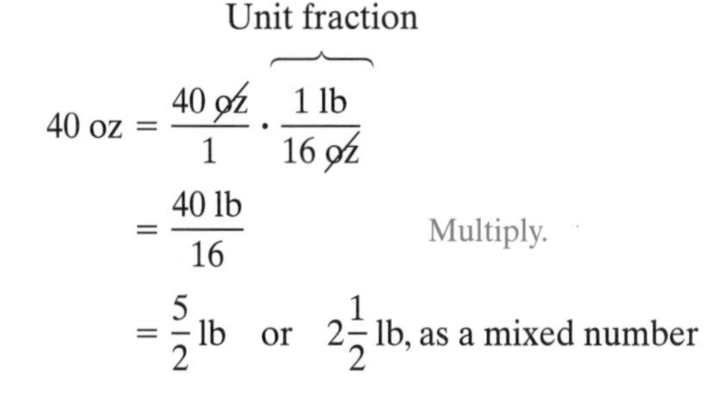

$$40\text{ oz} = \frac{40\text{ }\cancel{\text{oz}}}{1} \cdot \overbrace{\frac{1\text{ lb}}{16\text{ }\cancel{\text{oz}}}}^{\text{Unit fraction}}$$

$$= \frac{40\text{ lb}}{16} \qquad \text{Multiply.}$$

$$= \frac{5}{2}\text{ lb} \quad \text{or} \quad 2\frac{1}{2}\text{ lb, as a mixed number}$$

EXAMPLE 1

Convert 9000 pounds to tons.

Solution:

Multiply 9000 lb by the unit fraction $\frac{1\text{ ton}}{2000\text{ lb}}$. ← units converting to / ← original units

PRACTICE PROBLEM 1

Convert 4500 pounds to tons.

Answer:

1. $2\frac{1}{4}$ tons

$$9000 \text{ lb} = 9000 \cancel{\text{lb}} \cdot \frac{1 \text{ ton}}{2000 \cancel{\text{lb}}} = \frac{9000 \text{ tons}}{2000} = \frac{9}{2} \text{ tons or } 4\frac{1}{2} \text{ tons}$$

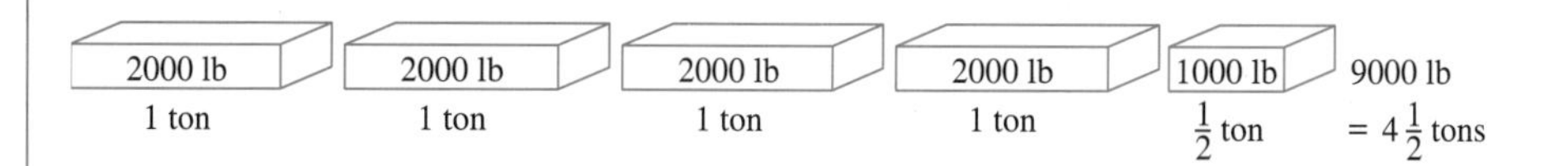

PRACTICE PROBLEM 2

Convert 56 ounces to pounds.

EXAMPLE 2

Convert 3 pounds to ounces.

Solution:

Multiply by the unit fraction $\frac{16 \text{ oz}}{1 \text{ lb}}$ to convert from pounds to ounces.

$$3 \text{ lb} = 3 \cancel{\text{lb}} \cdot \frac{16 \text{ oz}}{1 \cancel{\text{lb}}} = 3 \cdot 16 \text{ oz} = 48 \text{ oz}$$

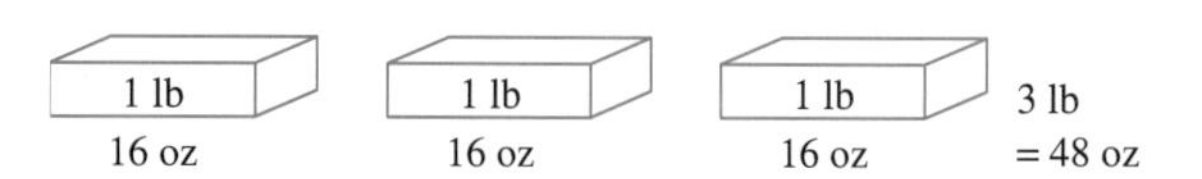

2 As with length, it is sometimes useful to simplify a measurement of weight by writing it in terms of mixed units.

33 ounces = _______ lb _______ oz

Because 16 oz = 1 lb, divide 16 into 33 to see how many pounds are in 33 ounces. The quotient is the number of pounds and the remainder is the number of ounces. To see why we divide 16 into 33, notice that

$$33 \text{ oz} = 33 \cancel{\text{oz}} \cdot \frac{1 \text{ lb}}{16 \cancel{\text{oz}}} = \frac{33}{16} \text{ lb}$$

$$\begin{array}{r} 2 \text{ lb } 1 \text{ oz} \\ 16\overline{)33} \\ -32 \\ \hline 1 \end{array}$$

Thus 33 ounces is the same as 2 lb 1 oz.

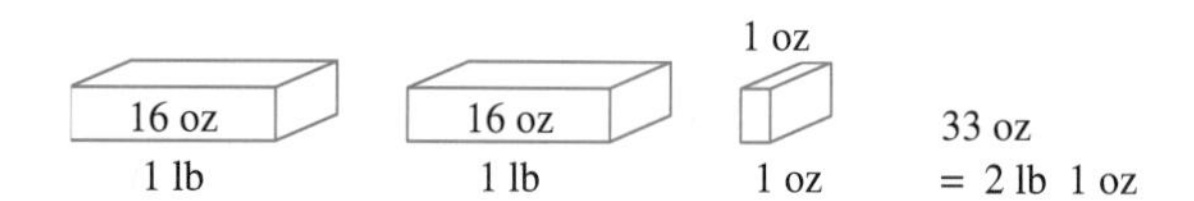

3 Performing arithmetic operations on units of weight works the same way as performing arithmetic operations on units of length.

Answer:

2. $3\frac{1}{2}$ pounds

EXAMPLE 3

Subtract 3 tons 1350 lb from 8 tons 1000 lb.

Solution:

To subtract, line up similar units.

$$\begin{array}{r} 8 \text{ tons } 1000 \text{ lb} \\ - \; 3 \text{ tons } 1350 \text{ lb} \\ \hline \end{array}$$

Since we cannot subtract 1350 lb from 1000 lb, we borrow 1 ton from the 8 tons. Write 1 ton as 2000 lb and combine it with the 1000 lb.

$$\begin{array}{r} 7 \text{ tons} + (1 \text{ ton}) \; 2000 \text{ lb} \\ \not{8} \text{ tons } 1000 \text{ lb} \\ - \; 3 \text{ tons } 1350 \text{ lb} \\ \hline \end{array} \quad \text{becomes} \quad \begin{array}{r} 7 \text{ tons } 3000 \text{ lb} \\ - \; 3 \text{ tons } 1350 \text{ lb} \\ \hline 4 \text{ tons } 1650 \text{ lb} \end{array}$$

To check, see that the sum of 4 tons 1650 lb and 3 tons 1350 lb is 8 tons 1000 lb.

PRACTICE PROBLEM 3

Subtract 5 tons 1200 lb from 8 tons 100 lb.

EXAMPLE 4

Multiply 5 lb 9 oz by 6.

Solution:

Multiply 5 lb by 6 and multiply 9 oz by 6.

$$\begin{array}{rr} 5 \text{ lb} & 9 \text{ oz} \\ \times & 6 \\ \hline 30 \text{ lb} & 54 \text{ oz} \end{array}$$

To write 54 oz as mixed units, divide by 16 (1 lb = 16 oz):

$$\begin{array}{r} 3 \text{ lb } 6 \text{ oz} \\ 16\overline{)54} \\ -48 \\ \hline 6 \end{array}$$

Thus,

$$30 \text{ lb } 54 \text{ oz} = 30 \text{ lb} + 3 \text{ lb } 6 \text{ oz} = 33 \text{ lb } 6 \text{ oz.}$$

PRACTICE PROBLEM 4

Multiply 4 lb 11 oz by 8.

EXAMPLE 5

Divide 9 lb 6 oz by 2.

Solution:

Divide each of the units by 2.

$$\begin{array}{rr} 4 \text{ lb} & 11 \text{ oz} \\ 2\overline{)9 \text{ lb}} & 6 \text{ oz} \\ -8 & \\ \hline 1 \text{ lb} = & 16 \text{ oz} \\ & \overline{22 \text{ oz}} \end{array}$$

Divide 2 into 22 oz to get 11 oz.

To check, multiply 4 pounds 11 ounces by 2. The result will be 9 pounds 6 ounces.

PRACTICE PROBLEM 5

Divide 5 lb 8 oz by 4.

Answers:

3. 2 tons 900 lb
4. 37 lb 8 oz
5. 1 lb 6 oz

PRACTICE PROBLEM 6

A 5-lb 14-oz batch of cookies is packed in a 6-oz container before it is mailed. Find the total weight.

EXAMPLE 6

Bryan weighed 8 lb 8 oz at birth. By the time he was 1 year old, he had gained 11 lb 14 oz. Find his weight at age 1 year.

Solution:

birth weight	→	8 lb 8 oz
+ weight gained	→	+ 11 lb 14 oz
total weight	→	19 lb 22 oz

Since 22 oz equals 1 lb 6 oz,

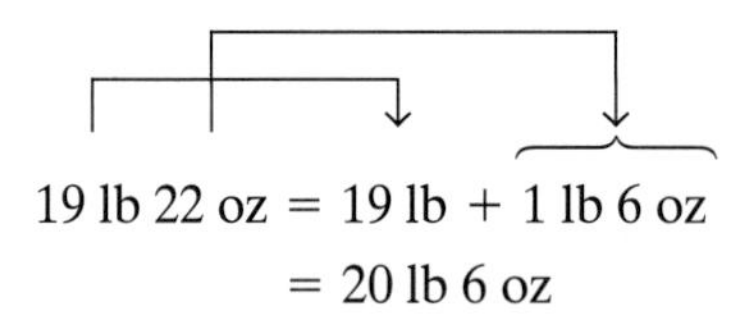

$$19 \text{ lb } 22 \text{ oz} = 19 \text{ lb} + 1 \text{ lb } 6 \text{ oz}$$
$$= 20 \text{ lb } 6 \text{ oz}$$

Bryan weighed 20 lb 6 oz on his first birthday.

MASS

In scientific and technical areas, a careful distinction is made between **weight** and **mass. Weight** is really a measure of the pull of gravity. The farther from Earth an object gets, the less it weighs. However, **mass** is a measure of the amount of substance in the object and does not change. Astronauts orbiting Earth weigh much less than they weigh on Earth, but they have the same mass in orbit as they do on Earth.

4 The basic unit of mass in the metric system is the **gram.** It is defined as the mass of water contained in a cube 1 centimeter (cm) on each side.

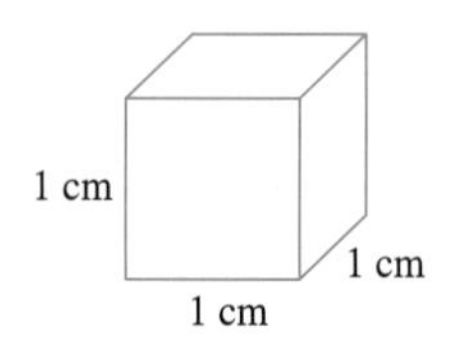

The following examples may help you get a feeling for metric masses.

A tablet contains 200 milligrams of ibuprofen.

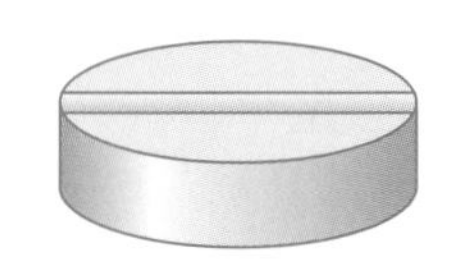

A large paper clip weighs approximately 1 gram.

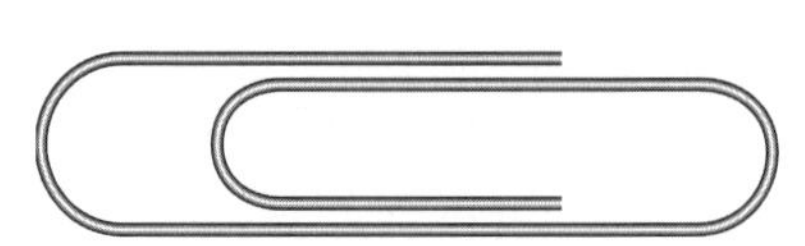

A box of crackers weighs 453 grams.

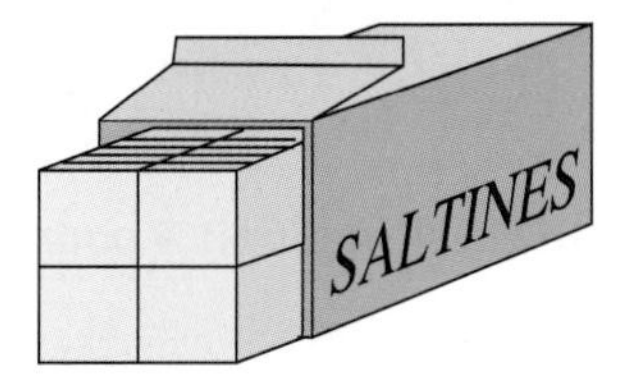

Answer:
6. 6 lb 4 oz

A kilogram is slightly over 2 pounds.
An adult woman may weigh 60 kilograms.

The prefixes for units of mass in the metric system are the same as for units of length, as shown in the following table.

PREFIX	MEANING	METRIC UNIT OF MASS
kilo	1000	1 kilogram (kg) = 1000 grams (g)
hecto	100	1 hectogram (hg) = 100 g
deka	10	1 dekagram (dag) = 10 g
		1 gram (g) = 1 g
deci	1/10	1 decigram (dg) = 1/10 g or 0.1 g
centi	1/100	1 centigram (cg) = 1/100 g or 0.01 g
milli	1/1000	1 milligram (mg) = 1/1000 g or 0.001 g

The **milligram,** the **gram,** and the **kilogram** are the three most commonly used units of mass in the metric system.

As with lengths, all units of mass are powers of 10 of the gram, so converting from one unit of mass to another only involves moving the decimal point. To convert from one unit of mass to another in the metric system, list the units of mass in order from largest to smallest.

Convert 4300 milligrams to grams. To convert from milligrams to grams, move along the table 3 units to the left.

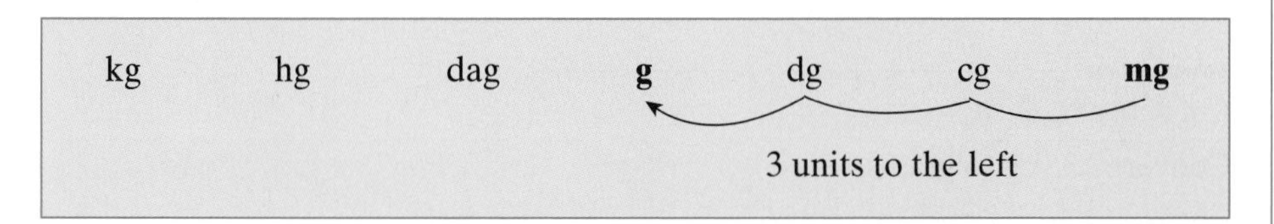

This means that we move the decimal point 3 places to the left to convert from milligrams to grams.

$$4300 \text{ mg} = 4.3 \text{ g}$$

The same conversion can be done with unit fractions.

$$4300 \text{ mg} = \frac{4300 \cancel{\text{mg}}}{1} \cdot \frac{0.001 \text{ g}}{1 \cancel{\text{mg}}}$$

$$= 4300 \cdot 0.001 \text{ g}$$

$$= 4.3 \text{ g}$$ To multiply by 0.001, move the decimal point 3 places to the left.

To see that this is reasonable, study the diagram on the next page.

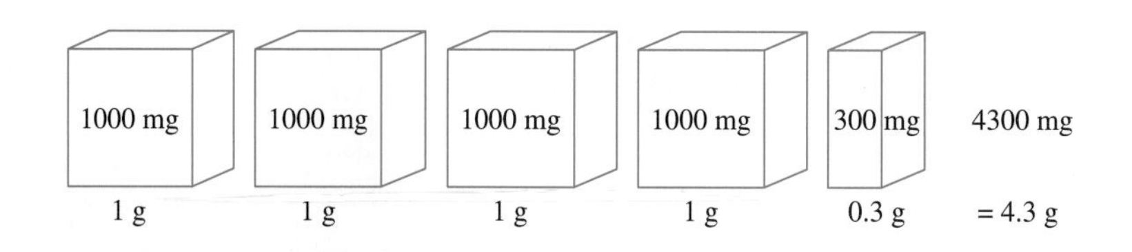

Thus 4300 mg = 4.3 g.

PRACTICE PROBLEM 7
Convert 3.41 g to milligrams.

EXAMPLE 7

Convert 3.2 kg to grams.

Solution:

$$\text{Unit fraction: } 3.2 \text{ kg} = 3.2 \cancel{\text{kg}} \cdot \frac{1000 \text{ g}}{1 \cancel{\text{kg}}} = 3200 \text{ g}$$

List the units of mass and move from kilograms to grams.

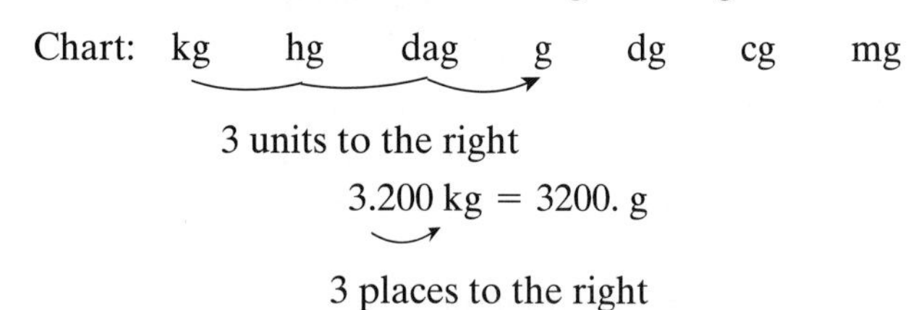

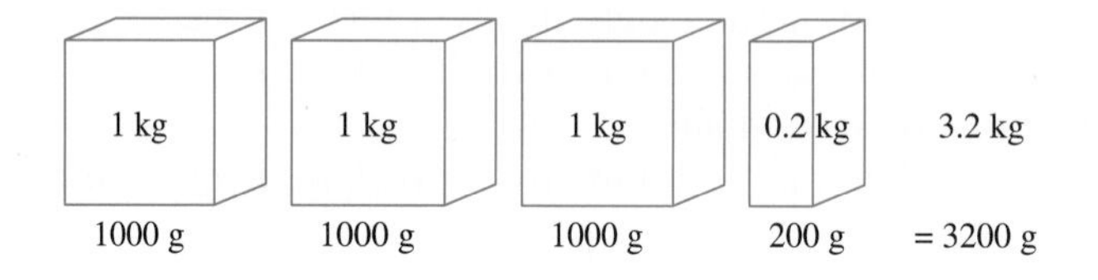

PRACTICE PROBLEM 8
Convert 56.2 cg to grams.

EXAMPLE 8

Convert 2.35 cg to grams.

Solution:

List the units of mass and move from centigrams to grams.

kg hg dag g dg cg mg

2 units to the left

02.35 cg = 0.0235 g

2 places to the left

5 Arithmetic operations can be performed with metric units of mass just as we performed operations with metric units of length. Convert each number to the same unit of mass and add, subtract, multiply, or divide as with decimals.

Answers:
7. 3410 mg
8. 0.562 g

EXAMPLE 9

Subtract 5.4 dg from 1.6 g.

Solution:

Convert both numbers to decigrams or to grams before subtracting.

$$5.4 \text{ dg} = 0.54 \text{ g} \quad \text{or} \quad 1.6 \text{ g} = 16 \text{ dg}$$

$$\begin{array}{r} 1.60 \text{ g} \\ -0.54 \text{ g} \\ \hline 1.06 \text{ g} \end{array} \qquad \begin{array}{r} 16.0 \text{ dg} \\ -\ 5.4 \text{ dg} \\ \hline 10.6 \text{ dg} \end{array}$$

The difference is 1.06 g or 10.6 dg.

PRACTICE PROBLEM 9

Subtract 3.1 dg from 2.5 g.

EXAMPLE 10

Multiply 15.4 kg by 5.

Solution:

Multiply the two numbers together.

$$\begin{array}{r} 15.4 \text{ kg} \\ \times \quad 5 \\ \hline 77.0 \text{ kg} \end{array}$$

The result is 77.0 kg.

PRACTICE PROBLEM 10

Multiply 12.6 kg by 4.

EXAMPLE 11

An elevator has a weight limit of 1400 kg. A sign posted in the elevator indicates that the maximum capacity of the elevator is 17 persons. What is the average allowable weight for each passenger, rounded to the nearest kilogram?

Solution:

To solve, notice that the total weight of
1400 kilograms $\div$ 17 = average weight

$$\begin{array}{r} 82.3 \text{ kg} \approx 82 \text{ kg} \\ 17\overline{)1400.0} \text{ kg} \\ -136 \\ \hline 40 \\ -34 \\ \hline 60 \\ -51 \\ \hline 9 \end{array}$$

Each passenger can weigh an average of 82 kg. (Recall that a kilogram is slightly over 2 pounds, so 82 kilograms is over 164 pounds. For a better approximation, see the calculator box in this section.)

PRACTICE PROBLEM 11

Twenty-four bags of cement weigh a total of 550 kg. Find the average weight of 1 bag, rounded to the nearest kilogram.

Answers:

9. 2.19 g or 21.9 dg
10. 50.4 kg
11. 23 kg

CALCULATOR EXPLORATIONS

METRIC TO U.S. CUSTOMARY CONVERSIONS IN WEIGHT

To convert between the two systems of measurement in weight, the following **approximations** can be used.

grams × 0.035 ≈ ounces	ounces × 28.35 ≈ grams
kilograms × 2.20 ≈ pounds	pounds × 0.454 ≈ kilograms
grams × 0.0022 ≈ pounds	pounds × 454 ≈ grams

EXAMPLE

A bulldog weighs 35 pounds. How many kilograms is this?

Solution:

From the above,

pounds × 0.454 ≈ kilograms
↓
35 × 0.454 ≈ kilograms

To multiply on your calculator, press the keys

[35] [×] [0.454] [=]

The display will read [15.89].

35 pounds ≈ 15.89 kilograms

Convert the following as indicated.

1. Convert 15 ounces to grams.
2. Convert 11.2 grams to ounces.
3. Convert 7 kilograms to pounds.
4. Convert 23 pounds to kilograms.
5. A piece of candy weighs 5 grams. How many ounces is this?
6. If a person weighs 82 kilograms, how many pounds is this?

MENTAL MATH

Convert the following mentally.

1. 16 ounces to pounds
2. 32 ounces to pounds
3. 1 ton to pounds
4. 3 tons to pounds
5. 1 pound to ounces
6. 3 pounds to ounces
7. 2000 pounds to tons
8. 4000 pounds to tons

Determine whether the measurement in each statement is reasonable.

9. The doctor prescribed a pill containing 2 kg of medication.
10. A full-grown cat weighs approximately 15 g.
11. A bag of flour weighs 4.5 kg.
12. A staple weighs 15 mg.
13. A professor weighs less than 150 g.
14. A car weighs 2000 mg.

EXERCISE SET 7.5

Convert the following as indicated. See Examples 1 and 2.

1. 2 pounds to ounces

2. 3 pounds to ounces

3. 5 tons to pounds

4. 3 tons to pounds

5. 12,000 pounds to tons

6. 32,000 pounds to tons

7. 60 ounces to pounds

8. 90 ounces to pounds

9. 3500 pounds to tons

10. 9000 pounds to tons

11. 16.25 pounds to ounces

12. 14.5 pounds to ounces

13. 4.9 tons to pounds

14. 8.3 tons to pounds

15. $4\frac{3}{4}$ pounds to ounces

16. $9\frac{1}{8}$ tons to pounds

ANSWERS

1. ______
2. ______
3. ______
4. ______
5. ______
6. ______
7. ______
8. ______
9. ______
10. ______
11. ______
12. ______
13. ______
14. ______
15. ______
16. ______

ANSWERS

17. ______
18. ______
19. ______
20. ______
21. ______
22. ______
23. ______
24. ______
25. ______
26. ______
27. ______
28. ______
29. ______
30. ______
31. ______
32. ______
33. ______
34. ______

17. 2950 pounds to the nearest tenth of a ton

18. 51 ounces to the nearest tenth of a pound

Perform the indicated operations. See Examples 3 through 5.

19. 34 lb 12 oz + 18 lb 14 oz

20. 6 lb 10 oz + 10 lb 8 oz

21. 6 tons 1540 lb + 2 tons 850 lb

22. 2 tons 1575 lb + 1 ton 480 lb

23. 5 tons 1050 lb − 2 tons 875 lb

24. 4 tons 850 lb − 1 ton 260 lb

25. 12 lb 4 oz − 3 lb 9 oz

26. 45 lb 6 oz − 26 lb 10 oz

27. 5 lb 3 oz × 6

28. 2 lb 5 oz × 5

29. 6 tons 1500 lb ÷ 5

30. 5 tons 400 lb ÷ 4

Convert the following as indicated. See Examples 7 and 8.

31. 500 g to kilograms

32. 650 g to kilograms

33. 4 g to milligrams

34. 9 g to milligrams

35. 25 kg to grams

36. 18 kg to grams

37. 48 mg to grams

38. 112 mg to grams

39. 6.3 g to kilograms

40. 4.9 g to kilograms

41. 15.14 g to milligrams

42. 16.23 g to milligrams

43. 4.01 kg to grams

44. 3.16 kg to grams

Perform the indicated operations. See Examples 9 and 10.

45. 3.8 mg + 9.7 mg

46. 41.6 g + 9.8 g

47. 205 mg + 5.61 g

48. 2.1 g + 153 mg

49. 9 g − 7150 mg

50. 4 kg − 2410 g

51. 1.61 kg − 250 g

52. 6.13 g − 418 mg

ANSWERS

35. ______
36. ______
37. ______
38. ______
39. ______
40. ______
41. ______
42. ______
43. ______
44. ______
45. ______
46. ______
47. ______
48. ______
49. ______
50. ______
51. ______
52. ______

ANSWERS

53. ______
54. ______
55. ______
56. ______
57. ______
58. ______
59. ______
60. ______
61. ______
62. ______
63. ______
64. ______
65. ______
66. ______
67. ______
68. ______
69. ______
70. ______
71. ______
72. ______

53. 5.2 kg $\times$ 2.6

54. 4.8 kg $\times$ 9.3

55. 17 kg $\div$ 8

56. 8.25 g $\div$ 6

Solve. See Examples 6 and 11.

57. A can of 7-Up weighs 336 grams. Find the weight in kilograms of 24 cans.

58. Guy normally weighs 73 kg, but he lost 2800 grams after being sick with the flu. Find Guy's new weight.

59. Sudafed is a decongestant that comes in two strengths. The regular strength contains 60 mg of medication. The extra strength contains 0.09 g of medication. How much extra medication is in the extra-strength tablet?

60. A small can of Planters Sunflower Nuts weighs 177 grams. If each can contains 6 servings, find the weight of one serving.

61. Doris has two open containers of Uncle Ben's rice. If she combines 1 lb 10 oz from one container with 3 lb 14 oz from the other container, how much total rice does she have?

62. Dru Mizel maintains the records of the amount of coal delivered to his department in the steel mill. In January, 3 tons 1500 lb were delivered. In February, 2 tons 1200 lb were delivered. Find the total amount delivered in these two months.

63. Carla Hamtini was amazed when she grew a 28 lb 10 oz zucchini in her garden, but later she learned that the heaviest zucchini ever grown weighed 64 lb. 8 oz. It was grown in Llanharry, Wales by B. Lavery in 1990. How far below the record weight was Carla's zucchini? (*Source: The Guinness Book of Records,* 1996)

64. The heaviest baby born in good health weighed an incredible 22 lb 8 oz. He was born in Italy in September, 1955. How much heavier is this than a 7 lb 12 oz baby? (*Source: The Guinness Book of Records,* 1996)

65. A package of Trailway's Gorp, a high-energy hiking trail mix, contains 0.3 kg of nuts, 0.15 kg of chocolate bits, and 400 g of raisins. Find the total weight of the package.

66. The manufacturer of Anacin wants to reduce the caffeine content of its aspirin by $\frac{1}{4}$. Currently, each regular tablet contains 32 mg of caffeine. How much caffeine should be removed from each tablet?

67. A regular-size bag of Lay's potato chips weighs 198 grams. Find the weight of a dozen bags, rounded to the nearest hundredth of a kilogram.

68. Clarence's cat weighs a hefty 9 kg. The vet has recommended that the cat lose 1500 grams. How much should the cat weigh?

69. One bag of Pepperidge Farm Bordeaux cookies weighs $6\frac{3}{4}$ ounces. How many pounds will a dozen bags weigh?

70. One can of Payless Red Beets weighs $8\frac{1}{2}$ ounces. How much will eight cans weigh?

71. A carton of 12 boxes of Quaker Oats Oatmeal weighs 6.432 kg. Each box includes 26 grams packaging material. What is the actual weight of the oatmeal in the carton?

72. The supermarket prepares hamburger in 3-lb 4-oz market packages. When Leo Gonzalas gets home, he divides the package in half before refrigerating the meat. How much will each package weigh?

EXAMPLE 4

Multiply 3 qt 1 pt by 3.

Solution:

Multiply each of the units of capacity by 3.

$$\begin{array}{r} 3 \text{ qt } 1 \text{ pt} \\ \times \quad\quad 3 \\ \hline 9 \text{ qt } 3 \text{ pt} \end{array}$$

Since 3 pints is the same as 1 quart and 1 pint,

$$9 \text{ qt } 3 \text{ pt} = 9 \text{ qt} + \overbrace{1 \text{ qt } 1 \text{ pt}} = 10 \text{ qt } 1 \text{ pt}$$

The 10 quarts can be changed to gallons by dividing by 4, since there are 4 quarts in a gallon. To see why we divide, notice that

$$10 \text{ qt} = 10 \text{ qt} \cdot \frac{1 \text{ gal}}{4 \text{ qt}} = \frac{10}{4} \text{ gal}$$

$$\begin{array}{r} 2 \text{ gal } 2 \text{ qt} \\ 4\overline{)10 \text{ qts}} \\ -8 \\ \hline 2 \end{array}$$

Then the product is 10 qt 1 pt or 2 gal 2 qt 1 pt.

PRACTICE PROBLEM 4

Multiply 2 gal 3 qt by 2.

EXAMPLE 5

Divide 3 gal 2 qt by 2.

Solution:

Divide each unit of capacity by 2.

$$\begin{array}{rr} 1 \text{ gal} & 3 \text{ qt} \\ 2\overline{)3 \text{ gal}} & 2 \text{ qt} \\ -2 & \\ \hline 1 \text{ gal} = & 4 \text{ qt} \\ & \underline{6 \text{ qt}} \end{array} \quad 6 \text{ qt} \div 2 = 3 \text{ qt}$$

PRACTICE PROBLEM 5

Divide 6 gal 1 qt by 2.

EXAMPLE 6

An aquarium contains 6 gal 3 qt of water. If 2 gal 2 qt of water is added, what is the total amount of water in the aquarium?

Solution:

$$\begin{array}{rcr} \text{beginning water} & \rightarrow & 6 \text{ gal } 3 \text{ qt} \\ + \quad \text{water added} & \rightarrow & + \ 2 \text{ gal } 2 \text{ qt} \\ \hline \text{total water} & \rightarrow & 8 \text{ gal } 5 \text{ qt} \end{array}$$

Since 5 qt = 1 gal 1 qt,

$$\begin{aligned} &8 \text{ gal} \quad \overbrace{5 \text{ qt}} \\ &= 8 \text{ gal} + 1 \text{ gal } 1 \text{ qt} \\ &= 9 \text{ gal } 1 \text{ qt} \end{aligned}$$

The total amount of water is 9 gallons 1 quart.

PRACTICE PROBLEM 6

A large oil drum contains 15 gal 3 qt of oil. How much will be in the drum if an additional 4 gal 3 qt of oil is poured into it?

Answers:
4. 5 gal 2 qt
5. 3 gal 1 pt
6. 20 gal 2 qt

CAPACITY: METRIC SYSTEM OF MEASUREMENT

3 Thus far, we know that the basic unit of length in the metric system is the meter and the basic unit of mass in the metric system is the gram. What is the basic unit of capacity? The **liter** is the basic unit of capacity in the metric system. By definition, a **liter** is the capacity or volume of a cube measuring 10 centimeters on each side.

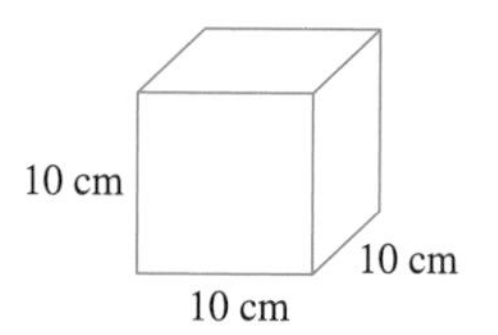

The following examples may help you get a feeling for metric capacities.

One liter of liquid is slightly more than one quart.

Many soft drinks are packaged in 2-liter bottles.

The metric system was designed to be a consistent system. Once again, the prefixes for metric units of capacity are the same as for metric units of length and mass, as summarized in the following table.

PREFIX	MEANING	METRIC UNIT OF CAPACITY
kilo	1000	1 kiloliter (kl) = 1000 liters (L)
hecto	100	1 hectoliter (hl) = 100 L
deka	10	1 dekaliter (dal) = 10 L
		1 liter (L) = 1 L
deci	1/10	1 deciliter (dl) = 1/10 L or 0.1 L
centi	1/100	1 centiliter (cl) = 1/100 L or 0.01 L
milli	1/1000	1 milliliter (ml) = 1/1000 L or 0.001 L

The **milliliter** and the **liter** are the two most commonly used metric units of capacity.

Converting from one unit of capacity to another involves multiplying by powers of 10 or moving the decimal point to the left or to the right. Listing units of capacity in order from largest to smallest helps to keep track of how many places to move the decimal point when converting.

Convert 2.6 liters to milliliters. To convert from liters to milliliters, we move along the table 3 units to the right.

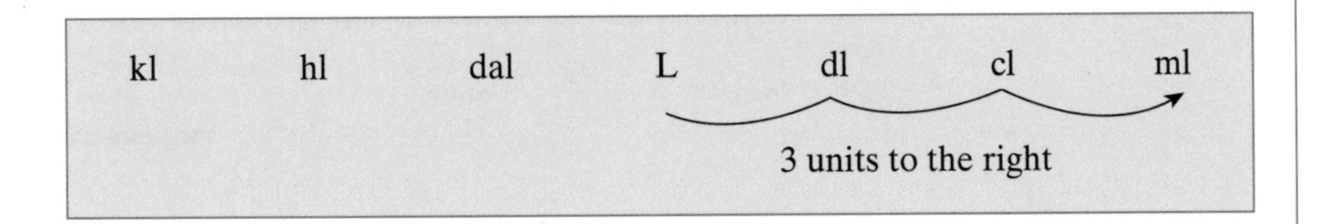

This means that we move the decimal point 3 places to the right to convert from liters to milliliters.

2.600 L = 2600. ml

This same conversion can be done with unit fractions.

$$2.6\text{ L} = \frac{2.6\text{ L}}{1} \cdot \frac{1000\text{ ml}}{1\text{ L}}$$
$$= 2.6 \cdot 1000\text{ ml}$$
$$= 2600\text{ ml}$$

To multiply by 1000, move the decimal point 3 places to the right.

To visualize the result, study the diagram below.

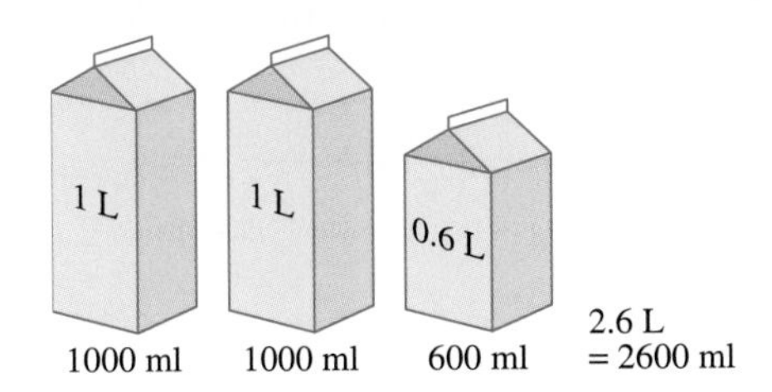

Thus 2.6 L = 2600 ml.

EXAMPLE 7

Convert 3210 ml to liters.

Solution:

Unit fraction: $3210\text{ ml} = 3210\text{ ml} \cdot \frac{1\text{ L}}{1000\text{ ml}} = 3.21\text{ L}$

List the unit measures and move from milliliters to liters.

Chart: kl hl dal L dl cl ml

3 units to the left

PRACTICE PROBLEM 7

Convert 2100 ml to liters.

Answer:

7. 2.1 L

3210 ml = 3.210 L

3 places to the left

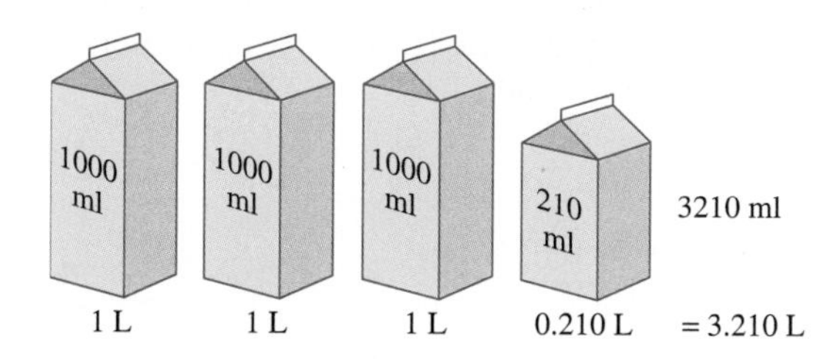

PRACTICE PROBLEM 8
Convert 2.13 dal to liters.

EXAMPLE 8

Convert 0.185 dl to milliliters.

Solution:

List the unit measures and move from deciliters to milliliters.

kl hl dal L dl cl ml

2 units to the right

0.185 dl = 18.5 ml

2 places to the right

4 As was true for length and weight, arithmetic operations involving metric units of capacity can also be performed. Make sure the metric units of capacity are the same before adding, subtracting, multiplying, or dividing.

PRACTICE PROBLEM 9
Add 1250 ml to 2.9 L.

EXAMPLE 9

Add 2400 ml to 8.9 L.

Solution:

Convert both to milliliters or both to liters before adding the capacities together.

2400 ml = 2.4 L or 8.9 L = 8900 ml

$$\begin{array}{r} 2.4\text{ L} \\ +\ 8.9\text{ L} \\ \hline 11.3\text{ L} \end{array} \qquad \begin{array}{r} 2400\text{ ml} \\ +\ 8900\text{ ml} \\ \hline 11{,}300\text{ ml} \end{array}$$

The total is 11.3 L or 11,300 ml. They both represent the same capacity.

PRACTICE PROBLEM 10
Divide 146.9 L by 13.

Answers:
8. 21.3 L
9. 4150 ml or 4.15 L
10. 11.3 L

EXAMPLE 10

Divide 18.08 ml by 16.

Solution:

Divide 16 into 18.08.

```
      1.13 ml
16)18.08 ml
  -16
    2 0
   -1 6
      48
     -48
       0
```

The solution is 1.13 ml.

EXAMPLE 11

A patient hooked up to an IV unit in the hospital is to receive 12.5 ml of medication every hour. How much medication does the patient receive in 3.5 hours?

Solution:

Multiply 12.5 ml by 3.5.

```
medication per hour  →       12.5 ml
×             hours  →   ×    3.5
-------------------         -----
   total medication           625
                             375
                          -------
                          43.75 ml
```

The patient receives 43.75 ml of medication.

PRACTICE PROBLEM 11

If 28.6 L of water can be pumped every minute, how much water can be pumped in 8.5 minutes?

Answer:

11. 243.1 L

CALCULATOR EXPLORATIONS

METRIC TO U.S. CUSTOMARY CONVERSIONS IN CAPACITY

To convert between the two systems of measurement in weight, the following **approximations** can be used.

liters $\times$ 1.06 $\approx$ quarts	quarts $\times$ 0.946 $\approx$ liters
liters $\times$ 0.264 $\approx$ gallons	gallons $\times$ 3.785 $\approx$ liters

EXAMPLE

How many quarts are there in a 2-liter bottle of cola?

Solution:

From the above.

liters $\times$ 1.06 $\approx$ quarts

↓

2 $\times$ 1.06 $\approx$ quarts

To multiply on your calculator, press the keys

[2] [×] [1.06] [=]

The display should read [2.12].

2 liters $\approx$ 2.12 quarts

Convert the following as indicated.

1. 5 quarts to liters

2. 26 gallons to liters

3. 17.5 liters to gallons

4. 7.8 liters to quarts

5. A 1-gallon container holds how many liters?

6. How many quarts are contained in a 3-liter bottle of cola?

MENTAL MATH

Convert the following mentally.

1. 2 c to pints
2. 4 c to pints
3. 4 qt to gallons
4. 8 qt to gallons
5. 2 pt to quarts
6. 6 pt to quarts
7. 8 fl oz to cups
8. 24 fl oz to cups
9. 1 pt to cups
10. 3 pt to cups
11. 1 gal to quarts
12. 2 gal to quarts

Determine whether the measurement in each statement is reasonable.

13. Clair took a dose of 2 L of cough medicine to cure her cough.

14. John drank 250 ml of milk for lunch.

15. Jeannie likes to relax in a tub filled with 3000 ml of hot water

16. Sarah pumped 20 L of gasoline into her car yesterday.

EXERCISE SET 7.6

Convert each measurement as indicated. See Examples 1 and 2.

1. 32 fluid ounces to cups

2. 16 quarts to gallons

3. 8 quarts to pints

4. 9 pints to quarts

5. 10 quarts to gallons

6. 15 cups to pints

7. 80 fluid ounces to pints

8. 18 pints to gallons

9. 2 quarts to cups

10. 3 pints to fluid ounces

11. 120 fluid ounces to quarts

12. 20 cups to gallons

13. 6 gallons to fluid ounces

14. 5 quarts to cups

15. $4\frac{1}{2}$ pints to cups

16. $6\frac{1}{2}$ gallons to quarts

ANSWERS

1. __________

2. __________

3. __________

4. __________

5. __________

6. __________

7. __________

8. __________

9. __________

10. __________

11. __________

12. __________

13. __________

14. __________

15. __________

16. __________

ANSWERS

17. ______
18. ______
19. ______
20. ______
21. ______
22. ______
23. ______
24. ______
25. ______
26. ______
27. ______
28. ______
29. ______
30. ______
31. ______
32. ______
33. ______
34. ______
35. ______
36. ______
37. ______
38. ______
39. ______
40. ______
41. ______

17. $2\frac{3}{4}$ gallons to pints

18. $3\frac{1}{4}$ quarts to cups

Perform the following arithmetic operations. See Examples 3 through 5.

19. 4 gal 3 qt + 5 gal 2 qt

20. 2 gal 3 qt + 8 gal 3 qt

21. 1 c 5 fl oz + 2 c 7 fl oz

22. 2 c 3 fl oz + 2 c 6 fl oz

23. 3 gal − 1 gal 3 qt

24. 2 pt − 1 pt 1 c

25. 3 gal 1 qt − 1 qt 1 pt

26. 3 qt 1 c − 1 c 4 fl oz

27. 1 pt 1 c × 3

28. 1 qt 1 pt × 2

29. 8 gal 2 qt × 2

30. 6 gal 1 pt × 2

31. 9 gal 2 qt ÷ 2

32. 5 gal 6 fl oz ÷ 2

Convert the following as indicated. See Examples 7 and 8.

33. 5 L to milliliters

34. 8 L to milliliters

35. 4500 ml to liters

36. 3100 ml to liters

37. 410 L to kiloliters

38. 250 L to kiloliters

39. 64 ml to liters

40. 39 ml to liters

41. 0.16 kl to liters

CHAPTER 7 REVIEW

(7.1) *Write each ratio as a fraction in lowest terms.*

1. 6000 people to 4800 people

2. 121 births to 143 births

3. $2\frac{1}{4}$ days to 10 days

4. 14 quarters to 5 quarters

5. 4 weeks to 15 weeks

6. 4 yards to 8 yards

7. $3\frac{1}{2}$ dollars to 7 dollars

8. 3.5 centimeters to 75 centimeters

Write each rate as a fraction in lowest terms.

9. 8 stillborn births to 1000 live births.

10. 6 professors for 20 graduate research assistants

11. 15 word-processing pages printed in 6 minutes

12. 8 computers assembled in 6 hours

Write each phrase as a unit rate.

13. 468 miles in 9 hours

14. 180 feet in 12 seconds

15. \$0.93 for 3 pears

16. \$6.96 for 4 diskettes

17. 260 kilometers in 4 hours

18. 8 gallons of pesticide for 6 acres of crops

19. \$184 for books for 5 college courses

20. 52 bushels of fruit from 4 trees

Compare the unit rates and decide which is the better buy. Round to the nearest cent.

21. Taco sauce: 8 ounces for \$0.99 or 12 ounces for \$1.69

22. Peanut butter: 18 ounces for \$1.49 or 28 ounces for \$2.39

23. 2% milk: 16 ounces for \$0.59, $\frac{1}{2}$ gallon for \$1.69, or 1 gallon for \$2.29 (1 gallon = 128 fluid ounces)

24. Coca-Cola: 12 ounces for \$0.59, 16 ounces for \$0.79, or 32 ounces for \$1.19

(7.2) *Write the sentence as a proportion.*

25. 20 men is to 14 women as 10 men is to 7 women.

26. 50 tries is to 4 successes as 25 tries is to 2 successes.

ANSWERS

1. ______
2. ______
3. ______
4. ______
5. ______
6. ______
7. ______
8. ______
9. ______
10. ______
11. ______
12. ______
13. ______
14. ______
15. ______
16. ______
17. ______
18. ______
19. ______
20. ______
21. ______
22. ______
23. ______
24. ______
25. ______
26. ______

ANSWERS

27. ______
28. ______
29. ______
30. ______
31. ______
32. ______
33. ______
34. ______
35. ______
36. ______
37. ______
38. ______
39. ______
40. ______
41. ______
42. ______
43. ______
44. ______
45. ______
46. ______
47. ______
48. ______
49. ______
50. ______
51. ______
52. ______

27. 16 sandwiches is to 8 players as 2 sandwiches is to 1 player.

28. 12 tires is to 3 cars as 4 tires is to 1 car.

Determine whether each proportion is a true statement.

29. $\frac{21}{8} = \frac{14}{6}$

30. $\frac{3}{5} = \frac{60}{100}$

31. $\frac{3.1}{6.2} = \frac{0.8}{0.16}$

32. $\frac{3.75}{3} = \frac{7.5}{6}$

Solve the proportion for the variable.

33. $\frac{x}{6} = \frac{15}{18}$

34. $\frac{y}{9} = \frac{5}{3}$

35. $\frac{4}{13} = \frac{10}{x}$

36. $\frac{8}{5} = \frac{9}{z}$

37. $\frac{16}{3} = \frac{y}{6}$

38. $\frac{x}{3} = \frac{9}{2}$

39. $\frac{x}{5} = \frac{27}{2\frac{1}{4}}$

40. $\frac{2\frac{1}{2}}{6} = \frac{3}{z}$

41. $\frac{x}{0.4} = \frac{4.7}{3}$. Round to the nearest hundredth.

42. $\frac{0.07}{0.3} = \frac{7.2}{n}$. Round to the nearest tenth.

(7.3) *Solve.*

The ratio of a quarterback's completed passes to attempted passes is 3 to 7.

43. If he attempts 32 passes, find how many passes he completed. Round to nearest whole pass.

44. If he completed 15 passes, find how many passes he attempted.

One bag of pesticide covers 4000 square feet of crops.

45. Find how many bags of pesticide should be purchased to cover a rectangular garden 180 feet by 175 feet.

46. Find how many bags of pesticide should be purchased to cover a square garden 250 feet on each side.

An owner of a Ford Escort can drive 420 miles on 11 gallons of gas.

47. If Tom Aloiso ran out of gas in an Escort and AAA comes to his rescue with $1\frac{1}{2}$ gallons of gas, determine whether Tom can then drive to a gas station 65 miles away.

48. Find how many gallons of gas Tom can expect to burn on a 3000-mile trip. Round to the nearest gallon.

Yearly homeowner property taxes are figured at a rate of $1.15 tax for every $100 of house value.

49. If a homeowner pays $627.90 in property taxes, find the value of his home.

50. Find the property taxes on a townhouse valued at $89,000.

On an architect's blueprint, 1 inch = 12 feet.

51. Find the length of a wall represented by a $3\frac{3}{8}$-inch line on the blueprint.

52. If an exterior wall is 99 feet long, find how long the blueprint measurement should be.

Given that the pairs of triangles are similar, find the unknown length x.

53.

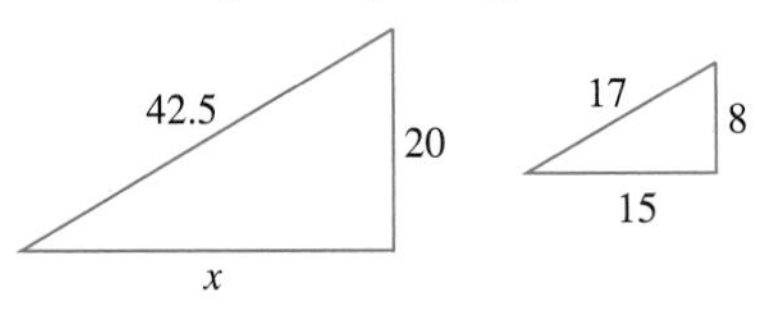

54.

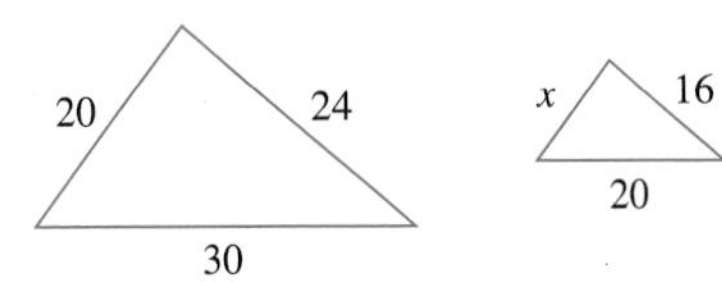

55.

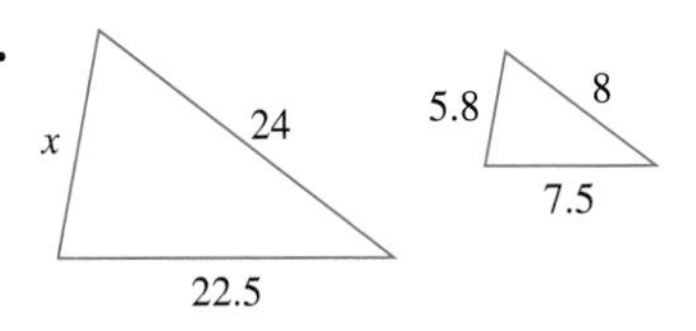

56.

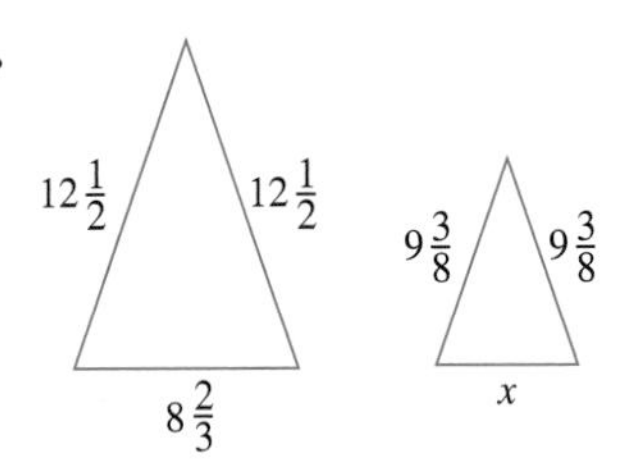

Solve.

57. A housepainter needs to estimate the height of a condominium. He estimates the length of his shadow to be 7 feet long and the length of the building's shadow to be 42 feet long. Find the height of the building if the housepainter is $5\frac{1}{2}$ feet tall.

58. Santa's elves are making a triangular sail for a toy sailboat. The toy sail is to be the same shape as a real sailboat's sail. Use the following diagram to find the unknown lengths x and y.

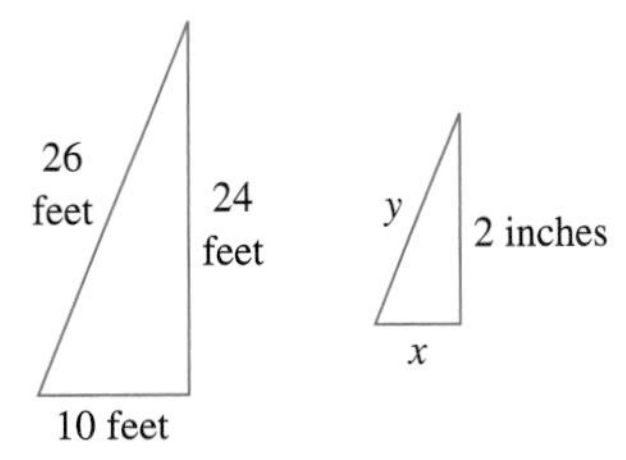

(7.4) *Convert.*

59. 108 in. to feet

60. 72 ft to yards

61. 2.5 mi to feet

62. 6.25 ft to inches

63. 52 ft = ____ yd ____ ft

64. 46 in. = ____ ft ____ in.

65. 42 m to centimeters

66. 82 cm to millimeters

67. 12.18 mm to meters

68. 2.31 m to kilometers

Perform the indicated arithmetic operations.

69. 4 yd 2 ft + 16 yd 2 ft

70. 12 ft 1 in. − 4 ft 8 in.

71. 8 ft 3 in. × 5

72. 7 ft 4 in. ÷ 2

73. 8 cm + 15 mm

74. 4 m + 126 cm

75. 9.3 km − 183 m

76. 4100 mm − 3 m

Solve.

77. A bolt of cloth contains 333 yd 1 ft of cotton ticking. Find the amount of material that remains after 163 yd 2 ft is removed from the bolt.

78. The local ambulance corps plans to award 20 framed certificates of valor to some of its outstanding members. If each frame requires 6 ft 4 in. of framing material, how much material is needed for all the frames?

ANSWERS

53. ____
54. ____
55. ____
56. ____
57. ____
58. ____
59. ____
60. ____
61. ____
62. ____
63. ____
64. ____
65. ____
66. ____
67. ____
68. ____
69. ____
70. ____
71. ____
72. ____
73. ____
74. ____
75. ____
76. ____
77. ____
78. ____

ANSWERS

79. ______
80. ______
81. ______
82. ______
83. ______
84. ______
85. ______
86. ______
87. ______
88. ______
89. ______
90. ______
91. ______
92. ______
93. ______
94. ______
95. ______
96. ______
97. ______
98. ______
99. ______
100. ______
101. ______
102. ______
103. ______
104. ______

79. The trip from Philadelphia to Washington, D.C., is 312.6 km. Four friends agree to share the driving equally. How far must each drive on this round-trip vacation?

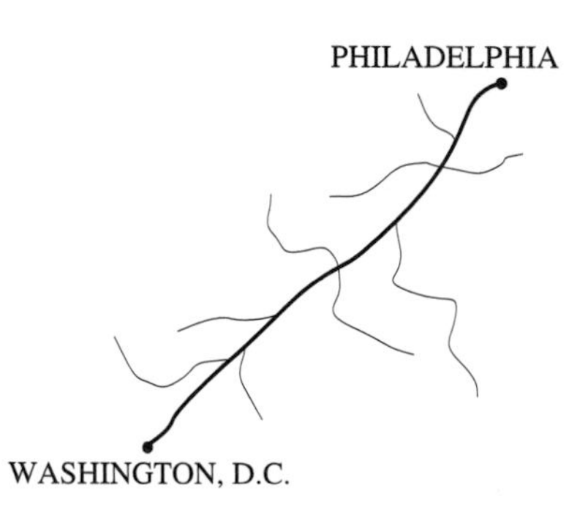

80. The college has ordered that NO SMOKING signs be placed above the doorway to each classroom. Each sign is 0.8 m long and 30 cm wide. Find the area of each sign.

(7.5) *Convert.*

81. 66 oz to pounds

82. 2.3 tons to pounds

83. 52 oz = ____ lb ____ oz

84. 8200 lb = ____ tons ____ lb

85. 1400 mg to grams

86. 40 kg to grams

87. 2.1 hg to dekagrams

88. 0.03 mg to decigrams

Perform the indicated arithmetic operations.

89. 6 lb 5 oz − 2 lb 12 oz

90. 5 tons 1600 lb + 4 tons 1200 lb

91. 6 tons 2250 lb ÷ 3

92. 8 lb 6 oz × 4

93. 1300 mg + 3.6 g

94. 4.8 kg + 4200 g

95. 9.3 g − 1200 mg

96. 6.3 kg × 8

Solve the following.

97. Lois ordered 1 lb 12 oz of soft-center candies and 2 lb 8 oz of chewy-center candies for her party. Find the total weight of the candy ordered.

98. Four local townships jointly purchase 38 tons 300 lb of cinders to spread on their roads during an ice storm. Determine the weight of the cinders each township receives if they share the purchase equally.

99. Linda ordered 8.3 kg of whole wheat flour from the health store, but she received 450 g less. How much flour did she actually receive?

100. Eight friends spent a weekend in the Poconos tapping maple trees and preparing 9.3 kg of maple syrup. Find the weight each friend receives if they share the syrup equally.

(7.6) *Convert.*

101. 16 pints to quarts

102. 40 fluid ounces to cups

103. 6.75 gallons to quarts

104. 8.5 pints to cups

105. 9 pt = ____ qt ____ pt

106. 15 qt = ____ gal ____ qt

107. 3.8 L to milliliters

108. 4.2 ml to deciliters

109. 14 hl to kiloliters

110. 30.6 L to centiliters

Perform the indicated arithmetic operations.

111. 1 qt 1 pt + 3 qt 1 pt

112. 3 gal 2 qt 1 pt × 2

113. 0.946 L − 210 ml

114. 6.1 L + 9400 ml

Solve.

115. Carlos prepares 4 gal 2 qt of iced tea for a block party. During the first 30 minutes of the party, 1 gal 3 qt of the tea is consumed. How much iced tea remains?

116. A recipe for soup stock calls for 1 c 4 fl oz of beef broth. How much should be used if the recipe is cut in half?

117. Each bottle of Kiwi liquid shoe polish holds 85 ml of the polish. Find the number of liters of shoe polish contained in 8 boxes if each box contains 16 bottles.

118. Ivan Miller wants to pour three separate containers of saline solution into a single vat with a capacity of 10 liters. Will 6 liters of solution in the first container combined with 1300 milliliters in the second container and 2.6 liters in the third container fit into the larger vat?

(7.7) *Convert. Round to the nearest tenth of a degree, if necessary.*

119. 245°C to degrees Fahrenheit

120. 160°C to degrees Fahrenheit

121. 42°C to degrees Fahrenheit

122. 86°C to degrees Fahrenheit

123. 93.2°F to degrees Celsius

124. 51.8°F to degrees Celsius

125. 41.3°F to degrees Celsius

126. 80°F to degrees Celsius

Solve. Round to the nearest tenth of a degree, if necessary.

127. A sharp dip in the jet stream caused the temperature in New Orleans to drop to 35°F. Find the corresponding temperature in degrees Celsius.

128. The recipe for meat loaf calls for a 165°C oven. Find the setting used if the oven has a Fahrenheit thermometer.

ANSWERS

105. ________
106. ________
107. ________
108. ________
109. ________
110. ________
111. ________
112. ________
113. ________
114. ________
115. ________
116. ________
117. ________
118. ________
119. ________
120. ________
121. ________
122. ________
123. ________
124. ________
125. ________
126. ________
127. ________
128. ________

CHAPTER 7 TEST

1. Write the ratio as a fraction in lowest terms: 4500 trees to 6500 trees.

2. On Monday the price of Zodiac stock was $8\frac{3}{4}$ dollars, but the stock rose to $10\frac{1}{2}$ dollars on Tuesday. Find the ratio of Monday's price to Tuesday's price.

Write each phrase as a unit rate.

3. 650 kilometers in 8 hours

4. 9 inches of rain in 30 days

Compare the unit rates and decide which is the better buy.

5. Steak sauce: 8 ounces for $1.19 or 12 ounces for $1.89

6. Jelly: 16 ounces for $1.49 or 24 ounces for $2.39

Solve the proportion for the variable.

7. $\frac{x}{3} = \frac{15}{9}$

8. $\frac{\frac{15}{12}}{\frac{3}{7}} = \frac{x}{\frac{4}{5}}$

9. $\frac{1.5}{5} = \frac{2.4}{z}$

Convert as indicated.

10. 280 in. to feet and inches

11. $2\frac{1}{2}$ gal to quarts

12. 30 oz to pounds

13. 2.8 tons to pounds

14. 38 pt to gallons

15. 40 mg to grams

16. 2.4 kg to grams

17. 3.6 cm to millimeters

18. 4.3 dg to grams

19. 0.83 L to milliliters

Perform the indicated operations.

20. 3 qt 1 pt + 2 qt 1 pt

21. 8 lb 6 oz − 4 lb 9 oz

22. 2 ft 9 in. × 3

23. 5 gal 2 qt ÷ 2

24. 8 cm − 14 mm

25. 1.8 km + 456 m

Convert. Round to the nearest tenth of a degree, if necessary.

26. Convert 84°F to degrees Celsius.

27. Convert 12.6°C to degrees Fahrenheit.

ANSWERS

1. ______
2. ______
3. ______
4. ______
5. ______
6. ______
7. ______
8. ______
9. ______
10. ______
11. ______
12. ______
13. ______
14. ______
15. ______
16. ______
17. ______
18. ______
19. ______
20. ______
21. ______
22. ______
23. ______
24. ______
25. ______
26. ______
27. ______

ANSWERS

28. ______________

29. ______________

30. ______________

31. ______________

32. ______________

33. ______________

34. ______________

Solve.

28. Given that the following triangles are similar, find the missing length.

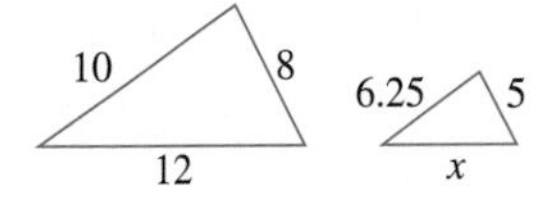

29. Tamara, a surveyor, needs to estimate the height of a tower. She estimates the length of her shadow to be 4 feet long and the length of the tower's shadow to be 48 feet long. Find the height of the tower if she is $5\frac{3}{4}$ feet tall.

30. The sugar maples in front of Bette's house are 8.4 meters tall. Because they interfere with the phone lines, the telephone company plans to remove the top third of the trees. How tall will the maples be after they are cut back?

31. A total of 15 gal 1 qt of oil has been removed from a 20-gallon drum. How much oil still remains in the container?

32. The doctors are quite concerned about Lucia, who is running a 41°C fever. Find Lucia's temperature in degrees Fahrenheit.

33. The engineer in charge of the bridge construction said that the span of the bridge would be 88 m, but the actual construction required it to be 340 cm longer. Find the span of the bridge.

34. If 2 ft 9 in. of material is used to manufacture one scarf, how much material is needed for 6 scarves?

57. $\frac{1}{50}$

58. $\frac{1}{25}$

59. 1.75

60. 1.25

61. $\frac{5}{7}$

62. $\frac{5}{14}$

63. 0.002

64. 0.009

65. $\frac{8}{15}$

66. $\frac{12}{19}$

Solve. See Examples 1, 7, and 8.

67. If 30 students are enrolled in a class and the attendance today is 100%, determine how many students attended.

68. A basketball player makes 81 out of 100 attempted free throws. Write the ratio of the number of free throws made to number of total free throws attempted as a percent.

69. A basketball player makes 45 out of 50 free throws. Write the ratio of the number of free throws made to number of total free throws attempted as a percent.

70. The sales tax in Slidell, Louisiana, is 8.25%. Write 8.25% as a decimal.

71. The Munoz family saves 0.10 of their take-home pay. Write 0.10 as a percent.

72. Less than 25% of a person's calories should come from fat. Write 25% as a decimal.

73. The cost of an item for sale is 0.7 of the sales price. Write 0.7 as a percent.

74. A real estate agent receives a commission of 3% of the sale price of a house. Write 3% as a decimal.

75. In 1994, the average monthly payment for financing new and used cars went up by 6.25%. Write 6.25% as a decimal. (*Source:* The American Automobile Manufacturers Association)

76. A health insurance company pays 80% of a person's medical costs. Write 80% as a decimal.

77. In 1994, 52.5% of the votes were cast for Fife Symington and he was elected governor of the State of Utah. Write 52.5% as a fraction. (*Source:* Elections Research Center)

78. 73.7% of the work force in the United States works 35 hours or more per week. Write 73.7% as a fraction. (*Source:* U.S. Bureau of Labor)

79. The Lemoine family spends $\frac{1}{5}$ of their take-home pay for food. Write $\frac{1}{5}$ as a percent.

80. A toaster is on sale for $\frac{4}{5}$ of the original price. Write $\frac{4}{5}$ as a percent.

81. The number of violent crimes in Stockholm is reduced 15%. Write a fraction equal to 15%.

82. Less than 25% of the calories in a diet should be from fat. Write a fraction equal to 25%.

ANSWERS

57. _____
58. _____
59. _____
60. _____
61. _____
62. _____
63. _____
64. _____
65. _____
66. _____
67. _____
68. _____
69. _____
70. _____
71. _____
72. _____
73. _____
74. _____
75. _____
76. _____
77. _____
78. _____
79. _____
80. _____
81. _____
82. _____

ANSWERS

83. ______

84. ______

85. ______

86. ______

89. ______

90. ______

91. ______

92. ______

The bar graph below shows the predicted fastest growing occupations. Use this graph to find the percent increase predicted for the occupations listed.

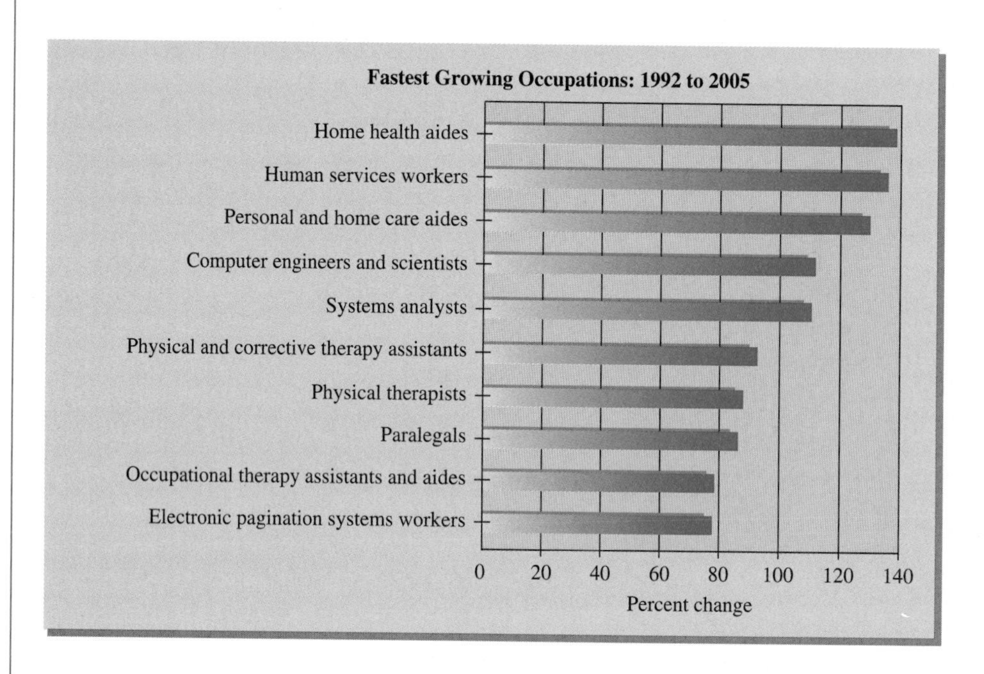

83. Home health aides

84. Computer engineers and scientists

85. Paralegals

86. Physical therapists

Complete the table.

87.

PERCENT	DECIMAL	FRACTION
35%		
		$\frac{1}{5}$
	.5	
70%		
		$\frac{3}{8}$

88.

PERCENT	DECIMAL	FRACTION
	0.525	
		$\frac{3}{4}$
$66\frac{2}{3}\%$		
		$\frac{5}{6}$
100%		

Write the fractions as a decimal and then write the decimal as a percent. Round the decimal to three decimal places and the percent to the nearest tenth of a percent.

89. $\frac{21}{79}$

90. $\frac{56}{102}$

91. $\frac{850}{736}$

92. $\frac{506}{248}$

Fill in the blanks.

93. A fraction written as a percent is greater than 100% when the numerator is ______ (greater/less) than the denominator.

94. A decimal written as a percent is less than 100% when the decimal is ______ (greater/less) than 1.

Solution:

1. UNDERSTAND. Notice that the nutrition label tells us that a serving of this food contains a total of 80 calories and 10 of these calories are from fat.
2. ASSIGN. Let x = percent of calories from fat.
3. TRANSLATE.

In words:	Number of fat calories	is	what percent	of	Total calories
	↓	↓	↓	↓	↓
Translate:	10	=	x	·	80

4. SOLVE.

$$10 = 80x$$

$$\frac{10}{80} = \frac{80x}{80} \quad \text{Divide both sides by 80.}$$

$$0.125 = x \quad \text{Simplify.}$$

$$12.5\% = x \quad \text{Write as a percent.}$$

5. CHECK. Check to see that 10 is 12.5% of 80.
6. STATE. This food contains 12.5% of its total calories from fat.

NUTRITION FACTS	
Serving Size 1 pouch (20g)	
Servings Per Container 6	
AMOUNT PER SERVING	
Calories	80
Calories from Fat	10
	% DAILY VALUE*
Total Fat 1g	2%
Sodium 45mg	2%
Total Carbohydrate 17g	6%
Sugars 9g	
Protein 0g	
Vitamin C	25%
Not a significant source of saturated fat, cholesterol, dietary fiber, vitamin A, calcium and iron.	
* Percent Daily Values are based on a 2,000 calorie diet.	

Fruit snacks nutrition label

EXERCISE SET 8.2

Write each of the following as an equation. Do not solve. See Examples 1 through 3.

1. 15% of 72 is what number?

2. What number is 25% of 55?

3. 30% of what number is 80?

4. 0.5 is 20% of what number?

5. What percent of 90 is 20?

6. 0.008 is 50% of what number?

7. 19 is 40% of what number?

8. 72% of 63 is what?

9. What number is 9% of 43?

10. 4.5 is what percent of 45?

Solve. See Examples 4 and 5.

11. 10% of 35 is what number?

12. 25% of 60 is what number?

13. What is 14% of 52?

14. What is 30% of 17?

Solve. See Examples 6 and 7.

15. 30 is 5% of what number?

16. 25 is 25% of what number?

17. 1.2 is 12% of what number?

18. 0.22 is 44% of what number?

Solve. If necessary, round to the nearest hundredth percent. See Examples 8 and 9.

19. 4 is what percent of 56?

20. 9 is what percent of 81?

21. 2.58 is what percent of 50?

22. 3.15 is what percent of 100?

Solve.

23. 0.1 is 10% of what number?

24. 0.5 is 5% of what number?

25. 35 is $16\frac{2}{3}\%$ of what number?

26. 7.2 is $6\frac{1}{4}\%$ of what number?

27. 125% of 36 is what number?

28. 200% of 13.5 is what number?

29. 126 is what percent of 31.5?

30. 264 is what percent of 33?

ANSWERS

1. ______
2. ______
3. ______
4. ______
5. ______
6. ______
7. ______
8. ______
9. ______
10. ______
11. ______
12. ______
13. ______
14. ______
15. ______
16. ______
17. ______
18. ______
19. ______
20. ______
21. ______
22. ______
23. ______
24. ______
25. ______
26. ______
27. ______
28. ______
29. ______
30. ______

ANSWERS

31. ______
32. ______
33. ______
34. ______
35. ______
36. ______
37. ______
38. ______
39. ______
40. ______
41. ______
42. ______
43. ______
44. ______
45. ______
46. ______
47. ______
48. ______
49. ______
50. ______

Solve. If necessary, round to the nearest hundredth. See Example 10.

31. A 5% tax is charged on the $450.00 purchase price of a TV. 5% of $450.00 is what number?

32. A VCR regularly sells for $250, but is on sale for $220. $220 is what percent of $250?

33. Liability insurance on a car costs $525 per year. The insurance company deducts 10% if the driver has not had an accident in the past three years. Find the amount deducted. (*Hint:* 10% of $525 is what number?)

34. A sale advertises you can save $66 or 44% on the purchase of a tweed coat. Find the original price of the tweed coat. (*Hint:* 44% of what number is 66?)

35. The interest paid on a home improvement loan was $6960. If this **amount is 11% of** the amount **borrowed,** find how much was borrowed.

36. Vera Faciane earns $1200 per month and budgets $75 per week for food. What **percent of** her monthly **income is** spent on **food?**

37. Last year, Mai Toberlan bought a share of stock for $83. She was paid a dividend of $5.13. Determine what **percent of** the stock **price is** the **dividend.**

38. The Hodder family paid 20% of the purchase price of a $75,000 home as a down payment. Determine the amount of the down payment.

39. The population of Orlando, Florida, was 128,000 in 1980. The population increased 36% over the next 12 years. What was the population increase of Orlando from 1980 to 1992? (*Source: 1980 Census of Population* and U.S. Bureau of the Census)

40. An owner of a repair service company estimates that, for every 40 hours a repairperson is on the job, he can only bill for 75% of the hours. The remaining hours, the repairperson is idle or driving to or from a job. Determine the number of hours per 40-hour week the owner can bill for a repairman.

41. The **assessed value** of a piece of property is set by government as a percent of the **market value** of the property. Find the assessed value of a piece of property whose market value is $87,500, if the assessed value rate is 35%.

42. **Property tax** due on a home is a percent of its **assessed value.** Find the property tax due on a home whose assessed value is $130,625 if the tax rate is 2.15%.

43. A **federal excise tax** is a tax charged on certain **commodities** or **services.** Suppose a federal excise tax is charged at the rate of 3% of long-distance telephone service. Determine the federal excise tax on a long-distance telephone bill of $37.68. Round to the nearest cent.

44. A state charges a tax of $0.18 on one pack of cigarettes. If a pack of cigarettes sells for $2.23, including the tax, find the rate of the tax.

45. Gold cost $385.20 per ounce on January 3. On April 21, it cost $337.60 per ounce. What percent of $385.20 is $337.60?

46. The social security taxes an employee pays are 15.02% of total wages. Find the amount of social security tax if wages are $725.

47. Next year's enrollment at a university is expected to increase by 8.3% over this year's enrollment, which is 12,465. Determine next year's increase in enrollment to the nearest whole.

48. A down payment of 20% of the purchase price of a house is often required in order to finance the remaining 80% of the price. A couple saved $12,385 as a down payment. Determine the price of the house the couple can finance.

49. 18% of Frank's gross pay is withheld for income tax, and 7.525% is withheld for social security tax. Find the **total** amount withheld from Frank's gross pay of $320.

50. A manufacturer of electronic components expects 1.04% of its product to be defective. Determine the number of defective components expected in a batch of 28,350 components.

Find what percent of total calories is from fat. If necessary round to the nearest tenth of a percent. See Example 11.

51.

NUTRITION FACTS	
Serving Size 18 crackers (29g)	
Servings Per Container About 9	
AMOUNT PER SERVING	
Calories 120 Calories from Fat 35	
	% DAILY VALUE
Total Fat 4g	**6%**
Saturated Fat 0.5g	**3%**
Polyunsaturated Fat 0g	
Monounsaturated Fat 1.5g	
Cholesterol 0mg	**0%**
Sodium 220mg	**9%**
Total Carbohydrate 21g	**7%**
Dietary Fiber 2g	**7%**
Sugars 3g	
Protein 2g	
Vitamin A 0% • Vitamin C 0%	
Calcium 2% • Iron 4% Phosphorus 10%	

Snack Crackers

52.

NUTRITION FACTS	
Serving Size 28 crackers (31g)	
Servings Per Container About 6	
AMOUNT PER SERVING	
Calories 130 Calories from Fat 35	
	% DAILY VALUE*
Total Fat 4g	**6%**
Saturated Fat 2g	**10%**
Polyunsaturated Fat 1g	
Monounsaturated Fat 1g	
Cholesterol 0mg	**0%**
Sodium 470mg	**20%**
Total Carbohydrate 23g	**8%**
Dietary Fiber 1g	**4%**
Sugars 4g	
Protein 2g	
Vitamin A 0% • Vitamin C 0%	
Calcium 0% • Iron 2%	

Snack Crackers

ANSWERS

51. ________

52. ________

ANSWERS

53. ______

54. ______

55. ______

56. ______

57. ______

58. ______

59. ______

60. ______

61. ______

62. ______

63. ______

64. ______

53.

NUTRITION FACTS	
Serving Size 2 waffles (72g)	
Servings Per Container 4	
AMOUNT PER SERVING	
Calories 190 Calories from Fat 50	
	% DAILY VALUE*
Total Fat 6g	**10%**
Saturated Fat 1g	**5%**
Cholesterol 10mg	**4%**
Sodium 540mg	**22%**
Potassium 115 mg	**3%**
Total Carbohydrate 28g	**9%**
Dietary Fiber 1g	**4%**
Sugars 6g	
Protein 5g	
Vitamin A	0%
Vitamin C	0%
Calcium	8%
Iron	20%
Thiamin	15%
Riboflavin	20%
Niacin	15%
Vitamin B_6	25%
Vitamin B_{12}	25%
Phosphorus	25%

Waffles

54.

NUTRITION FACTS	
Serving Size $\frac{2}{3}$ cup (29g)	
Servings Per Container 1.5	
AMOUNT PER SERVING	
Calories 150 Calories from Fat 80	
	% DAILY VALUE*
Total Fat 9g	**14%**
Saturated Fat 3g	**16%**
Cholesterol 0mg	**0%**
Sodium 90mg	**4%**
Total Carbohydrate 16g	**5%**
Dietary Fiber 2g	**8%**
Sugars 0g	
Protein 2g	
Vitamin A 0% • Vitamin C 15%	
Calcium 0% • Iron 2%	
*Percent Daily Values are based on a 2,000 calorie diet.	

Potato Sticks

55. Which snack cracker is lower in fat, the snack cracker in Exercise 51 or the snack cracker in Exercise 52?

56. Which is lower in fat, the potato sticks in Exercise 54 or the snack cracker in Exercise 52?

Review Exercises

Solve the proportion for x. See Section 7.2.

57. $\frac{1}{6} = \frac{x}{54}$

58. $\frac{2}{5} = \frac{8}{x}$

59. $\frac{x}{7} = \frac{4}{21}$

60. $\frac{4}{x} = \frac{12}{5}$

61. $\frac{5}{8} = \frac{x}{3}$

62. $\frac{16}{x} = \frac{48}{120}$

63. $\frac{10}{3} = \frac{150}{x}$

64. $\frac{x}{24} = \frac{14}{3}$

8.3

INTEREST

TAPE PA 8.3

OBJECTIVES

1. Calculate simple interest.
2. Use a compound interest table to calculate compound interest.
3. Find monthly payments on loans.

1 **Interest** is money charged for using other people's money. When you borrow money, you pay interest. When you loan or invest money, you earn interest. The money borrowed, loaned, or invested is called the **principal amount,** or simply **principal.** Interest is normally stated in terms of a percent of the principal for a given period of time. The **interest rate** is the percent used in computing the interest. Unless stated otherwise, *the rate is understood to be per year.* When the interest is computed on the original principal, it is called **simple interest.** The simple interest formula is as follows:

SIMPLE INTEREST FORMULA

Simple **I**nterest = **P**rincipal · **R**ate · **T**ime

or

$$I = P \cdot R \cdot T$$

where the rate is understood to be per year.

EXAMPLE 1

Find the simple interest after 2 years on \$500 at an interest rate of 12%.

Solution:

In this example, $P = \$500$, $R = 12\%$, and $T = 2$ years. Replace the variables by values in the formula $I = P \cdot R \cdot T$.

$$I = P \cdot R \cdot T$$

$I = \$500 \cdot 12\% \cdot 2$ Let $P = \$500$, $R = 12\%$, and $T = 2$.

$= \$500 \cdot (0.12) \cdot 2$ Write 12% as a decimal.

$= \$120$ Multiply.

The simple interest is \$120.

PRACTICE PROBLEM 1

Find the simple interest after 3 years on \$750 at an interest rate of 8%.

REMINDER Remember that, unless stated otherwise, the interest rate given is **per year.** A time period given other than years must be converted to years. For example, if the time period is in months, divide by 12 to write it as a fraction of a year. To see why, study the examples below.

$$8 \text{ months} = 8 \not{\text{mo}} \cdot \frac{1 \text{ yr}}{12 \not{\text{mo}}} = \frac{8}{12} \text{ year} = \frac{2}{3} \text{ year}$$

$$18 \text{ months} = 18 \not{\text{mo}} \cdot \frac{1 \text{ yr}}{12 \not{\text{mo}}} = \frac{18}{12} \text{ years} = \frac{3}{2} \text{ years} \quad \text{or} \quad 1\frac{1}{2} \text{ years}$$

Answer:
1. \$180

PRACTICE PROBLEM 2

Juanita Lopez borrowed $800 for 9 months at a simple interest rate of 20%. How much interest did she pay?

EXAMPLE 2

Ivan Borski borrowed $2400 at 10% simple interest for 8 months to buy a used Chevy S-10. Find the simple interest he paid.

Solution:

$P = \$2400$, $R = 10\%$ or 0.10, and $T = \frac{8}{12}$ year or $\frac{2}{3}$ year.

$$\begin{aligned} I &= P \cdot R \cdot T \\ &= \$2400 \cdot (0.10) \cdot \frac{2}{3} \\ &= \$160 \end{aligned}$$

The interest on Ivan's loan is $160.

When money is borrowed, the borrower pays the original amount borrowed or the principal, as well as the interest. When money is invested, the investor receives the original amount invested, or the principal as well as the interest. In either case, the **total amount** is the sum of the principal and the interest.

> Total amount (paid or received) = principal + interest

PRACTICE PROBLEM 3

If $500 is borrowed at a simple interest rate of 12% for 6 months, find the total amount paid.

EXAMPLE 3

An accountant invested $2000 at a simple interest rate of 10% for 2 years. What total amount of money will she have from her investment in 2 years?

Solution:

First, find her interest.

$$\begin{aligned} I &= P \cdot R \cdot T \\ &= \$2000 \cdot (0.10) \cdot 2 \\ &= \$400 \end{aligned}$$

Next add the interest to the principal.

$$\begin{aligned} \text{total amount} &= \text{principal} + \text{interest} \\ &= \$2000 + \$400 \\ &= \$2400 \end{aligned}$$

After 2 years, she has a total amount of $2400.

2 Recall that simple interest depends on the original principal only. Another type of interest is **compound interest. Compound interest** is computed on not only the principal, but also on the interest already earned in previous compounding periods. Compound interest is used more often than simple interest.

Let's see how compound interest differs from simple interest. Suppose that $2000 is invested at 7% interest **compounded annually** for 3 years. This means that interest is added to the principal at the end of each year and next year's interest is computed on this new amount. In this section, we will round dollar amounts to the nearest cent.

Answers:
2. $120
3. $530

23. \$10,000 is compounded semiannually at a rate of 9% for 20 years.

24. \$3500 is compounded daily at a rate of 8% for 10 years.

Find the total amount of compound interest earned. See Example 5.

25. \$2675 is compounded annually at a rate of 9% for 1 year.

26. \$6375 is compounded semiannually at a rate of 10% for 1 year.

27. \$10,000 is compounded quarterly at a rate of 15% for 5 years.

28. \$11,500 is compounded semiannually at a rate of 16% for 5 years.

29. \$2050 is compounded daily at a rate of 18% for 5 years.

30. \$1100 is compounded quarterly at a rate of 13% for 10 years.

31. \$2000 is compounded annually at a rate of 8% for 5 years.

32. \$2000 is compounded semiannually at a rate of 8% for 5 years.

33. \$2000 is compounded quarterly at a rate of 8% for 5 years.

34. \$2000 is compounded daily at a rate of 8% for 5 years.

ANSWERS

23. ______

24. ______

25. ______

26. ______

27. ______

28. ______

29. ______

30. ______

31. ______

32. ______

33. ______

34. ______

ANSWERS

35. ______

36. ______

37. ______

38. ______

39. ______

40. ______

41. ______

42. ______

43. ______

44. ______

45. ______

46. ______

47. ______

48. ______

49. ______

Solve. See Example 6.

35. A college student borrows \$1500 for 6 months to pay for a semester of school. If the interest is \$61.88, find the monthly payment.

36. Jim Tillman borrows \$1800 for 9 months. If the interest is \$148.90, find his monthly payment.

37. \$20,000 is borrowed for 4 years. If the interest on the loan is \$10,588.70, find the monthly payment.

38. \$105,000 is borrowed for 15 years. If the interest on the loan is \$181,125.00, find the monthly payment.

39. Explain how to look up the compound interest factor in the Compound Interest Table.

40. Explain how to find the amount of interest on a compounded account.

41. Compare the following accounts:

Account 1: \$1000 is invested for 10 years at a simple interest rate of 6%.

Account 2: \$1000 is compounded semiannually at a rate of 6% for 10 years.

Discuss how the interest is computed for each account. Determine which account earns more interest. Why?

Review Exercises

Perform the indicated operation. See Sections 5.2 to 5.4.

42. $\begin{array}{r} 0.82 \\ +\ 0.23 \\ \hline \end{array}$

43. $\begin{array}{r} 6.28 \\ -1.19 \\ \hline \end{array}$

44. $\begin{array}{r} 1.3 \\ \times\ 0.7 \\ \hline \end{array}$

45. $\begin{array}{r} 2.04 \\ \times\ 0.22 \\ \hline \end{array}$

46. (0.2)(1.56)

47. (0.83)(3.01)

48. $76.2 \div 40$

49. $6.66 \div 0.03$

TAPE PA 8.4

8.4

SALES TAX AND WAGES

OBJECTIVES

1. Calculate sales tax.
2. Calculate gross pay for hourly wages, salaries, and commissions.
3. Calculate net pay.

1 Percents are frequently used in the retail trade. For example, most states levy a tax on certain items when purchased as a source of revenue. This tax is called a **sales tax,** and retail stores collect it for the state. Sales tax is almost always stated as a percent of the purchase price.

A 6% sales tax rate on a purchase of a $10.00 item yields a sales tax of

$$\text{sales tax} = 6\% \text{ of } \$10 = 0.06 \cdot \$10.00 = \$0.60$$

The total price to the customer would be

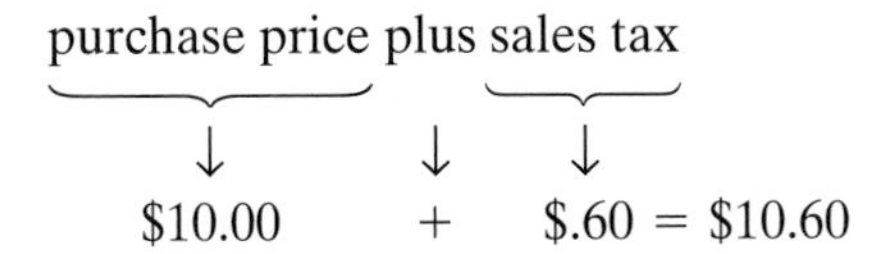

This example suggests the following formulas for sales tax.

> **SALES TAX FORMULAS**
>
> sales tax = tax rate · purchase price
>
> total price = purchase price + sales tax

In this section we will round dollar amounts to the nearest cent.

EXAMPLE 1

Find the sales tax and the total price on a purchase of an $85.50 trench coat in a city where the sales tax rate is 7.5%.

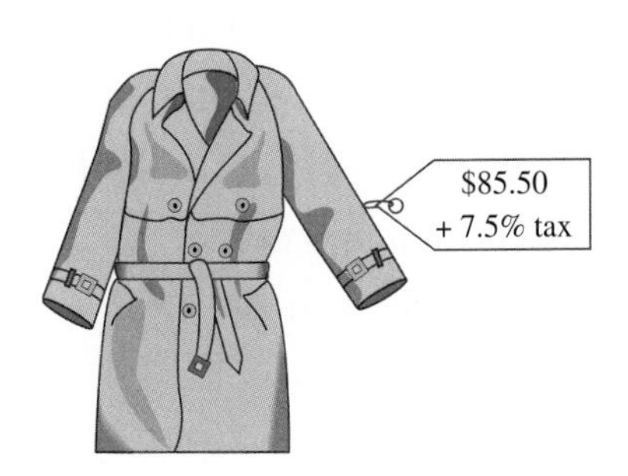

Solution:

The purchase price is $85.50 and the tax rate is 7.5%.

PRACTICE PROBLEM 1

If the percent sales tax is 6%, what is the sales tax and the total amount due on a $29.90 Goodgrip tire?

1. tax, $1.79; total, $31.69

In words: sales tax = tax rate · purchase price

Translate: sales tax = 7.5% · \$85.50

= (0.075) · \$85.5 Write 7.5% as a decimal.

≈ \$6.41 Rounded to the nearest cent.

Thus

In words: total price = purchase price + sales tax

Translate: total price = \$85.50 + \$6.41

= \$91.91.

The sales tax on \$85.50 is \$6.41 and the total price is \$91.91.

2 A **wage** is payment for performing work. Hourly wage, commissions, and salary are some of the ways wages can be paid. **Gross pay** is the wage before deductions such as federal income tax, state income tax, or insurance are taken out. **Net pay** is the wage after deductions have been taken out of the gross pay. Many employees are paid an **hourly wage,** which means they are paid a fixed wage for each hour worked.

PRACTICE PROBLEM 2

Jane Chin worked 40 hours last week. Her hourly rate is \$18.52 per hour. Calculate Jane's gross pay.

EXAMPLE 2

Brenda Owens is an electrician for Central Power and Light Electric Company. She worked 40 hours last week. Calculate her gross pay for last week if her hourly wage is \$13.88.

Solution:

Brenda worked 40 hours last week to be paid at the hourly rate of \$13.88. To find her gross pay, multiply her number of hours worked by her hourly wage.

In words: gross pay = number of hours worked · hourly wage

Translate: gross pay = 40(\$13.88)

= \$555.20

Brenda's gross pay for the week is \$555.20.

An employee who is paid a **salary** receives a fixed wage for each work period, such as each week, every two weeks, or each month. Salaried employees work a certain number of hours in their work period, but they receive no extra pay for extra hours worked.

Answer:

2. \$740.80

17. Before taking a typing course, Geoffry could type 32 words per minute. By the end of the course, he was able to type 76 words per minute. Find the percent increase.

18. By better defining the potential market, Zip Company was able to increase last year's sales 13%, to \$225,000 this year. Find last year's sales.

19. The price of a loaf of bread decreased from \$1.09 to \$0.89. Find the percent decrease.

20. From 1995 to 2004, the number of doctorates awarded to women is projected to increase 15%. The number of women who received doctorates in 1995 was 17,000. Find the predicted number of women to be awarded doctorates in 2004. (*Source:* U.S. National Center for Education Statistics)

21. The nation's Hispanic population is projected to increase 33% during the 10-year period from 1995 to 2005. The number of Hispanics in 1995 was 26.8 million. Find the increase to the nearest hundredth of a million. (*Source:* U.S. Bureau of the Census, *U.S. Census of Population:* 1970.)

22. The population of Tokyo is expected to increase from 26,518 in 1994 to 28,700 in 2015. Find this percent increase. (*Source:* United Nations, Dept. for Economic and Social Information and Policy Analysis)

Review Exercises

Find the simple interest using the formula $I = PRT$. See Section 7.3.

23. Find the simple interest if \$5000 is invested at 3% for 2 years.

24. Find the simple interest if \$6500 is invested at 12% for 4 years.

25. Find the simple interest if \$4500 is invested at 10% for 3 years.

26. Find the simple interest if \$95,000 is invested at $10\frac{1}{2}\%$ for $2\frac{1}{2}$ years.

27. Find the simple interest if \$30,000 is invested at 8.6% for 6 months.

28. Find the simple interest if \$1600 is invested at 8.9% for 6 months.

ANSWERS

17. ________

18. ________

19. ________

20. ________

21. ________

22. ________

23. ________

24. ________

25. ________

26. ________

27. ________

28. ________

GROUP ACTIVITY

INVESTIGATING PROBABILITY

MATERIALS:

- paper or foam cup, 30 thumbtacks

The measure of the chance of an event occurring is its probability. In this activity you will investigate one way that probabilities can be estimated.

1. Place the thumbtacks in the cup. Shake the cup and toss out the thumbtacks onto a flat surface.

NUMBER OF TACKS	NUMBER OF TACKS LANDING POINT UP	FRACTION OF POINT-UP TACKS	PERCENT OF POINT-UP TACKS
30			
60			
90			
120			
150			

Count the number of tacks that land point up, and record this number in the table. Complete the table. (*Hint:* For 60 thumbtacks, count the number of tacks landing point up in two tosses of the 30 thumbtacks, etc.)

2. For each row of the table, find the fraction of tacks that landed point up. Express this fraction as a percent. Add these values to the table in columns labeled "Fraction of Point-Up Tacks" and "Percent of Point-Up Tacks."

3. Each of the percents you computed in question 2 is an *estimate* of the probability that a single thumbtack will land point up when tossed. When you estimate a probability experimentally, the larger the number of trials used (i.e., number of tacks tossed), the better the estimate of the actual probability. What do you suppose the value of the actual probability is? Explain your reasoning.

4. Combine your results for all 450 of your tack tosses recorded in the table with the other groups' results. Of this total number of tacks, compute the percent of tacks that landed point up. This is your best estimate of the probability that a tack will land point up when tossed.

5. If you tossed 200 thumbtacks, what percent would you expect to land point up? How many tacks would you expect to land point up? Use the percent (probability) you computed in question 4 to make this calculation. What if you tossed 300 thumbtacks?

CHAPTER 8 HIGHLIGHTS

DEFINITIONS AND CONCEPTS	EXAMPLES
SECTION 8.1 PERCENTS, DECIMALS, AND FRACTIONS	
SUMMARY OF CONVERTING PERCENTS, DECIMALS, AND FRACTIONS	
To write a percent as a decimal, drop the % symbol and move the decimal point two places to the left.	$25\% = 25.\% = 0.25$
To write a decimal as a percent, move the decimal point two places to the right and attach %.	$0.7 = 0.70 = 70\%$
To write a percent as a fraction, drop the % symbol and write the number over 100.	$13\% = \frac{13}{100}$
To write a fraction as a percent, multiply the fraction by 100%.	$\frac{1}{3} = \frac{1}{3} \cdot 100\% = \frac{100}{3}\% = 33\frac{1}{3}\%$
SECTION 8.2 PERCENT EQUATIONS AND PROBLEM SOLVING	
Three key words in the statement of a percent problem are **of,** which means multiplication (·), **is,** which means equals (=), and **what** (or some equivalent word or phrase), which stands for the unknown.	Find the following. Six is 12% of what number? ↓ ↓ ↓ ↓ ↓ $6 = 12\% \cdot x$ $6 = 0.12x$ Write 12% as a decimal. $\frac{6}{0.12} = \frac{0.12x}{0.12}$ Divide by 0.12. $50 = x$
SECTION 8.3 INTEREST	
SIMPLE INTEREST FORMULA Interest = Principal · Rate · Time $I = P \cdot R \cdot T$ where the rate is understood to be per year.	Find the simple interest after 3 years on \$800 at an interest rate of 5%. $I = P \cdot R \cdot T$ $I = \$800 \cdot 5\% \cdot 3$ Let $P = \$800$, $R = 5\%$, and $T = 3$. $I = \$800 \cdot 0.05 \cdot 3$ Write 5% as 0.05. $I = \$120$ Multiply. The interest is \$120.
Compound interest is computed not only on the principal, but also on interest already earned in previous compounding periods. (See Appendix D.)	\$800 is invested at 5% compounded quarterly for 10 years. Find the total amount at the end of 10 years. total amount = original principal · compound interest factor $= \$800 \cdot 1.64362$ $\approx \$1314.90$

(8.2) *Write each of the following as an equation and solve for the variable. If necessary, round to the nearest tenth.*

47. 1250 is 1.25% of what?

48. What is $33\frac{1}{3}\%$ of 24,000?

49. 2502 is what percent of 3456?

50. 22.9 is 20% of what?

51. What is 40% of 7500?

52. 693 is what percent of 462?

53. 104.5 is 25% of what?

54. 50.4 is 5.25% of what?

55. What is $83\frac{1}{3}\%$ of 42.06?

56. 383 is what percent of 875?

57. In a survey of 2000 people, it was found that 1320 have a microwave oven. Find the percent of people who own microwaves.

58. A sales tax of $1.53 is added to an item's price of $152.99. Find the sales tax rate.

59. Of the 12,365 freshmen entering County College, 2000 are enrolled in Intermediate Algebra. Find the percent of entering freshmen who are enrolled in Intermediate Algebra.

(8.3) *Solve.*

60. Find the simple interest due on $4000 loaned for 3 months at 12% interest.

61. Find how long it takes $2000 to grow to $2210 if it is invested at a simple interest rate of 7%.

62. Find the total amount due on an 8-month loan of $1200 at a simple interest rate of 15%.

63. Find the simple interest on $600 for 2 months at 6%.

ANSWERS

47. ______

48. ______

49. ______

50. ______

51. ______

52. ______

53. ______

54. ______

55. ______

56. ______

57. ______

58. ______

59. ______

60. ______

61. ______

62. ______

63. ______

ANSWERS

64. ______

65. ______

66. ______

67. ______

68. ______

69. ______

70. ______

71. ______

72. ______

73. ______

74. ______

75. ______

76. ______

Solve.

64. Find the total amount in an account if \$5500 is compounded annually at 12% for 15 years.

65. Find the compound interest earned if \$6000 is compounded semiannually at 11% for 10 years.

66. Find the compound interest earned if \$100 is compounded quarterly at 12% for 5 years.

67. Find the amount of interest earned on an account if \$300 is compounded daily at a rate of 17% for 1 year.

68. Find the compound interest earned if \$1000 is compounded quarterly at 18% for 20 years.

(8.4) *Solve. Round dollar amounts to the nearest cent.*

69. Find the sales tax paid on a \$25.50 purchase if the sales tax rate is 4.5%.

70. If a sales tax rate of 4.5% is applied to only nonfood items, find the sales tax on the purchase of 2 gallons of milk at \$2.55 each and a magazine at \$2.50.

71. If the sales tax rate is 5.5%, what is the total amount charged for a \$250 coat?

72. Russ James is a sales representative for a chemical company and is paid a commission rate of 5% on all sales. Find his commission if he sold \$100,000 worth of chemicals last month.

73. Carol Sell is a sales clerk in a clothing store. She receives a commission of 7.5% on all sales. Find her gross pay for the week if her sales for the week were \$4005.68.

74. A brick layer is paid \$15.50 per hour. Find the brick layer's gross pay if he worked 43 hours.

75. A cashier for a grocery store is paid \$8.00 per hour. Find his gross pay if he worked 40 hours.

76. A sales clerk is paid \$9.35 per hour. Find his gross pay if he worked 40 hours.

77. A teacher is paid a monthly salary of \$2450. Find his yearly salary if he works the whole year.

78. A piano teacher at Werlein's Music is paid a salary of \$1300 per month. Find her yearly salary.

79. A window designer for a department store is paid \$21,500 per year. What is his monthly salary?

(8.5) *Solve.*

80. A Bose radio priced at \$236 is increased by 25%. Find the new price.

81. A retail electronic store bought 50 telephones for resale at a cost of \$24 each. The store increases the price 100% and then sells them. Find the new price.

82. The current charge for dumping waste in a local landfill is \$16 per cubic foot. To cover new environmental costs, the charge will increase to \$33 per cubic foot. Find the percent increase.

83. The number of violent crimes in a city increased from 534 to 675. Find the percent increase. Round to the nearest tenth.

84. This year the fund drive for a charity collected \$215,000. Next year, a 4% decrease is expected. Find how much is expected to be collected in next year's drive.

85. A local union negotiated a new contract that increases the hourly pay 15% over last year's pay. The old hourly rate was \$11.50. Find the new hourly rate rounded to the nearest cent.

ANSWERS

77. ______________

78. ______________

79. ______________

80. ______________

81. ______________

82. ______________

83. ______________

84. ______________

85. ______________

CHAPTER

9

POLYNOMIALS

BUSINESS ANALYSIS

Suppose you own a business that manufactures baskets. How many baskets should you make? What price should you charge for each basket? What kind of profit can you expect from your basket sales? These kinds of questions can be answered with the help of a business analyst.

IN THE CHAPTER GROUP ACTIVITY ON PAGE 653, YOU WILL HAVE THE OPPORTUNITY TO ANSWER THESE QUESTIONS ABOUT A BASKET MANUFACTURING BUSINESS.

TAPE PA 9.1

9.1

ADDING AND SUBTRACTING POLYNOMIALS

OBJECTIVES

1. Add polynomials.
2. Subtract polynomials.
3. Evaluate polynomials at given replacement values.

Before we add and subtract polynomials, let's first review some definitions presented in Section 3.1. Recall that the *addends* of an algebraic expression are the *terms* of the expression.

EXPRESSIONS

$3x + 5$ — 2 terms

$7y^2 + (-6y) + 4$ — 3 terms

Also, recall that *like terms* can be added or subtracted by using the distributive property. For example,

$$7x + 3x = (7 + 3)x = 10x$$

1 Some terms are also called **monomials.** A monomial is a term that contains only whole number exponents and no variable in the denominator.

MONOMIALS	NOT MONOMIALS	
$3x^2$	$\frac{2}{y}$	Variable in denominator.
$-\frac{1}{2}a^2bc^3$	$-2x^{-5}$	Not a whole number exponent.
7		

A sum or difference of monomials is called a **polynomial.**

> A **polynomial** is a monomial or a sum or difference of monomials.

EXAMPLES OF POLYNOMIALS

$$5x^3 - 6x^2 + 2x - 10, \quad -1.2y^3 + 0.7y, \quad z, \quad \frac{1}{3}r - \frac{1}{2}, \quad 0$$

To add polynomials, we use the commutative and associative properties to rearrange and group like terms.

Then we combine like terms.

> **TO ADD POLYNOMIALS**
> To add polynomials, combine like terms.

EXAMPLE 1

Add: $(3x - 1) + (-6x + 2)$.

Solution:

$$\begin{aligned}(3x - 1) + (-6x + 2) &= (3x - 6x) + (-1 + 2) && \text{Group like terms.}\\ &= (-3x) + (1) && \text{Combine like terms.}\\ &= -3x + 1\end{aligned}$$

PRACTICE PROBLEM 1

Add: $(2y + 7) + (9y - 14)$.

EXAMPLE 2

Add: $(9y^2 - 6y) + (7y^2 + 10y + 2)$.

Solution:

$$\begin{aligned}(9y^2 - 6y) + (7y^2 + 10y + 2) &= \underbrace{9y^2 + 7y^2}\underbrace{- 6y + 10y} + 2 && \text{Group like terms.}\\ &= 16y^2 + 4y + 2\end{aligned}$$

PRACTICE PROBLEM 2

Add: $(5x^2 + 4x - 3) + (x^2 - 6x)$.

EXAMPLE 3

Find the sum of $(-y^2 + 2y + 1.7)$ and $(12y^2 - 6y - 3.6)$.

Solution:

Recall that "sum" means addition.

$$\begin{aligned}&(-y^2 + 2y + 1.7) + (12y^2 - 6y - 3.6)\\ &= \underbrace{-y^2 + 12y^2} + \underbrace{2y - 6y} + \underbrace{1.7 - 3.6} && \text{Group like terms.}\\ &= 11y^2 - 4y - 1.9 && \text{Combine like terms.}\end{aligned}$$

PRACTICE PROBLEM 3

Find the sum of $(7z^2 - 4.2z + 11)$ and $(-9z^2 - 1.9z + 4)$.

Polynomials can also be added vertically. To do so, line up like terms underneath one another. Let's vertically add the polynomials in Example 3.

EXAMPLE 4

Find the sum of $(-y^2 + 2y + 1.7)$ and $(12y^2 - 6y - 3.6)$. Use a vertical format.

Solution:

Line up like terms underneath one another.

$$\begin{array}{r}-y^2 + 2y + 1.7\\ +\ 12y^2 - 6y - 3.6\\ \hline 11y^2 - 4y - 1.9\end{array}$$

PRACTICE PROBLEM 4

Add the polynomials in Practice Problem 3 vertically.

Notice that we are finding the same sum in Example 3 as in Example 4. Of course, the results are the same.

2 To subtract one polynomial from another, recall how we subtract numbers. To subtract a number, we add its opposite: $a - b = a + (-b)$.

For example,

$$\begin{aligned}7 - 10 &= 7 + (-10)\\ &= -3\end{aligned}$$

To subtract a polynomial, we also add its opposite. Just as the opposite of 3 is -3, the opposite of $(2x^2 - 5x + 1)$ is $-(2x^2 - 5x + 1)$. Let's practice simplifying the opposite of a polynomial.

Answers:
1. $11y - 7$
2. $6x^2 - 2x - 3$
3. and **4.** $-2z^2 - 6.1z + 15$

PRACTICE PROBLEM 5
Simplify $-(7y^2 + 4y - 6)$.

EXAMPLE 5
Simplify $-(2x^2 - 5x + 1)$.

Solution:
Rewrite $-(2x^2 - 5x + 1)$ as $-1(2x^2 - 5x + 1)$ and use the distributive property.

$$\begin{aligned} -(2x^2 - 5x + 1) &= -1(2x^2 - 5x + 1) \\ &= -1(2x^2) + (-1)(-5x) + (-1)(1) \\ &= -2x^2 + 5x - 1 \end{aligned}$$

Notice the result of Example 5.

$$-(2x^2 - 5x + 1) = -2x^2 + 5x - 1$$

This means that the opposite of a polynomial can be found by changing the signs of the terms of the polynomial. This leads to the following.

> **TO SUBTRACT POLYNOMIALS**
> To subtract polynomials, change the signs of the terms of the polynomial being subtracted, then add.

PRACTICE PROBLEM 6
Subtract: $(3b - 2) - (7b + 23)$.

EXAMPLE 6
Subtract: $(5a + 7) - (2a - 10)$.

Solution:

$$\begin{aligned} (5a + 7) - (2a - 10) &= (5a + 7) + (-2a + 10) && \text{Add the opposite of } 2a - 10. \\ &= 5a - 2a + 7 + 10 && \text{Group like terms.} \\ &= 3a + 17 \end{aligned}$$

PRACTICE PROBLEM 7
Subtract $(3x^2 - 12x)$ from $(-4x^2 + 20x + 17)$.

EXAMPLE 7
Subtract $(-6z^2 - 2z + 13)$ from $(4z^2 - 20z)$.

Solution:
Be careful when arranging the polynomials in this example.

$$\begin{aligned} (4z^2 - 20z) - (-6z^2 - 2z + 13) &= (4z^2 - 20z) + (6z^2 + 2z - 13) \\ &= 4z^2 + 6z^2 - 20z + 2z - 13 && \text{Group like terms.} \\ &= 10z^2 - 18z - 13 \end{aligned}$$

Just as with adding polynomials, we can subtract polynomials using a vertical format. Let's subtract the polynomials in Example 7 using a vertical format.

PRACTICE PROBLEM 8
Subtract $(3x^2 - 12x)$ from $(-4x^2 + 20x + 17)$. Use a vertical format.

EXAMPLE 8
Subtract $(-6z^2 - 2z + 13)$ from $(4z^2 - 20z)$. Use a vertical format.

Solution:
Line up like terms underneath one another.

$$\begin{array}{r} 4z^2 - 20z \\ -\,(-6z^2 - 2z + 13) \\ \hline \end{array} \qquad \begin{array}{r} 4z^2 - 20z \\ 6z^2 + 2z - 13 \\ \hline 10z^2 - 18z - 13 \end{array}$$

Answers:
5. $-7y^2 - 4y + 6$
6. $-4b - 25$
7. and **8.** $-7x^2 + 32x + 17$

Review Exercises

Evaluate. See Sections 1.7 and 2.4.

51. 3^4 **52.** $(-2)^5$ **53.** $(-5)^2$ **54.** 4^3

Write using exponential notation. See Section 1.8.

55. $x \cdot x \cdot x$ **56.** $y \cdot y \cdot y \cdot y \cdot y$ **57.** $2 \cdot 2 \cdot a \cdot a \cdot a \cdot a$ **58.** $5 \cdot 5 \cdot 5 \cdot b \cdot b$

ANSWERS

51. __________

52. __________

53. __________

54. __________

55. __________

56. __________

57. __________

58. __________

9.2

MULTIPLICATION PROPERTIES OF EXPONENTS

OBJECTIVES

1. Use the product rule for exponents.
2. Use the power of a power rule for exponents.
3. Use the power of a product rule for exponents.

TAPE PA 9.2

1 Recall from Section 1.8 that an exponent has the same meaning whether the base is a number or a variable. For example,

$$5^3 = \underbrace{5 \cdot 5 \cdot 5}_{\text{3 factors of 5}} \quad \text{and} \quad x^3 = \underbrace{x \cdot x \cdot x}_{\text{3 factors of } x}$$

We can use this definition of an exponent to discover properties that will help us to simplify products and powers of exponential expressions.

For example, let's use the definition of an exponent to find the product of x^3 and x^4.

$$\begin{aligned} x^3 \cdot x^4 &= (x \cdot x \cdot x)(x \cdot x \cdot x \cdot x) \\ &= \underbrace{x \cdot x \cdot x \cdot x \cdot x \cdot x \cdot x}_{\text{7 factors of } x} \\ &= x^7 \end{aligned}$$

Notice that the result is the same if we add the exponents.

$$x^3 \cdot x^4 = x^{3+4} = x^7$$

This suggests the following product property for exponents.

> **PRODUCT PROPERTY FOR EXPONENTS**
> If m and n are positive integers and a is a real number, then
>
> $$a^m \cdot a^n = a^{m+n}$$

In other words, to multiply two exponential expressions with the same base, keep the base and add the exponents.

EXAMPLE 1

Multiply: $y^7 \cdot y^2$.

Solution:

$$\begin{aligned} y^7 \cdot y^2 &= y^{7+2} && \text{Use the product property for exponents.} \\ &= y^9 && \text{Simplify.} \end{aligned}$$

PRACTICE PROBLEM 1

Multiply: $z^4 \cdot z^8$.

Answer:

1. z^{12}

PRACTICE PROBLEM 2
Multiply: $7y^5 \cdot 3y^9$.

EXAMPLE 2
Multiply: $3x^5 \cdot 6x^3$.

Solution:

$$\begin{aligned} 3x^5 \cdot 6x^3 &= (3 \cdot 6)(x^5 \cdot x^3) && \text{Apply the commutative and associative properties.} \\ &= 18x^{5+3} && \text{Use the product property for exponents.} \\ &= 18x^8 && \text{Simplify.} \end{aligned}$$

PRACTICE PROBLEM 3
Multiply: $(-7r^6s^2)(-3r^2s^5)$.

EXAMPLE 3
Multiply: $(-2a^4b^{10})(9a^5b^3)$.

Solution:
Use properties of multiplication to group numbers and like variables together.

$$\begin{aligned} (-2a^4b^{10})(9a^5b^3) &= (-2 \cdot 9)(a^4 \cdot a^5)(b^{10} \cdot b^3) \\ &= -18a^{4+5}b^{10+3} \\ &= -18a^9b^{13} \end{aligned}$$

PRACTICE PROBLEM 4
Multiply: $9y^4 \cdot 3y^2 \cdot y$. (Recall that $y = y^1$.)

EXAMPLE 4
Multiply: $2x^3 \cdot 3x \cdot 5x^6$.

Solution:
First notice the factor $3x$. Since there is one factor of x in $3x$, it can also be written as $3x^1$.

$$\begin{aligned} 2x^3 \cdot 3x^1 \cdot 5x^6 &= (2 \cdot 3 \cdot 5)(x^3 \cdot x^1 \cdot x^6) \\ &= 30x^{10} \end{aligned}$$

2 Next suppose that we want to simplify an exponential expression raised to a power. To see how we simplify $(x^2)^3$, we again use the definition of an exponent.

$$\begin{aligned} (x^2)^3 &= \underbrace{(x^2) \cdot (x^2) \cdot (x^2)}_{3 \text{ factors of } x^2} && \text{Apply the definition of an exponent.} \\ &= x^{2+2+2} && \text{Use the product property for exponents.} \\ &= x^6 && \text{Simplify.} \end{aligned}$$

Notice the result is exactly the same if we multiply the exponents.

$$(x^2)^3 = x^{2 \cdot 3} = x^6$$

This suggests the following power of a power property for exponents.

POWER OF A POWER PROPERTY FOR EXPONENTS
If m and n are positive integers and a is a real number, then

$$(a^m)^n = a^{m \cdot n}$$

In other words, to raise a power to a power, keep the base and multiply the exponents.

Answers:
2. $21y^{14}$
3. $21r^8s^7$
4. $27y^7$

9.3

MULTIPLYING POLYNOMIALS

TAPE PA 9.3

OBJECTIVES

1. Multiply a monomial and any polynomial.
2. Multiply two binomials.
3. Raise a binomial to a power of two.
4. Multiply any two polynomials.

1 Recall that a polynomial that consists of one term is called a **monomial.** For example, $5x$ is a monomial. To multiply a monomial and any polynomial, we use the distributive property and properties of exponents.

Recall the distributive property $a(b + c) = a \cdot b + a \cdot c$.

EXAMPLE 1

Multiply: $5x(3x^2 + 2)$.

Solution:

$$5x(3x^2 + 2) = 5x \cdot 3x^2 + 5x \cdot 2 \quad \text{Apply the distributive property.}$$
$$= 15x^3 + 10x$$

PRACTICE PROBLEM 1

Multiply: $3y(7y^2 + 5)$.

EXAMPLE 2

Multiply: $2z(4z^2 + 6z - 9)$.

Solution:

$$2z(4z^2 + 6z - 9) = 2z \cdot 4z^2 + 2z \cdot 6z + 2z(-9)$$
$$= 8z^3 + 12z^2 - 18z$$

PRACTICE PROBLEM 2

Multiply: $5r(8r^2 - r + 11)$.

To visualize multiplication by a monomial, study the rectangle below.

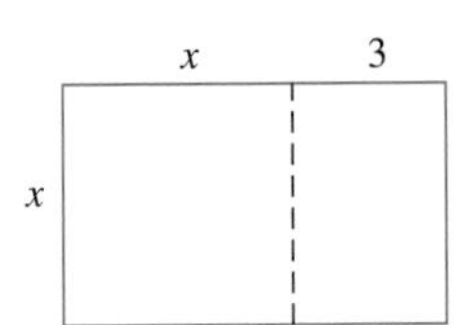

The width of the rectangle is x and its length is $x + 3$. One way to calculate the area of the rectangle is

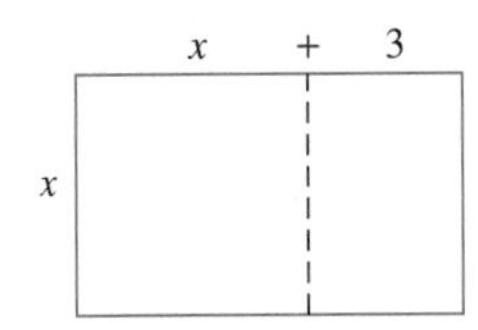

Area = width · length
$= x(x + 3)$

Another way to calculate the area of the rectangle is to find the sum of the areas of the smaller figures.

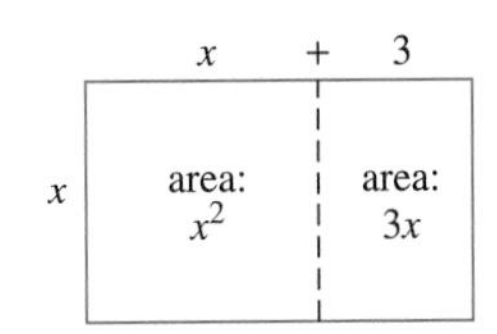

Area $= x^2 + 3x$

Answers:
1. $21y^3 + 15y$
2. $40r^3 - 5r^2 + 55r$

Since the areas must be equal, we have that

$$x(x + 3) = x^2 + 3x$$ As expected by the distributive property.

2 A polynomial that consists of exactly two terms is called a **binomial.** To multiply two binomials, we use a version of the distributive property:

$$(b + c)a = b \cdot a + c \cdot a$$

PRACTICE PROBLEM 3
Multiply: $(b + 7)(b + 5)$.

EXAMPLE 3

Multiply: $(x + 2)(x + 3)$.

Solution:

$(x + 2)(x + 3) = x(x + 3) + 2(x + 3)$	Apply the distributive property.
$= x \cdot x + x \cdot 3 + 2 \cdot x + 2 \cdot 3$	Apply the distributive property.
$= x^2 + 3x + 2x + 6$	Multiply.
$= x^2 + 5x + 6$	Combine like terms.

PRACTICE PROBLEM 4
Multiply: $(5x - 1)(5x + 4)$.

EXAMPLE 4

Multiply: $(4y + 9)(3y - 2)$.

Solution:

$(4y + 9)(3y - 2) = 4y(3y - 2) + 9(3y - 2)$	Apply the distributive property.
$= 4y \cdot 3y + 4y(-2) + 9 \cdot 3y + 9(-2)$	Apply the distributive property.
$= 12y^2 - 8y + 27y - 18$	Multiply.
$= 12y^2 + 19y - 18$	Combine like terms.

3 To raise a binomial to the power of 2, we use the definition of an exponent, and then multiply.

PRACTICE PROBLEM 5
Multiply: $(6y - 1)^2$.

EXAMPLE 5

Multiply: $(2x + 1)^2$.

Solution:

$(2x + 1)^2 = (2x + 1)(2x + 1)$	Apply the definition of an exponent.
$= 2x(2x + 1) + 1(2x + 1)$	Apply the distributive property.
$= 2x \cdot 2x + 2x \cdot 1 + 1 \cdot 2x + 1 \cdot 1$	Apply the distributive property.
$= 4x^2 + 2x + 2x + 1$	Multiply.
$= 4x^2 + 4x + 1$	Combine like terms.

Answers:
3. $b^2 + 12b + 35$
4. $25x^2 + 15x - 4$
5. $36y^2 - 12y + 1$

4 A polynomial that consists of exactly three terms is called a **trinomial.** Next, we multiply a binomial by a trinomial.

EXAMPLE 6

Multiply: $(3a + 2)(a^2 - 6a + 3)$.

Solution:

Use the distributive property to multiply $3a$ by the trinomial $(a^2 - 6a + 3)$ and then 2 by the trinomial.

$(3a + 2)(a^2 - 6a + 3) = 3a(a^2 - 6a + 3) + 2(a^2 - 6a + 3)$	Apply the distributive property.
$= 3a \cdot a^2 + 3a(-6a) + 3a \cdot 3 + 2 \cdot a^2 + 2(-6a) + 2 \cdot 3$	Apply the distributive property.
$= 3a^3 - 18a^2 + 9a + 2a^2 - 12a + 6$	Multiply.
$= 3a^3 - 16a^2 - 3a + 6$	Combine like terms.

PRACTICE PROBLEM 6

Multiply: $(2x + 5)(x^2 + 4x - 1)$.

In general, we have the following:

TO MULTIPLY TWO POLYNOMIALS

Multiply each term of the first polynomial by each term of the second polynomial, and then combine like terms.

A convenient method of multiplying polynomials is to use a vertical format similar to multiplying real numbers.

EXAMPLE 7

Find the product of $(a^2 - 6a + 3)$ and $(3a + 2)$ vertically.

Solution:

$$\begin{array}{rl} a^2 - 6a + 3 & \\ \times \qquad\qquad 3a + 2 & \\ \hline 2a^2 - 12a + 6 & \text{Multiply } a^2 - 6a + 3 \text{ by } 2. \\ 3a^3 - 18a^2 + 9a \qquad & \text{Multiply } a^2 - 6a + 3 \text{ by } 3a. \text{ Line up like terms.} \\ \hline 3a^3 - 16a^2 - 3a + 6 & \text{Combine like terms.} \end{array}$$

Notice that this example is the same as Example 6 and of course the products are the same.

PRACTICE PROBLEM 7

Multiply $(x^2 + 4x - 1)$ and $(2x + 5)$ vertically.

Answers:

6. $2x^3 + 13x^2 + 18x - 5$

7. $2x^3 + 13x^2 + 18x - 5$

EXERCISE SET 9.3

Multiply. See Examples 1 and 2.

1. $3x(9x^2 - 3)$

2. $4y(10y^3 + 2y)$

3. $-5a(4a^2 - 6a + 1)$

4. $-2b(3b^2 - 2b + 5)$

5. $7x^2(6x^2 - 5x + 7)$

6. $6z^2(-3z^2 - z + 4)$

Multiply. See Examples 3–5.

7. $(x + 3)(x + 10)$

8. $(y + 5)(y + 9)$

9. $(2x - 6)(x + 4)$

10. $(7z + 1)(z - 6)$

11. $(6a + 4)^2$

12. $(8b - 3)^2$

Multiply. See Examples 6 and 7.

13. $(a + 6)(a^2 - 6a + 3)$

14. $(y + 4)(y^2 + 8y - 2)$

15. $(4x - 5)(2x^2 + 3x - 10)$

16. $(9z - 2)(2z^2 + z + 1)$

17. $(x^3 + 2x + x^2)(3x + 1 + x^2)$

18. $(y^2 - 2y + 5)(y^3 + 2 + y)$

Multiply.

19. $10r(-3r + 2)$

20. $5x(4x^2 + 5)$

21. $-2y^2(3y + y^2 - 6)$

22. $3z^3(4z^4 - 2z + z^3)$

23. $(x + 2)(x + 12)$

24. $(y + 7)(y - 7)$

25. $(2a + 3)(2a - 3)$

26. $(6s + 1)(3s - 1)$

27. $(x + 5)^2$

28. $(x + 3)^2$

ANSWERS

1. ______
2. ______
3. ______
4. ______
5. ______
6. ______
7. ______
8. ______
9. ______
10. ______
11. ______
12. ______
13. ______
14. ______
15. ______
16. ______
17. ______
18. ______
19. ______
20. ______
21. ______
22. ______
23. ______
24. ______
25. ______
26. ______
27. ______
28. ______

ANSWERS

29. ______
30. ______
31. ______
32. ______
33. ______
34. ______
35. ______
36. ______
37. ______
38. ______
39. ______
40. ______
41. ______
42. ______
43. ______
44. ______
45. ______
46. ______
47. ______
48. ______
49. ______
50. ______
51. ______
52. ______

29. $\left(b + \frac{3}{5}\right)\left(b + \frac{4}{5}\right)$

30. $\left(a - \frac{7}{10}\right)\left(a + \frac{3}{10}\right)$

31. $(6x + 1)(x^2 + 4x + 1)$

32. $(9y - 1)(y^2 + 3y - 5)$

33. $(7x + 5)^2$

34. $(5x + 9)^2$

35. $(2x - 1)^2$

36. $(4a - 3)^2$

37. $(2x^2 - 3)(4x^3 + 2x - 3)$

38. $(3y^2 + 2)(5y^2 - y + 2)$

39. $(x^3 + x^2 + x)(x^2 + x + 1)$

40. $(a^4 + a^2 + 1)(a^4 + a^2 - 1)$

41. $(2z^2 - z + 1)(5z^2 + z - 2)$

42. $(2b^2 - 4b + 3)(b^2 - b + 2)$

Find the area of each figure.

❑ **43.**

$(y - 6)$ feet

$(y^2 + 3y + 2)$ feet

❑ **44.**

$(2x + 11)$ centimeters

Find the area of the shaded figure. To do so, subtract the area of the smaller square from the area of the larger geometric figure.

❑ **45.**

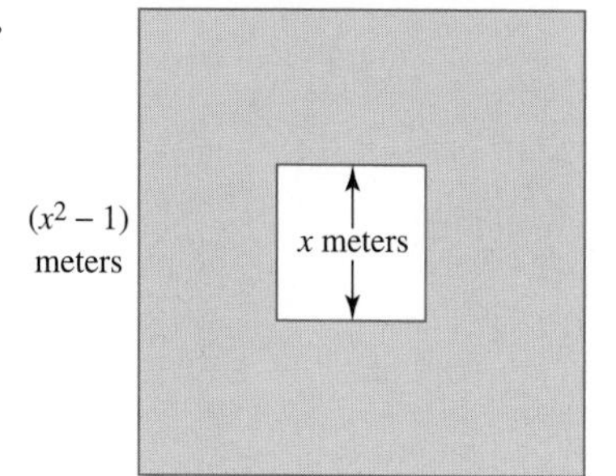

❑ **46.**

$(3x + 5)$ miles

$2x$ miles

$(3x - 5)$ miles

Review Exercises

Write each number as a product of prime numbers. See Section 4.2.

47. 50

48. 48

49. 72

50. 36

51. 200

52. 300

TAPE PA 9.4

9.4

INTRODUCTION TO FACTORING POLYNOMIALS

OBJECTIVES

1. Find the greatest common factor of a list of integers.
2. Find the greatest common factor of a list of terms.
3. Factor the greatest common factor from the terms of a polynomial.

Recall that when an integer is written as the product of 2 or more integers, each of these integers is called a *factor* of the product. This is true of polynomials also. When a polynomial is written as the product of two or more other polynomials, each of these polynomials is called a factor of the product.

-2	$\cdot$	4	$=$	-8	x^3	$\cdot$	x^7	$=$	x^{10}	$5(x + 2)$	$=$	$5x + 10$
factor		factor		product	factor		factor		product	factor factor		product

The process of writing a polynomial as a product is called **factoring.** Notice that factoring is the reverse process of multiplying.

factoring →

$$5x + 10 = 5(x + 2)$$

← multiplying

1 Before we factor polynomials, let's practice finding the greatest common factor of a list of integers. The **greatest common factor (GCF)** of a list of integers is the largest integer that is a factor of all the integers in the list. For example,

the GCF of 30 and 18 is 6

because 6 is the largest integer that is a factor of both 30 and 18.

If the GCF cannot be found by inspection, the following steps can be used.

TO FIND THE GCF OF A LIST OF INTEGERS

Step 1. Write each number as a product of prime numbers.

Step 2. Identify the common prime factors.

Step 3. The product of all common prime factors found in *step 2* is the greatest common factor. If there are no common prime factors, the greatest common factor is 1.

Recall from Section 4.2 that a prime number is a whole number other than 1, whose only factors are 1 and itself.

PRACTICE PROBLEM 1

Find the GCF of 42 and 28.

EXAMPLE 1

Find the GCF of 12 and 20.

Solution:

Step 1. Write each number as a product of primes.

$12 = 2 \cdot 2 \cdot 3$

$20 = 2 \cdot 2 \cdot 5$

Step 2. $12 = \boxed{2} \cdot \boxed{2} \cdot 3$

$20 = \boxed{2} \cdot \boxed{2} \cdot 5$

$\downarrow \quad \downarrow$

$2 \cdot 2$ Identify the common factors.

Step 3. The GCF is $2 \cdot 2 = 4$

2 How do we find the GCF of a list of variables raised to powers? For example, what is the GCF of y^3, y^5, and y^{10}? Notice that each variable term contains a factor of y^3 and no higher power of y is a factor of each term.

$y^3 = y^3$

$y^5 = y^3 \cdot y^2$ Recall the product rule for exponents.

$y^{10} = y^3 \cdot y^7$

The GCF of y^3, y^5, and y^{10} is y^3. From this example, we can see that **the GCF of a list of variables raised to powers is the variable raised to the smallest exponent in the list.**

PRACTICE PROBLEM 2

Find the GCF of z^7, z^8, and z.

EXAMPLE 2

Find the GCF of x^{11}, x^4, and x^6.

Solution:

The GCF is x^4, since 4 is the smallest exponent to which x is raised.

In general, **the GCF of a list of terms is the product of all common factors.**

PRACTICE PROBLEM 3

Find the GCF of the terms $6a^4$, $3a^5$, and $15a^2$.

EXAMPLE 3

Find the GCF of the list of terms $4x^3$, $12x$, and $10x^5$.

Solution:

The GCF of 4, 12, and 10 is 2.

The GCF of x^3, x^1, and x^5 is x^1.

Thus, the GCF of $4x^3$, $12x$, and $10x^5$ is $2x^1$ or $2x$.

3 Next, we practice factoring a polynomial by factoring the GCF from its terms. To do so, write each term of the polynomial as a product of the GCF and another factor, then apply the distributive property.

Answers:

1. 14

2. z

3. $3a^2$

EXAMPLE 4

Factor $7x^3 + 14x^2$.

Solution:

The GCF of $7x^3$ and $14x^2$ is $7x^2$.

$$7x^3 + 14x^2 = 7x^2 \cdot x^1 + 7x^2 \cdot 2$$
$$= 7x^2(x + 2) \quad \text{Apply the distributive property.}$$

Notice in Example 4 that we factored $7x^3 + 14x^2$ by writing it as the product $7x^2(x + 2)$.

Also notice that to check factoring we multiply

$$7x^2(x + 2) = 7x^2 \cdot x + 7x^2 \cdot 2$$
$$= 7x^3 + 14x^2$$

which is the original binomial.

PRACTICE PROBLEM 4

Factor $10y^7 + 5y^9$.

EXAMPLE 5

Factor $6x^2 - 24x + 6$.

Solution:

The GCF of the terms is 6.

$$6x^2 - 24x + 6 = 6 \cdot x^2 - 6 \cdot 4x + 6 \cdot 1$$
$$= 6(x^2 - 4x + 1)$$

PRACTICE PROBLEM 5

Factor $4z^2 - 12z + 2$.

EXAMPLE 6

Factor $-2a + 20b - 4b^2$.

Solution:

$$-2a + 20b - 4b^2 = 2 \cdot -a + 2 \cdot 10b - 2 \cdot 2b^2$$
$$= 2(-a + 10b - 2b^2)$$

When the coefficient of the first term is a negative number, we often factor out a negative common factor.

$$-2a + 20b - 4b^2 = (-2) \cdot a + (-2)(-10b) + (-2)(2b^2)$$
$$= -2(a - 10b + 2b^2)$$

Both $2(-a + 10b - 2b^2)$ and $-2(a - 10b + 2b^2)$ are factorizations of $-2a + 20b - 4b^2$.

PRACTICE PROBLEM 6

Factor $-3y^2 - 9y + 15x^2$.

Answers:

4. $5y^7(2 + y^2)$
5. $2(2z^2 - 6z + 1)$
6. $-3(y^2 + 3y - 5x^2)$

EXERCISE SET 9.4

Find the greatest common factor of each list of numbers. See Example 1.

1. 48 and 15

2. 36 and 20

3. 60 and 72

4. 96 and 45

5. 12, 20, and 36

6. 18, 24, and 60

7. 8, 32, and 100

8. 30 , 50, and 200

Find the greatest common factor of each list of terms. See Examples 2 and 3.

9. y^7, y^2, y^{10}

10. x^3, x, x^5

11. a^5, a^5, a^5

12. b^6, b^6, b^4

13. x^3y^2, xy^2, x^4y^2

14. a^5b^3, a^5b^2, a^5b

15. $3x^4, 5x^7, 10x$

16. $9z^6, z^5, 2z^3$

17. $2z^3, 14z^5, 18z^3$

18. $6y^7, 9y^6, 15y^5$

Factor. Check by multiplying. See Examples 4–6.

19. $3y^2 + 18y$

20. $2x^2 + 18x$

21. $10a^6 - 5a^8$

22. $21y^5 + y^{10}$

23. $4x^3 + 12x^2 + 20x$

24. $9b^3 - 54b^2 + 9b$

25. $z^7 - 6z^5$

26. $y^{10} + 4y^5$

ANSWERS

1. ______
2. ______
3. ______
4. ______
5. ______
6. ______
7. ______
8. ______
9. ______
10. ______
11. ______
12. ______
13. ______
14. ______
15. ______
16. ______
17. ______
18. ______
19. ______
20. ______
21. ______
22. ______
23. ______
24. ______
25. ______
26. ______

ANSWERS

27. ______________

28. ______________

29. ______________

30. ______________

31. ______________

32. ______________

33. ______________

34. ______________

35. ______________

36. ______________

37. ______________

38. ______________

27. $-35 + 14y - 7y^2$

28. $-20x + 4x^2 - 2$

29. $12a^5 - 36a^6$

30. $25z^3 - 20z^2$

31. The area of the larger rectangle below is $x(x + 2)$. Find another expression for the area by writing the sum of the areas of the smaller rectangles.

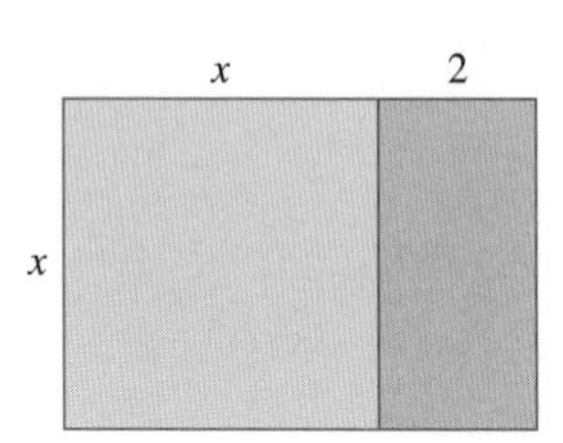

32. Write an expression for the area of the larger rectangle in two different ways.

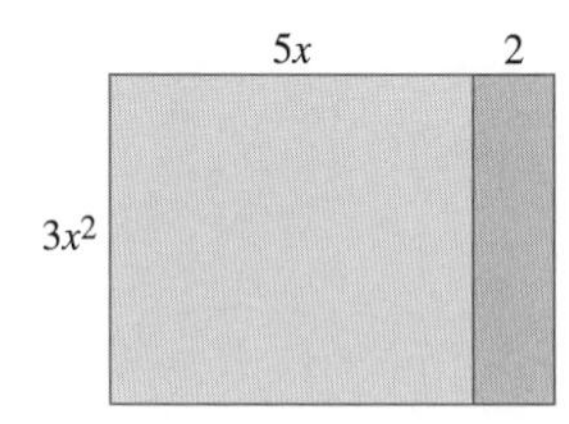

Review Exercises

Solve. See Sections 8.1 and 8.2.

33. Find 30% of 120.

34. Find 45% of 265.

35. Write 80% as a fraction.

36. Write 65% as a fraction.

37. Write $\frac{3}{8}$ as a percent.

38. Write $\frac{3}{4}$ as a percent.

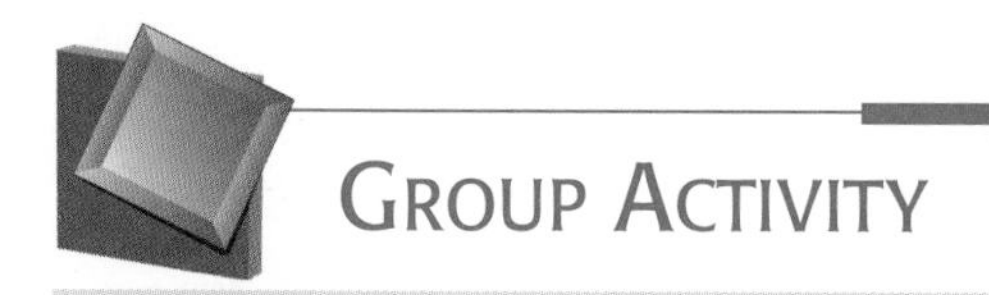

GROUP ACTIVITY

BUSINESS ANALYSIS

Suppose you own a business that manufactures baskets. You need to decide how many baskets to make. The more you make, the lower the price you charge will have to be, to try to sell them all. Naturally, each extra basket you make costs you money, as you have to buy materials.

	DESCRIPTION	ALGEBRAIC EXPRESSION
Number of baskets	Unknown	x
Total expenses for manufacturing the baskets	It will cost \$100 to buy special equipment to manufacture the baskets in addition to basket materials costing \$0.50 per basket	$100 + 0.50x$
Price charged per basket	For each additional basket produced, the price that must be charged per basket decreases from \$40 by an additional \$0.05	$40 - 0.05x$

1. Revenue is the amount of money collected from sales of the baskets. Revenue can be found by multiplying the price charged per basket by the number of baskets sold. Use the algebraic expressions given in the table to find a polynomial that represents the revenue from sales of baskets.
2. Profit is the amount of money you earn from selling the baskets after deducting the expenses for making the baskets. Profit can be found by subtracting total expenses from revenue. Find a polynomial that represents the profit from sales of baskets.
3. Complete the following table by evaluating the polynomial for profit that you found in question 2 for each of the numbers of baskets given in the table.

NUMBER OF BASKETS x	200	300	400	500	600
Total Expenses					
Revenue					
Profit					

4. Which number of baskets in the table will give you the largest profit from making and selling baskets?

CHAPTER 9 HIGHLIGHTS

DEFINITIONS AND CONCEPTS	EXAMPLES
SECTION 9.1 ADDING AND SUBTRACTING POLYNOMIALS	
A **polynomial** is a monomial or a sum or difference of monomials.	Polynomials $5x^2 - 6x + 2, \quad -\frac{9}{10}y, \quad 7$
To add polynomials, combine like terms.	Add: $(7z^2 - 6z + 2) + (5z^2 - 4z + 5)$. $(7z^2 - 6z + 2) + (5z^2 - 4z + 5)$ $= \underbrace{7z^2 + 5z^2} \underbrace{- 6z - 4z} + \underbrace{2 + 5}$ Group like terms. $= 12z^2 - 10z + 7$ Combine like terms.
To subtract polynomials, change the signs of the terms being subtracted, then add.	Subtract: $(20x - 6) - (30x - 6)$. $(20x - 6) - (30x - 6) = (20x - 6) + (-30x + 6)$ $= \underbrace{20x - 30x} - 6 + 6$ Group like terms. $= -10x$ Combine like terms.
SECTION 9.2 MULTIPLICATION PROPERTIES OF EXPONENTS	
Product property for exponents $a^m \cdot a^n = a^{m+n}$	$x^3 \cdot x^{11} = x^{3+11} = x^{14}$
Power of a power property for exponents $(a^m)^n = a^{m \cdot n}$	$(y^5)^3 = y^{5 \cdot 3} = y^{15}$
Power of a product property for exponents $(ab)^n = a^n b^n$	$(2z^5)^4 = 2^4(z^5)^4 = 16z^{20}$
SECTION 9.3 MULTIPLYING POLYNOMIALS	
A **monomial** is a polynomial with 1 term. A **binomial** is a polynomial with 2 terms. A **trinomial** is a polynomial with 3 terms.	Monomial: $-2x^2y^3$ Binomial: $5x - y$ Trinomial: $7z^3 + 0.5z + 1$
To multiply two polynomials, multiply each term of the first polynomial by each term of the second polynomial, and then combine like terms.	$(x + 2)(x^2 + 5x - 1)$ $= x(x^2 + 5x - 1) + 2(x^2 + 5x - 1)$ $= x \cdot x^2 + x \cdot 5x + x(-1) + 2 \cdot x^2 + 2 \cdot 5x + 2(-1)$ $= x^3 + 5x^2 - x + 2x^2 + 10x - 2$ $= x^3 + 7x^2 + 9x - 2$

DEFINITIONS AND CONCEPTS	EXAMPLES
SECTION 9.4 INTRODUCTION TO FACTORING POLYNOMIALS	
To Find the Greatest Common Factor of a List of Integers *Step 1.* Write each number as a product of prime numbers. *Step 2.* Identify the common prime factors. *Step 3.* The product of all common prime factors found in *step 2* is the greatest common factor. If there are no common prime factors, the greatest common factor is 1.	Find the GCF of 18 and 30. $18 = 2 \cdot 3 \cdot 3$ $30 = 2 \cdot 3 \cdot 5$ $\downarrow\ \downarrow$ GCF is $2 \cdot 3$ or 6.
The **GCF of a list of variables** raised to powers is the variable raised to the smallest exponent in the list.	The GCF of x^6, x^8, and x^3 is x^3.
The **GCF of a list of terms** is the product of all common factors.	Find the GCF of $6y^3$, $12y$, and $4y^7$. the GCF of 6, 12, and 4 is 2 the GCF of y^3, y, and y^7 is y the GCF of $6y^3$, $12y$, and $4y^7$ is $2y$
To factor the GCF from the terms of a polynomial, write each term as a product of the GCF and another factor, then apply the distributive property.	Factor $4y^6 + 6y^5$. The GCF of $4y^6$ and $6y^5$ is $2y^5$. $4y^6 + 6y^5 = 2y^5 \cdot 2y + 2y^5 \cdot 3$ $= 2y^5(2y + 3)$

APPENDIX

COMPOUND INTEREST TABLE

COMPOUNDED ANNUALLY	5%	6%	7%	8%	9%	10%	11%
1 year	1.05000	1.06000	1.07000	1.08000	1.09000	1.10000	1.11000
5 years	1.27628	1.33823	1.40255	1.46933	1.53862	1.61051	1.68506
10 years	1.62889	1.79085	1.96715	2.15892	2.36736	2.59374	2.83942
15 years	2.07893	2.39656	2.75903	3.17217	3.64248	4.17725	4.78459
20 years	2.65330	3.20714	3.86968	4.66096	5.60441	6.72750	8.06231

COMPOUNDED SEMIANNUALLY	5%	6%	7%	8%	9%	10%	11%
1 year	1.05063	1.06090	1.07123	1.08160	1.09203	1.10250	1.11303
5 years	1.28008	1.34392	1.41060	1.48024	1.55297	1.62889	1.70814
10 years	1.63862	1.80611	1.98979	2.19112	2.41171	2.65330	2.91776
15 years	2.09757	2.42726	2.80679	3.24340	3.74532	4.32194	4.98395
20 years	2.68506	3.26204	3.95926	4.80102	5.81636	7.03999	8.51331

COMPOUNDED QUARTERLY	5%	6%	7%	8%	9%	10%	11%
1 year	1.05095	1.06136	1.07186	1.08243	1.09308	1.10381	1.11462
5 years	1.28204	1.34686	1.41478	1.48595	1.56051	1.63862	1.72043
10 years	1.64362	1.81402	2.00160	2.20804	2.43519	2.68506	2.95987
15 years	2.10718	2.44322	2.83182	3.28103	3.80013	4.39979	5.09225
20 years	2.70148	3.29066	4.00639	4.87544	5.93015	7.20957	8.76085

COMPOUNDED DAILY	5%	6%	7%	8%	9%	10%	11%
1 year	1.05127	1.06183	1.07250	1.08328	1.09416	1.10516	1.11626
5 years	1.28400	1.34983	1.41902	1.49176	1.56823	1.64861	1.73311
10 years	1.64866	1.82203	2.01362	2.22535	2.45933	2.71791	3.00367
15 years	2.11689	2.45942	2.85736	3.31968	3.85678	4.48077	5.20569
20 years	2.71810	3.31979	4.05466	4.95216	6.04831	7.38703	9.02202

APPENDIX D COMPOUND INTEREST TABLE (CONTINUED)

COMPOUNDED ANNUALLY

	12%	13%	14%	15%	16%	17%	18%
1 year	1.12000	1.13000	1.14000	1.15000	1.16000	1.17000	1.18000
5 years	1.76234	1.84244	1.92541	2.01136	2.10034	2.19245	2.28776
10 years	3.10585	3.39457	3.70722	4.04556	4.41144	4.80683	5.23384
15 years	5.47357	6.25427	7.13794	8.13706	9.26552	10.53872	11.97375
20 years	9.64629	11.52309	13.74349	16.36654	19.46076	23.10560	27.39303

COMPOUNDED SEMIANNUALLY

	12%	13%	14%	15%	16%	17%	18%
1 year	1.12360	1.13423	1.14490	1.15563	1.16640	1.17723	1.18810
5 years	1.79085	1.87714	1.96715	2.06103	2.15892	2.26098	2.36736
10 years	3.20714	3.52365	3.86968	4.24785	4.66096	5.11205	5.60441
15 years	5.74349	6.61437	7.61226	8.75496	10.06266	11.55825	13.26768
20 years	10.28572	12.41607	14.97446	18.04424	21.72452	26.13302	31.40942

COMPOUNDED QUARTERLY

	12%	13%	14%	15%	16%	17%	18%
1 year	1.12551	1.13648	1.14752	1.15865	1.16986	1.18115	1.19252
5 years	1.80611	1.89584	1.98979	2.08815	2.19112	2.29891	2.41171
10 years	3.26204	3.59420	3.95926	4.36038	4.80102	5.28497	5.81636
15 years	5.89160	6.81402	7.87809	9.10513	10.51963	12.14965	14.02741
20 years	10.64089	12.91828	15.67574	19.01290	23.04980	27.93091	33.83010

COMPOUNDED DAILY

	12%	13%	14%	15%	16%	17%	18%
1 year	1.12747	1.13880	1.15024	1.16180	1.17347	1.18526	1.19716
5 years	1.82194	1.91532	2.01348	2.11667	2.22515	2.33918	2.45906
10 years	3.31946	3.66845	4.05411	4.48031	4.95130	5.47178	6.04696
15 years	6.04786	7.02625	8.16288	9.48335	11.01738	12.79950	14.86983
20 years	11.01883	13.45751	16.43582	20.07316	24.51533	29.94039	36.56577

APPENDIX E

REVIEW OF GEOMETRIC FIGURES

Name	Description	Figure
	PLANE FIGURES HAVE LENGTH AND WIDTH BUT NO THICKNESS OR DEPTH.	
POLYGON	Union of three or more coplanar line segments that intersect with each other only at each end point, with each end point shared by two segments.	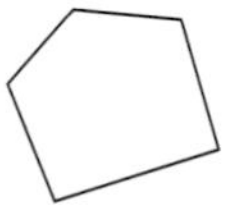
TRIANGLE	Polygon with three sides (sum of measures of three angles is 180°).	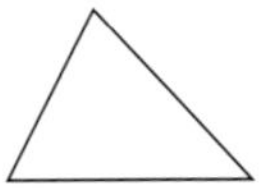
SCALENE TRIANGLE	Triangle with no sides of equal length.	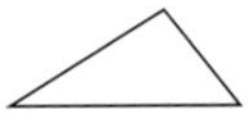
ISOSCELES TRIANGLE	Triangle with two sides of equal length.	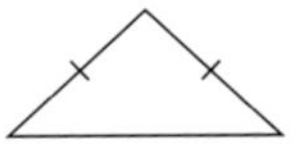
EQUILATERAL TRIANGLE	Triangle with all sides of equal length.	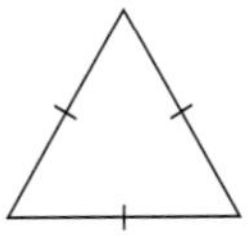
RIGHT TRIANGLE	Triangle that contains a right angle.	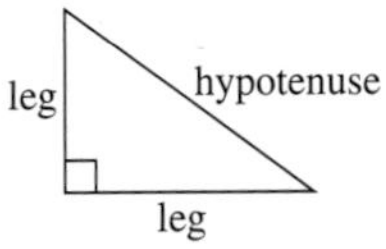 leg, hypotenuse, leg
QUADRILATERAL	Polygon with four sides (sum of measures of four angles is 360°).	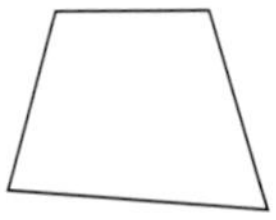

PLANE FIGURES HAVE LENGTH AND WIDTH BUT NO THICKNESS OR DEPTH.		
NAME	**DESCRIPTION**	**FIGURE**
TRAPEZOID	Quadrilateral with exactly one pair of opposite sides parallel.	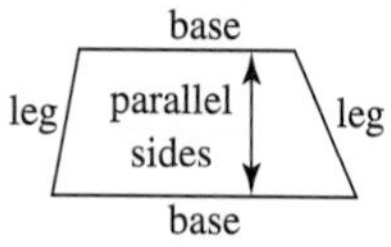 base; leg; parallel sides; leg; base
ISOSCELES TRAPEZOID	Trapezoid with legs of equal length.	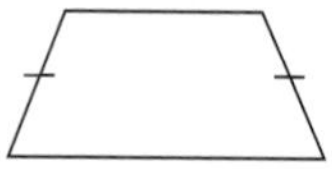
PARALLELOGRAM	Quadrilateral with both pairs of opposite sides parallel.	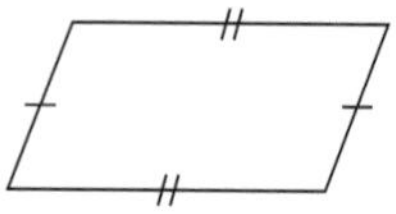
RHOMBUS	Parallelogram with all sides of equal length.	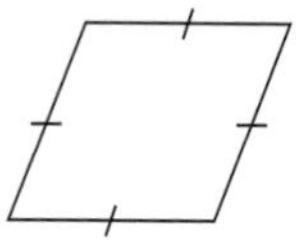
RECTANGLE	Parallelogram with four right angles.	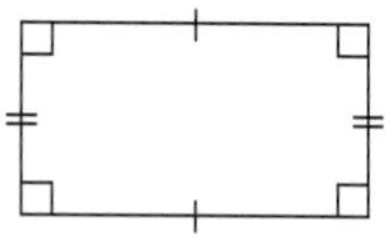
SQUARE	Rectangle with all sides of equal length.	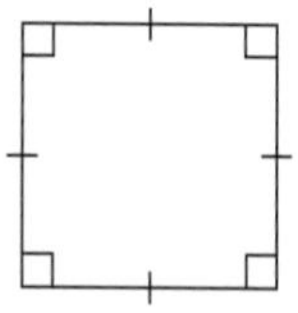
CIRCLE	All points in a plane the same distance from a fixed point called the **center.**	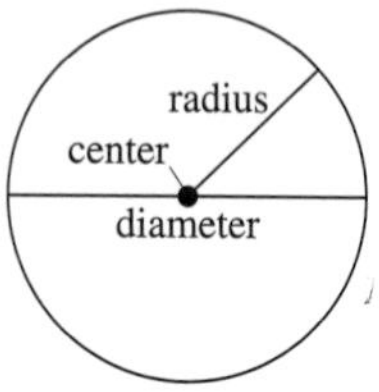 radius; center; diameter

SOLID FIGURES HAVE LENGTH, WIDTH, AND HEIGHT OR DEPTH.		
NAME	DESCRIPTION	FIGURE
RECTANGULAR SOLID	A solid with six sides, all of which are rectangles.	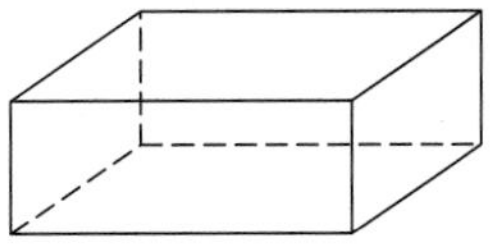
CUBE	A rectangular solid whose six sides are squares.	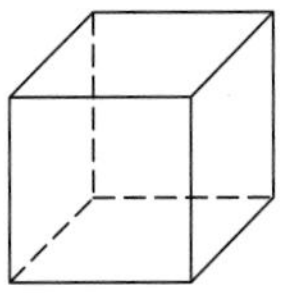
SPHERE	All points the same distance from a fixed point, called the **center.**	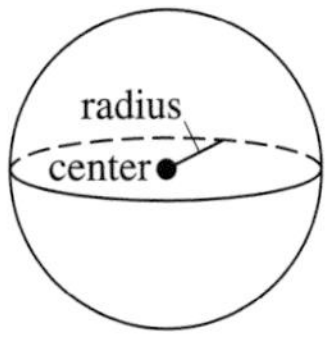 radius, center
RIGHT CIRCULAR CYLINDER	A cylinder consisting of two circular bases that are perpendicular to its altitude.	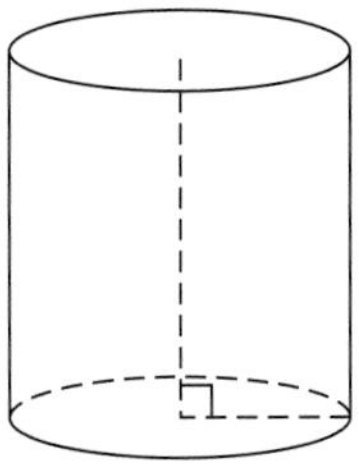
RIGHT CIRCULAR CONE	A cone with a circular base that is perpendicular to its altitude.	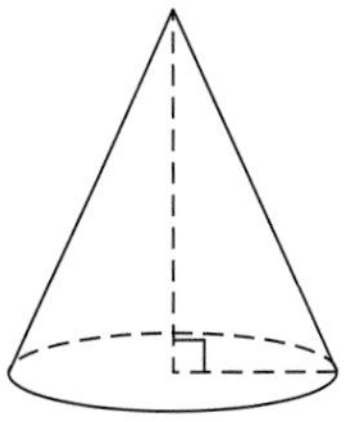

APPENDIX

F

Review of Angles, Lines, and Special Triangles

The word **geometry** is formed from the Greek words, **geo,** meaning earth, and **metron,** meaning measure. Geometry literally means to measure the earth.

This appendix contains a review of some basic geometric ideas. It will be assumed that fundamental ideas of geometry such as point, line, ray, and angle are known. In this appendix, the notation $\angle 1$ is read "angle 1" and the notation $m\,\angle 1$ is read "the measure of angle 1."

We first review types of angles.

Angles

An angle whose measure is more than 0° but but less than 90° is called an **acute angle.**

A **right angle** is an angle whose measure is 90°. A right angle can be indicated by a square drawn at the vertex of the angle, as shown below.

An angle whose measure is greater than 90° but less than 180° is called an **obtuse angle.**

An angle whose measure is 180° is called a **straight angle.**

Two angles are said to be **complementary** if the sum of their measures is 90°.

Each angle is called the **complement** of the other.

Two angles are said to be **supplementary** if the sum of their measures is 180°.

Each angle is called the **supplement** of the other.

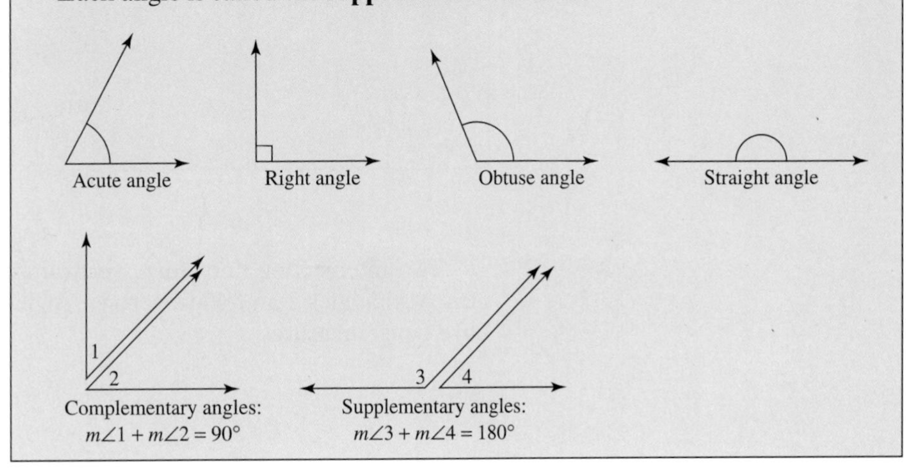

EXAMPLE 1

If an angle measures 28°, find its complement.

Solution:

Two angles are complementary if the sum of their measures is 90°. The complement of a 28° angle is an angle whose measure is $90° - 28° = 62°$. To check, notice that $28° + 62° = 90°$.

Plane is an undefined term that we will describe. A plane can be thought of as a flat surface with infinite length and width, but no thickness. A plane is two dimensional. The arrows in the following diagram indicate that a plane extends indefinitely and has no boundaries.

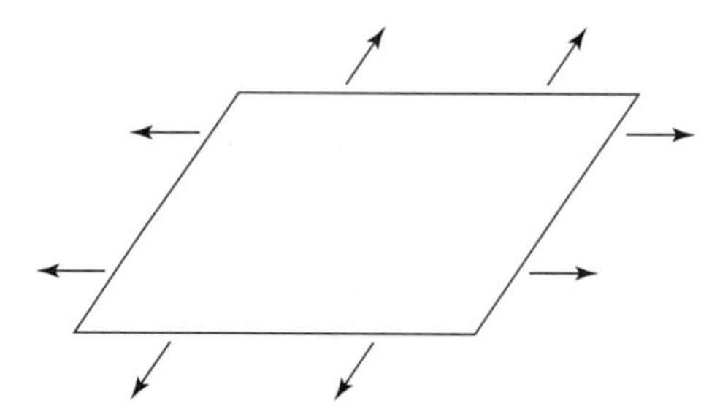

Figures that lie on a plane are called **plane figures.** (See the description of common plane figures in Appendix E.) Lines that lie in the same plane are called **coplanar.**

LINES

Two lines are **parallel** if they lie in the same plane but never meet.
Intersecting lines meet or cross in one point.
Two lines that form right angles when they intersect are said to be **perpendicular.**

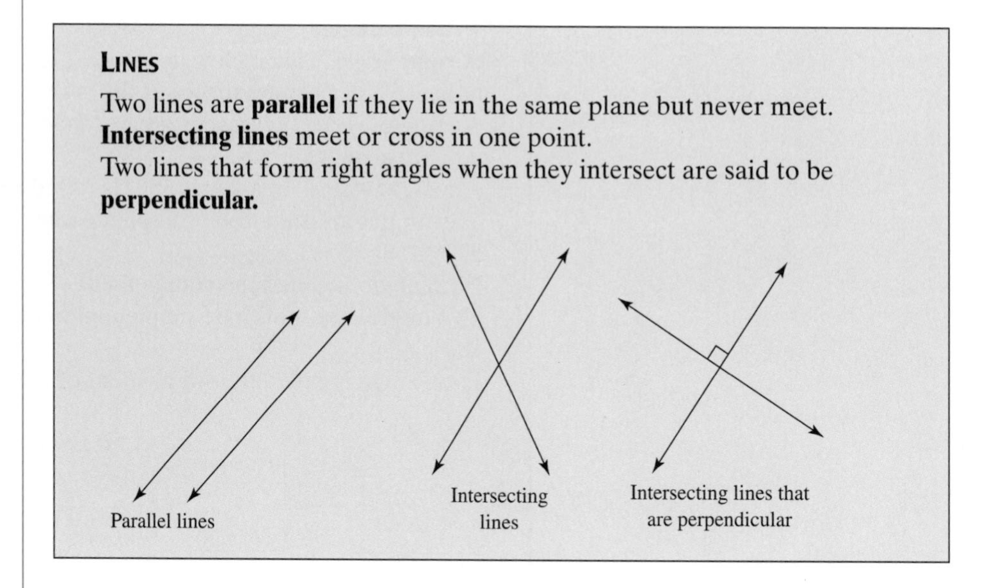

Two intersecting lines form **vertical angles.** Angles 1 and 3 are vertical angles. Also angles 2 and 4 are vertical angles. It can be shown that **vertical angles have equal measures.**

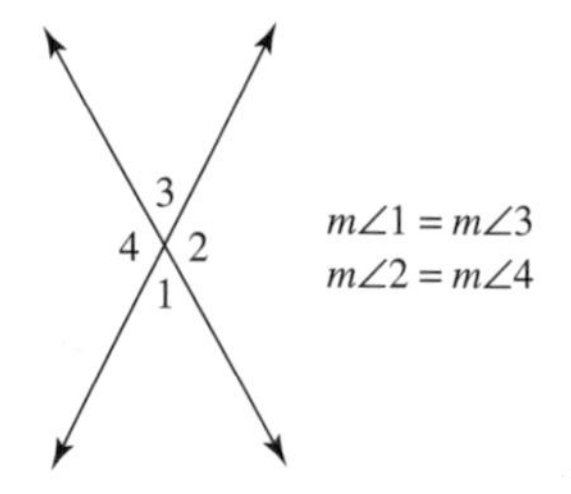

Adjacent angles have the same vertex and share a side. Angles 1 and 2 are adjacent angles. Other pairs of adjacent angles are angles 2 and 4, angles 3 and 4, and angles 3 and 1.

A **transversal** is a line that intersects two or more lines in the same plane. Line l is a transversal that intersects lines m and n. The eight angles formed are numbered and certain pairs of these angles are given special names.

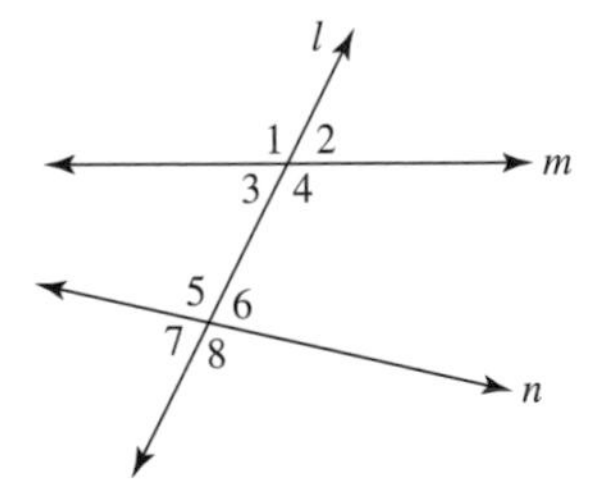

Corresponding angles: $\angle 1$ and $\angle 5$, $\angle 3$ and $\angle 7$, $\angle 2$ and $\angle 6$, and $\angle 4$, and $\angle 8$.
Exterior angles: $\angle 1$, $\angle 2$, $\angle 7$, and $\angle 8$.
Interior angles: $\angle 3$, $\angle 4$, $\angle 5$, and $\angle 6$.
Alternate interior angles: $\angle 3$ and $\angle 6$, $\angle 4$ and $\angle 5$.

These angles and parallel lines are related in the following manner.

PARALLEL LINES CUT BY A TRANSVERSAL

1. If two parallel lines are cut by a transversal, then
   a. corresponding angles are equal and
   b. alternate interior angles are equal.
2. If corresponding angles formed by two lines and a transversal are equal, then the lines are parallel.
3. If alternate interior angles formed by two lines and a transversal are equal, then the lines are parallel.

EXAMPLE 2

Given that lines m and n are parallel and that the measure of angle 1 is 100°, find the measures of angles 2, 3, and 4.

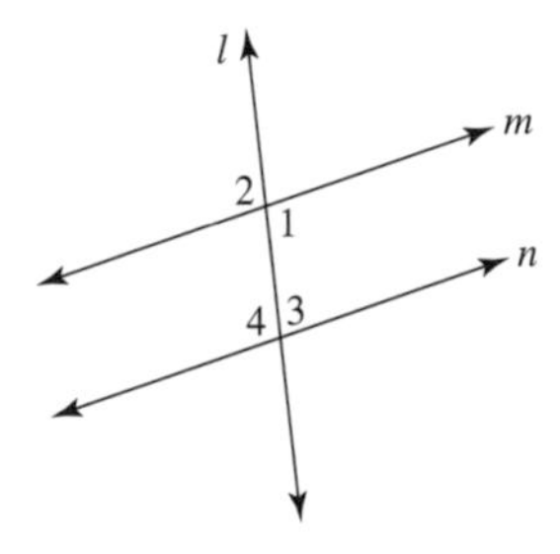

Solution:

$m\angle 2 = 100°$,
$m\angle 4 = 100°$,
$m\angle 3 = 180° - 100° = 80°$,

A **polygon** is the union of three or more coplanar line segments that intersect each other only at each end point, with each end point shared by exactly two segments.

A **triangle** is a polygon with three sides. The sum of the measures of the three angles of a triangle is 180°. In the following figure, $m\angle 1 + m\angle 2 + m\angle 3 = 180°$.

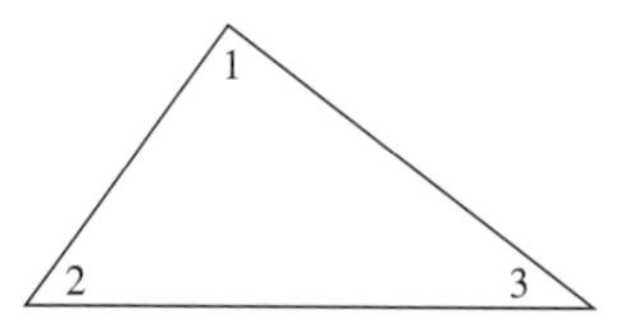

EXAMPLE 3

Find the measure of the third angle of the triangle shown.

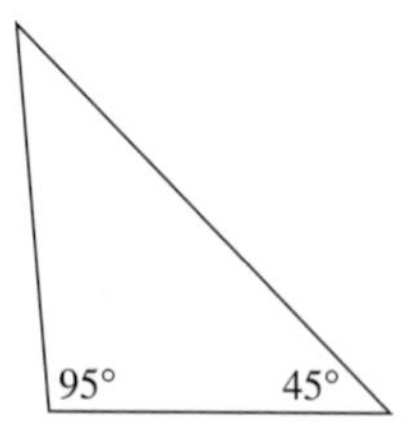

Solution:

The sum of the measures of the angles of a triangle is 180°. Since one angle measures 45° and the other angle measures 95°, the third angle measures $180° - 45° - 95° = 40°$.

Two triangles are **congruent** if they have the same size and the same shape. In congruent triangles, the measures of corresponding angles are equal and the lengths of corresponding sides are equal. The following triangles are congruent.

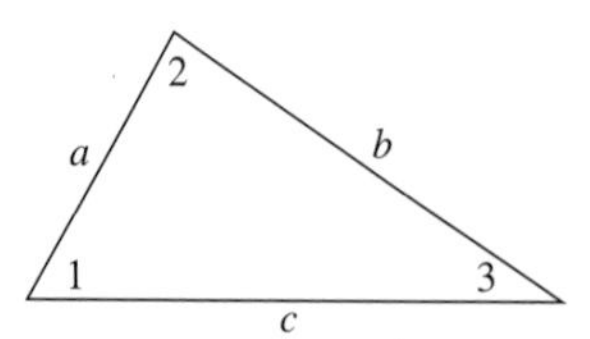

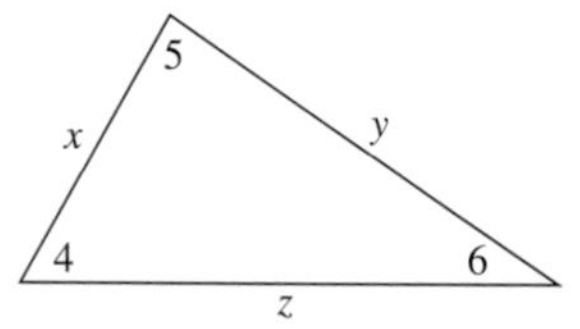

Corresponding angles are equal: $m\angle 1 = m\angle 4$, $m\angle 2 = m\angle 5$, and $m\angle 3 = m\angle 6$. Also, lengths of corresponding sides are equal: $a = x$, $b = y$, and $c = z$.

Any one of the following may be used to determine whether two triangles are congruent.

Congruent Triangles

1. If the measures of two angles of a triangle equal the measures of two angles of another triangle and the lengths of the sides between each pair of angles are equal, the triangles are congruent.

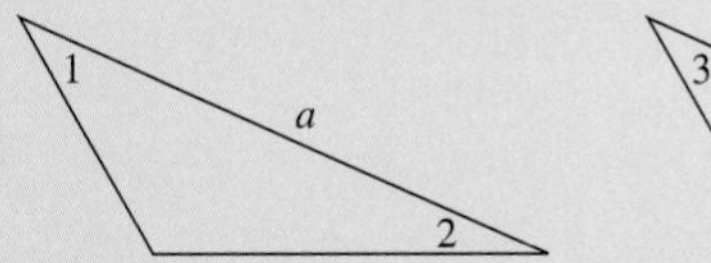

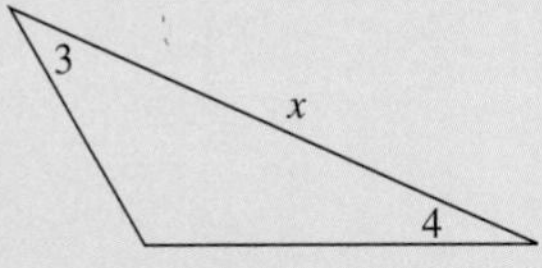

$m\angle 1 = m\angle 3$
$m\angle 2 = m\angle 4$
and
$a = x$

2. If the lengths of the three sides of a triangle equal the lengths of corresponding sides of another triangle, the triangles are congruent.

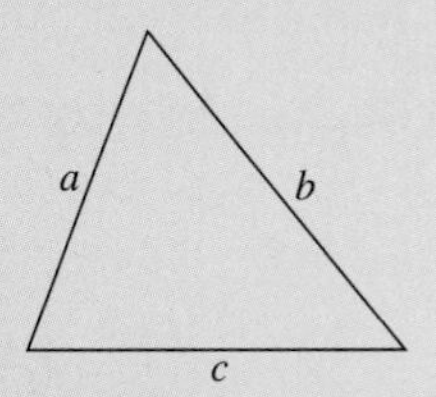

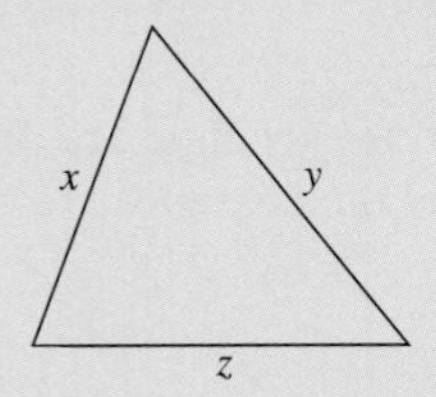

$a = x$
$b = y$
and
$c = z$

3. If the lengths of two sides of a triangle equal the lengths of corresponding sides of another triangle, and the measures of the angles between each pair of sides are equal, the triangles are congruent.

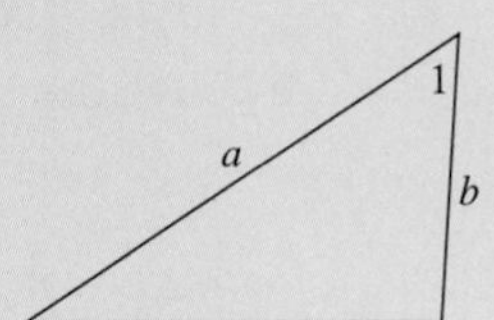

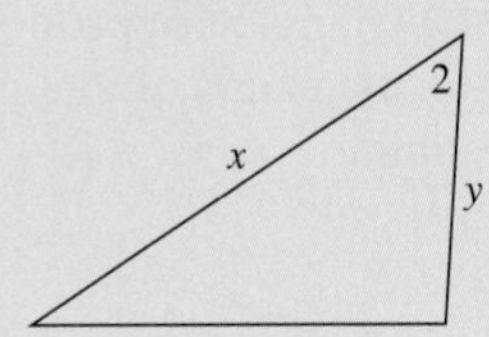

$a = x$
$b = y$
and
$m\angle 1 = m\angle 2$

Two triangles are similar if they have the same shape but not necessarily the same size. In similar triangles, the measures of corresponding angles are equal and corresponding sides are in proportion. The following triangles are similar. (All similar triangles drawn in this appendix will be oriented the same.)

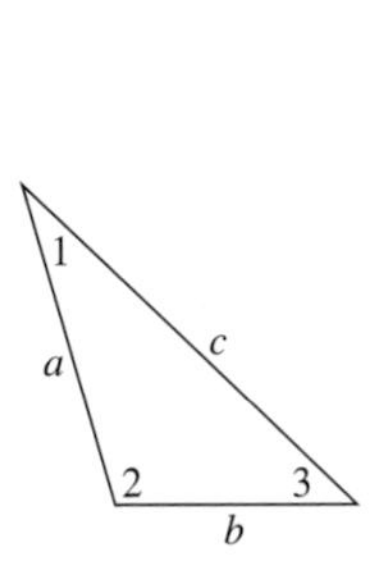

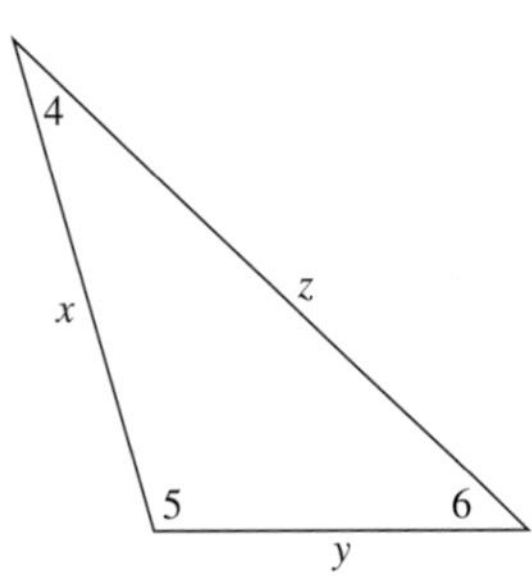

Corresponding angles are equal: $m\angle 1 = m\angle 4, m\angle 2 = m\angle 5$, and $m\angle 3 = m\angle 6$.

Also, corresponding sides are proportional: $\frac{a}{x} = \frac{b}{y} = \frac{c}{z}$.

Any one of the following may be used to determine whether two triangles are similar.

SIMILAR TRIANGLES

1. If the measures of two angles of a triangle equal the measures of two angles of another triangle, the triangles are similar.

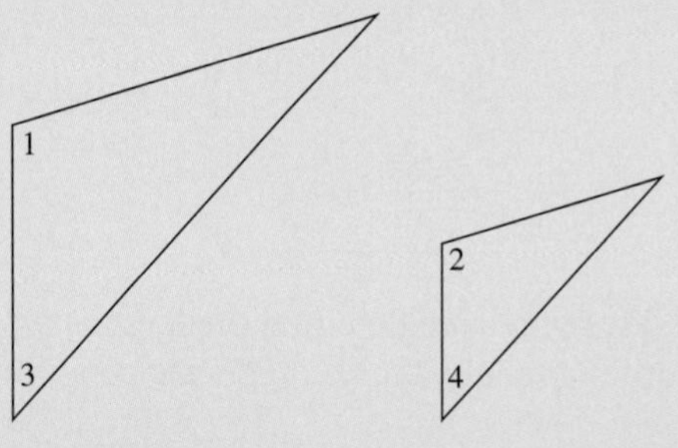

$$m\angle 1 = m\angle 2$$
and
$$m\angle 3 = m\angle 4$$

2. If three sides of one triangle are proportional to three sides of another triangle, the triangles are similar.

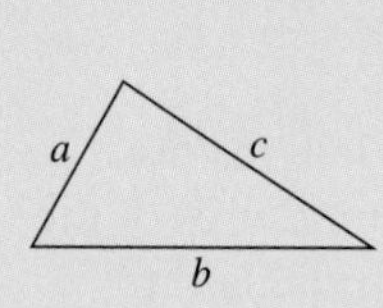

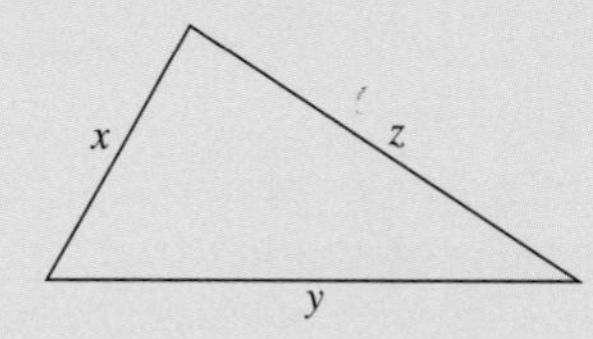

$$\frac{a}{x} = \frac{b}{y} = \frac{c}{z}$$

3. If two sides of a triangle are proportional to two sides of another triangle and the measures of the included angles are equal, the triangles are similar.

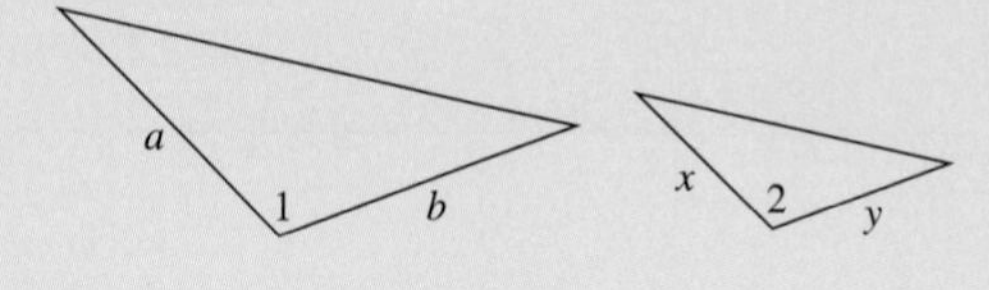

$$m\angle 1 = m\angle 2$$
and
$$\frac{a}{x} = \frac{b}{y}$$

EXAMPLE 4

Given that the following triangles are similar, find the missing length x.

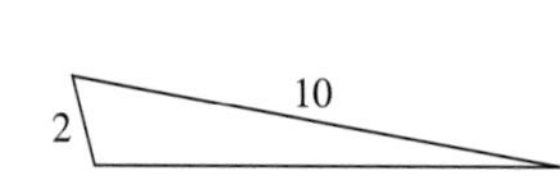

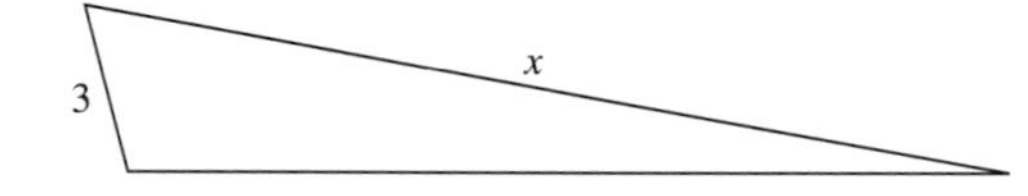

Solution:

Since the triangles are similar, corresponding sides are in proportion. Thus, $\frac{2}{3} = \frac{10}{x}$. To solve this equation for x, we cross multiply.

$$\frac{2}{3} = \frac{10}{x}$$

$$2x = 30$$

$$x = 15$$

The missing length is 15 units.

A **right triangle** contains a right angle. The side opposite the right angle is called the **hypotenuse,** and the other two sides are called the **legs.** The **Pythagorean theorem** gives a formula that relates the lengths of the three sides of a right triangle.

THE PYTHAGOREAN THEOREM

If a and b are the lengths of the legs of a right triangle, and c is the length of the hypotenuse, then $a^2 + b^2 = c^2$.

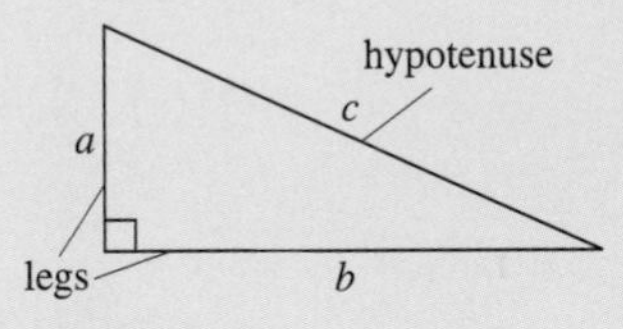

EXAMPLE 5

Find the length of the hypotenuse of a right triangle whose legs have lengths of 3 centimeters and 4 centimeters.

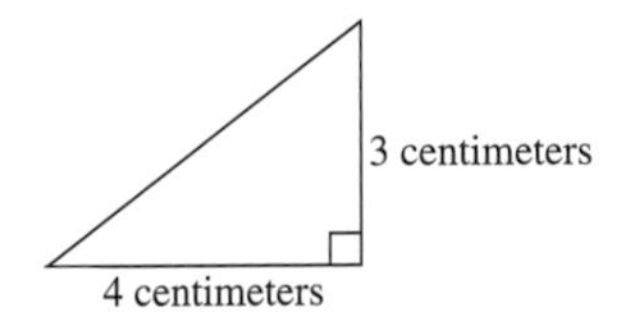

Solution:

Because we have a right triangle, we use the Pythagorean theorem. The legs are 3 centimeters and 4 centimeters, so let $a = 3$ and $b = 4$ in the formula.

$$a^2 + b^2 = c^2$$

$$3^2 + 4^2 = c^2$$

$$9 + 16 = c^2$$

$$25 = c^2$$

Since c represents a length, we assume that c is positive. Thus, if c^2 is 25, c must be 5. The hypotenuse has a length of 5 centimeters.

APPENDIX F EXERCISE SET

Find the complement of each angle. See Example 1.

1. $19°$ **2.** $65°$ **3.** $70.8°$

4. $45\frac{2}{3}°$ **5.** $11\frac{1}{4}°$ **6.** $19.6°$

Find the supplement of each angle.

7. $150°$ **8.** $90°$ **9.** $30.2°$

10. $81.9°$ **11.** $79\frac{1}{2}°$ **12.** $165\frac{8}{9}°$

13. If lines m and n are parallel, find the measures of angles 1 through 7. See Example 2.

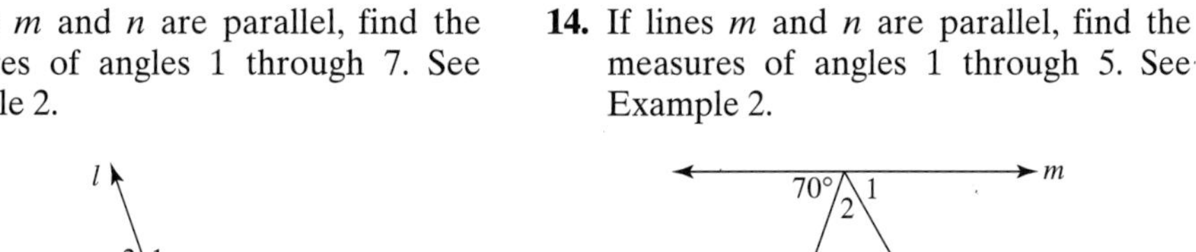

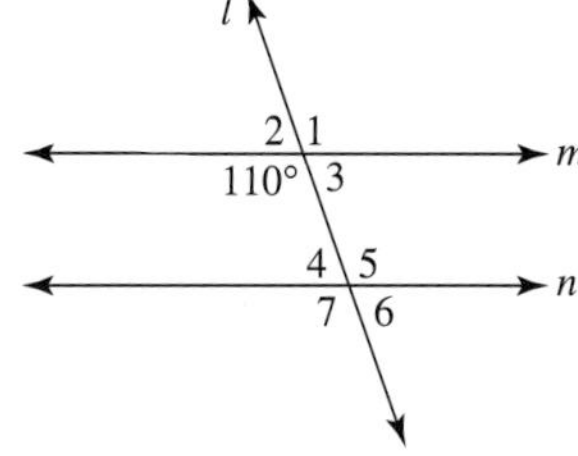

14. If lines m and n are parallel, find the measures of angles 1 through 5. See Example 2.

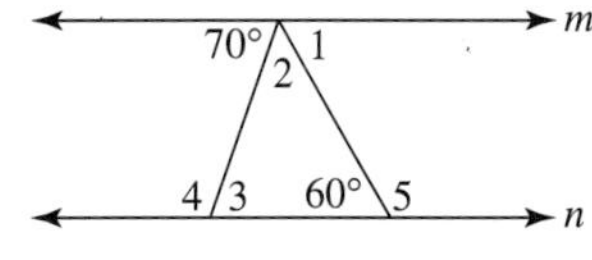

In each of the following, the measures of two angles of a triangle are given. Find the measure of the third angle. See Example 3.

15. $11°, 79°$ **16.** $8°, 102°$ **17.** $25°, 65°$

18. $44°, 19°$ **19.** $30°, 60°$ **20.** $67°, 23°$

In each of the following, the measure of one angle of a right triangle is given. Find the measures of the other two angles.

21. $45°$ **22.** $60°$ **23.** $17°$

24. $30°$ **25.** $39\frac{3}{4}°$ **26.** $72.6°$

ANSWERS

1. ______
2. ______
3. ______
4. ______
5. ______
6. ______
7. ______
8. ______
9. ______
10. ______
11. ______
12. ______
13. ______
14. ______
15. ______
16. ______
17. ______
18. ______
19. ______
20. ______
21. ______
22. ______
23. ______
24. ______
25. ______
26. ______

ANSWERS

27. ____________

28. ____________

29. ____________

30. ____________

31. ____________

32. ____________

33. ____________

34. ____________

Given that each of the following pairs of triangles is similar, find the missing length x. See Example 4.

27.

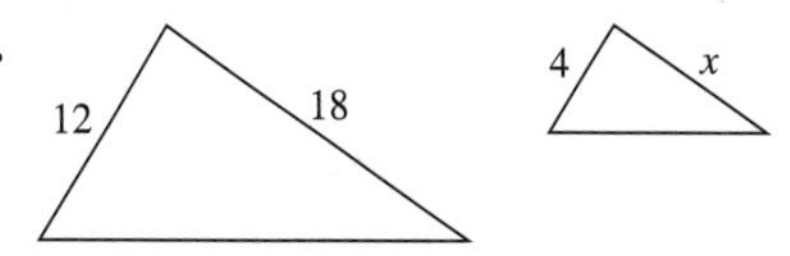

28.

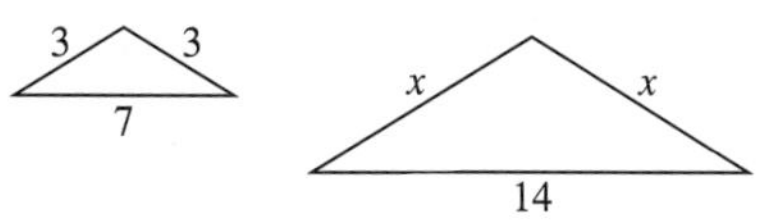

29.

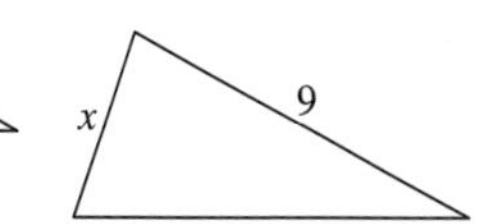

30.

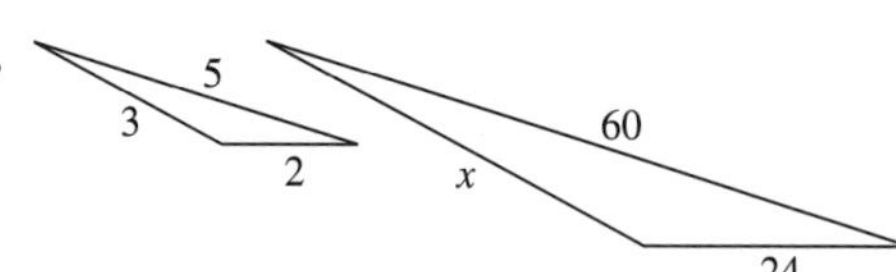

Use the Pythagorean theorem to find the missing lengths in the right triangles. See Example 5.

31.

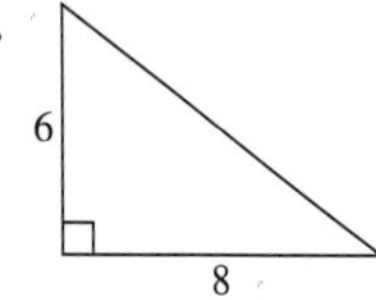

32.

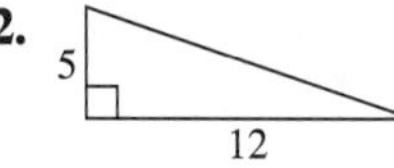

33.

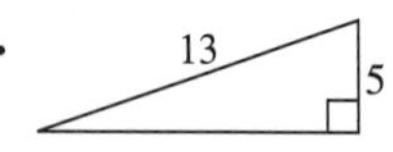

34.

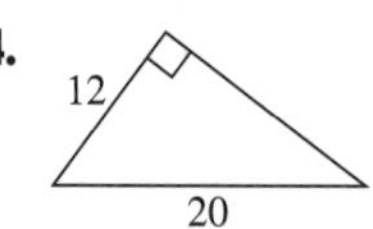

APPENDIX

G ANSWERS TO SELECTED EXERCISES

CHAPTER 1

Whole Numbers and Introduction to Algebra

Exercise Set 1.1

1. tens **3.** thousands **5.** hundred-thousands **7.** millions **9.** hundred-millions **11.** five thousand, four hundred twenty **13.** twenty-six thousand, nine hundred ninety **15.** one million, six hundred twenty thousand **17.** fifty-three million, five hundred twenty thousand, one hundred seventy **19.** ninety-seven billion, sixty million, five hundred fifty thousand **21.** five million, two hundred forty-eight thousand, four hundred one **23.** five hundred seventy thousand, one hundred sixty-two **25.** 6,508 **27.** 29,900 **29.** 6,504,019 **31.** 3,000,014 **33.** 15,000,040,016 **35.** 1821 **37.** 63,100,000 **39.** 400 + 6 **41.** 5000 + 200 + 90 **43.** 60,000 + 2000 + 40 + 7 **45.** 30,000 + 600 + 80 **47.** 30,000,000 + 9,000,000 + 600,000 + 80,000 **49.** $<$ **51.** $>$ **53.** $>$ **55.** $>$ **57.** four thousand **59.** 4000 + 100 + 40 + 5 **61.** Nile **63.** 218 **65.** USA and USSR **67.** 55,543 **69.** answers will vary

Calculator Explorations, Sec. 1.2

1. 134 **3.** 340 **5.** 2834

Mental Math, Sec. 1.2

1. 12 **3.** 9000 **5.** 1620

Exercise Set 1.2

1. 36 **3.** 92 **5.** 49 **7.** 5399 **9.** 117 **11.** 71 **13.** 117 **15.** 4 + 9 **17.** 13 + 2 **19.** (3 + 2) + 9 **21.** 4 + (1 + 10) **23.** (8 + 4) + 6 **25.** 25 **27.** 62 **29.** 212 **31.** 94 **33.** 910 **35.** 8273 **37.** 11,926 **39.** 1884 **41.** 16,717 **43.** 1110 **45.** 8999 **47.** 35,901 **49.** 612,389 **51.** 29 in. **53.** 25 ft **55.** 24 in. **57.** 8 yd **59.** $25,751 **61.** $30,448 **63.** 383 mi **65.** 361 ft **67.** no, $2648 < 2699$ **69.** 108,577,007 patties **71.** 2425 ft **73.** answers will vary **75.** 166,510,192

Calculator Explorations, Sec. 1.3

1. 770 **3.** 109 **5.** 8978

Mental Math, Sec. 1.3

1. 7 **3.** 5 **5.** 0 **7.** 400 **9.** 500

Exercise Set 1.3

1. 44 **3.** 60 **5.** 265 **7.** 254 **9.** 545 **11.** 600 **13.** 25 **15.** 45 **17.** 146 **19.** 288 **21.** 168 **23.** 6 **25.** 447 **27.** 5723 **29.** 504 **31.** 89 **33.** 79 **35.** 39,914 **37.** 32,711 **39.** 5041 **41.** 31,213 **43.** 4 **45.** 20 **47.** 7 **49.** 56 crawfish **51.** 264 pages **53.** 6065 ft **55.** $175 **57.** 358 miles **59.** $389 **61.** $448 **63.** 173 men **65.** Marilyn with 1018 votes **67.** Chicago O'Hare **69.** 165 thousand or 165,000 **71.** Procter & Gamble, General Motors, Philip Morris **73.** $804,040,000 **75.** $8,767,930,000

77. $\begin{array}{r} 5269 \\ -\ 2385 \\ \hline 2884 \end{array}$

Exercise Set 1.4

1. 630 **3.** 640 **5.** 790 **7.** 400 **9.** 1100 **11.** 43,000 **13.** 248,700 **15.** 36,000 **17.** 100,000 **19.** 60,000,000 **21.** ten: 5,280; hundred: 5,300; thousand: 5,000 **23.** ten: 9,440; hundred: 9,400; thousand: 9,000 **25.** ten: 14,880; hundred: 14,900; thousand: 15,000 **27.** 12,000

29. $\begin{array}{r} 30 \\ 40 \\ 40 \\ +\ 20 \\ \hline 130 \end{array}$ **31.** $\begin{array}{r} 650 \\ -\ 270 \\ \hline 380 \end{array}$ **33.** $\begin{array}{r} 730 \\ 630 \\ +\ 830 \\ \hline 2190 \end{array}$ **35.** $\begin{array}{r} 810 \\ -\ 800 \\ \hline 10 \end{array}$ **37.** $\begin{array}{r} 160 \\ 300 \\ +\ 150 \\ \hline 610 \end{array}$ **39.** $\begin{array}{r} 1800 \\ 1800 \\ +\ 1900 \\ \hline 5500 \end{array}$

41. $\begin{array}{r} 1800 \\ -\ 1500 \\ \hline 300 \end{array}$ **43.** $\begin{array}{r} 3000 \\ 1600 \\ +\ 3900 \\ \hline 8500 \end{array}$ **45.** correct **47.** incorrect; 1309 **49.** correct

51. correct **53.** $\begin{array}{r} \$800 \\ 1300 \\ +\ 1000 \\ \hline \$3100 \end{array}$ **55.** $\begin{array}{r} \$900 \\ 700 \\ 400 \\ 800 \\ +\ 400 \\ \hline \$3200 \end{array}$ **57.** $20 + 30 + 30 = 80$ miles

59. $\begin{array}{r} 29{,}000 \\ -\ 4{,}000 \\ \hline 25{,}000 \text{ feet} \end{array}$ **61.** $\begin{array}{r} 7{,}300{,}000 \\ -\ 1{,}600{,}000 \\ \hline 5{,}700{,}000 \end{array}$ **63.** $\begin{array}{r} 41{,}000{,}000 \text{ votes} \\ -\ 27{,}000{,}000 \text{ votes} \\ \hline 14{,}000{,}000 \text{ votes} \end{array}$ **65.** $\begin{array}{r} \$30{,}000 \\ 30{,}000 \\ 30{,}000 \\ 20{,}000 \\ +\ 30{,}000 \\ \hline \$140{,}000 \end{array}$

67. $1,400,000,000 **69.** $920,000,000 **71.** answers will vary

Calculator Explorations, Sec. 1.5

1. 3456 **3.** 15,322 **5.** 272,291

Mental Math, Sec. 1.5

1. 24 **3.** 0 **5.** 0 **7.** 87

Exercise Set 1.5

1. $9 \cdot 6$ **3.** $4 \cdot (8 \cdot 10)$ **5.** $(5 \cdot 7) \cdot 12$ **7.** $89 \cdot 32$ **9.** $4 \cdot 3 + 4 \cdot 9$ **11.** $2 \cdot 4 + 2 \cdot 6$ **13.** $10 \cdot 11 + 10 \cdot 7$ **15.** 252 **17.** 1872 **19.** 1662 **21.** 5310 **23.** 4172 **25.** 10,857 **27.** 11,326 **29.** 24,800 **31.** 0 **33.** 5900 **35.** 59,232 **37.** 142,506 **39.** 1,821,204 **41.** 456,135 **43.** 64,790 **45.** $600 \times 400 = 240{,}000$ **47.** $600 \times 500 = 300{,}000$ **49.** 63 sq. m **51.** 390 sq. ft **53.** 375 calories

55. \$1295 **57.** 192 cans **59.** 9900 sq. ft **61.** 42 keys **63.** 555 troops **65.** 3696 hits **67.** 495,864 sq. m
69. 58,380 pages **71.** 5828 pixels **73.** 1500 characters **75.** 71,343 mi **77.** 22,464 acres **79.** 50 students
81. apple and orange **83.** $\begin{array}{r} 42 \\ \times\ 93 \\ \hline 126 \\ 3780 \\ \hline 3906 \end{array}$ **85.** answers will vary **87.** answers will vary

Calculator Explorations, Sec. 1.6

1. 53 **3.** 62 **5.** 261 **7.** 0

Mental Math, Sec. 1.6

1. 5 **3.** 9 **5.** 0 **7.** 9 **9.** 1 **11.** 5 **13.** undefined **15.** 7 **17.** 0 **19.** 8

Exercise Set 1.6

1. 12 **3.** 37 **5.** 338 **7.** 16 R2 **9.** 563 R1 **11.** 37 R1 **13.** 265 R1 **15.** 49 **17.** 13 **19.** 97 R40
21. 206 **23.** 506 **25.** 202 R7 **27.** 45 **29.** 98 R100 **31.** 202 R15 **33.** 202 **35.** 26 **37.** 498
39. 79 **41.** 58 students **43.** \$252,000 **45.** 415 bushels **47.** 88 bridges
49. Yes, she needs 176 feet; she has 9 feet left over. **51.** 13 paychecks **53.** 16 ft **55.** 45 degrees
57. \$554,273,000 **59.** decrease **61.** 10 ft

Calculator Explorations, Sec. 1.7

1. 729 **3.** 1024 **5.** 2048 **7.** 2526 **9.** 4295 **11.** 8

Exercise Set 1.7

1. 3^4 **3.** 7^8 **5.** 12^3 **7.** $6^2 \cdot 5^3$ **9.** $9^3 \cdot 8$ **11.** $3 \cdot 2^5$ **13.** $3 \cdot 2^2 \cdot 5^3$ **15.** 25 **17.** 125 **19.** 64
21. 1024 **23.** 7 **25.** 243 **27.** 256 **29.** 64 **31.** 81 **33.** 729 **35.** 100 **37.** 10,000 **39.** 10
41. 1920 **43.** 729 **45.** 21 **47.** 8 **49.** 29 **51.** 4 **53.** 17 **55.** 46 **57.** 28 **59.** 10 **61.** 7
63. 4 **65.** 14 **67.** 72 **69.** 2 **71.** 35 **73.** 4 **75.** undefined **77.** 52 **79.** 44 **81.** 12 **83.** 13
85. 400 sq. m **87.** 64 sq. cm **89.** 10,000 sq. m **91.** $(2 + 3) \cdot 6 - 2$ **93.** $24 \div (3 \cdot 2) + 2 \cdot 5$ **95.** 1260 feet
97. 6,384,814

Exercise Set 1.8

1. 9 **3.** 26 **5.** 6 **7.** 3 **9.** 117 **11.** 94 **13.** 5 **15.** 626 **17.** 20 **19.** 4 **21.** 4 **23.** 0
25. 33 **27.** 121 **29.** 121 **31.** 100 **33.** 60 **35.** 4

37.

t	1	2	3	4
$16t^2$	16	64	144	256

39. $x + 5$ **41.** $x + 8$ **43.** $20 - x$ **45.** $512x$ **47.** $\frac{x}{2}$ **49.** $5x + (17 + x)$ **51.** $5x$ **53.** $11 - x$
55. $50 - 8x$ **57.** 274,657 **59.** 777 **61.** $5x$ **63.** increases

Chapter 1 Review

1. hundreds **3.** five thousand, four hundred eighty **5.** $6000 + 200 + 70 + 9$ **7.** 59,800 **9.** $<$ **11.** $<$
13. 1,595,138 **15.** 193,699 **17.** 13 **19.** 3 **21.** 38 **23.** 56 **25.** 110 **27.** 950 **29.** 1711

31. 197,699 **33.** 276 feet **35.** 33 **37.** 14 **39.** 362 **41.** 304 **43.** 2114 **45.** $15,626 **47.** May **49.** July, August, September **51.** 90 **53.** 470 **55.** 4800 **57.** 50,000,000 **59.** 4900 + 600 + 1900 = 7400 **61.** 10,200 **63.** 42 **65.** 0 **67.** 1410 **69.** 800 **71.** 3696 **73.** 0 **75.** 16,994 **77.** 113,634 **79.** 411,426 **81.** 50 · 30 = 1500 **83.** 5200 · 200 = 1,040,000 **85.** $4,897,341 **87.** 60 sq. m **89.** 3 **91.** 6 **93.** 5 R2 **95.** undefined **97.** 1 **99.** 15 **101.** 24 R2 **103.** 1 R17 **105.** 35 R15 **107.** 500 **109.** 506 **111.** 199 R8 **113.** 458 ft **115.** 7^4 **117.** $4 \cdot 2^3 \cdot 3^2$ **119.** 49 **121.** 1125 **123.** 13 **125.** 3 **127.** 32 **129.** 49 sq. m **131.** 5 **133.** undefined **135.** 121 **137.** 4 **139.** $x - 5$ **141.** $10 \div (x + 1)$

143.

x	0	1	2	3
$8x^2$	0	8	32	72

Chapter 1 Test

1. 141 **2.** 113 **3.** 14,880 **4.** 766 R42 **5.** 200 **6.** 48 **7.** 98 **8.** 0 **9.** undefined **10.** 33 **11.** 21 **12.** 36 **13.** 52,000 **14.** 6300 + 5400 + 2000 + 13,700 **15.** 4300 − 2700 = 1600 **16.** $17 **17.** $119 **18.** $126 **19.** 30 **20.** 1 **21.** **(a)** $17x$; **(b)** $20 - 2x$ **22.** 20 cm; 25 sq. cm **23.** 60 yd; 200 sq. yd **24.** 430 **25.** 21

CHAPTER 2
Integers

Exercise Set 2.1

1. −1445 **3.** +14,494 **5.** −15 **7.**

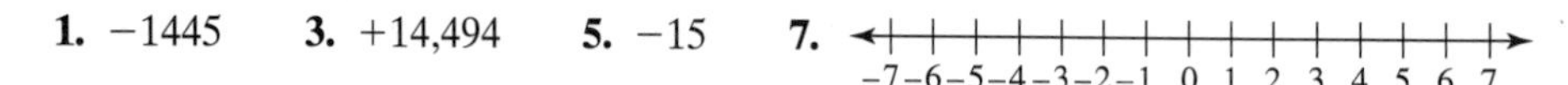

9.

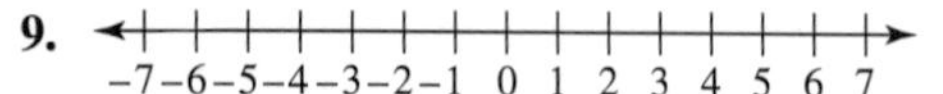

11. −7 −6 −5 −4 −3 −2 −1 0 1 2 3 4 5 6 7

13. −7 −6 −5 −4 −3 −2 −1 0 1 2 3 4 5 6 7

15. −7 −6 −5 −4 −3 −2 −1 0 1 2 3 4 5 6 7

17. < **19.** > **21.** > **23.** > **25.** < **27.** 5 **29.** 8 **31.** 0 **33.** 5 **35.** −5 **37.** 4 **39.** −23 **41.** 10 **43.** 7 **45.** −20 **47.** −3 **49.** 8 **51.** 14 **53.** 29 **55.** < **57.** > **59.** < **61.** = **63.** < **65.** > **67.** < **69.** = **71.** < **73.** < **75.** < **77.** −317 **79.** cooler **81.** false **83.** true **85.** false **87.** answers will vary **89.** 13 **91.** 35 **93.** 360

Mental Math, Sec. 2.2

1. 5 **3.** −35

Exercise Set 2.2

1.

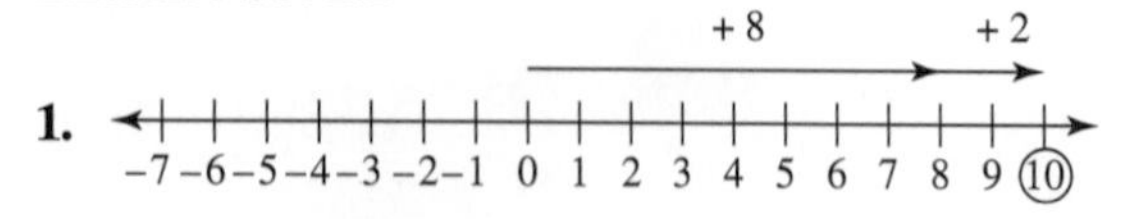

3.

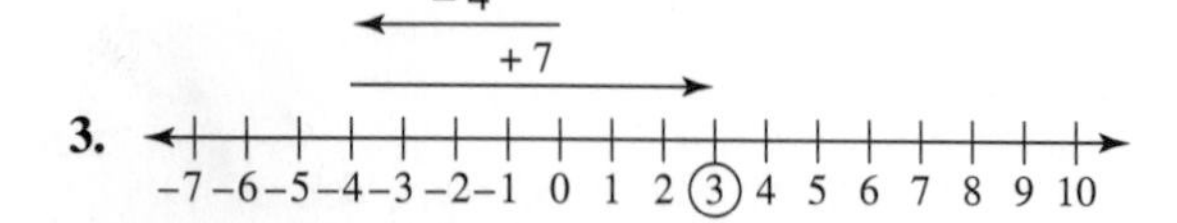

5. −13; +7; −13 −12 −11 −10 −9 −8 −7 (−6) −5 −4 −3 −2 −1 0 1 2 3 4 5 6 7

7. 35 **9.** −8 **11.** 0 **13.** 4 **15.** 2 **17.** −2 **19.** −7 **21.** −23 **23.** 16 **25.** 1 **27.** −70 **29.** −27 **31.** −11 **33.** −24 **35.** −57 **37.** −223 **39.** 0 **41.** 7 **43.** −3 **45.** −9 **47.** 30 **49.** 20 **51.** 51 **53.** −33 **55.** 21 **57.** 13 **59.** −33 **61.** −20 **63.** −125 **65.** 9 **67.** −100 **69.** 0 **71.** 16 **73.** 0 **75.** −55 **77.** −102 **79.** 2°C **81.** $-165 + (-16) = -181$; 181 ft below the surface **83.** Team 1, 7; Team 2, 6; winning team, Team 1 **85.** true **87.** false **89.** answers will vary **91.** 44 **93.** 0 **95.** 28

Exercise Set 2.3

1. 0 **3.** 5 **5.** −5 **7.** 14 **9.** 3 **11.** −18 **13.** −14 **15.** −38 **17.** −17 **19.** 13 **21.** 2 **23.** 0 **25.** −1 **27.** −27 **29.** 40 **31.** −8 **33.** 36 **35.** 0 **37.** 19 **39.** 0 **41.** 0 **43.** −4 **45.** −15 **47.** −14 **49.** −1100 **51.** 10 **53.** −22 **55.** −8 **57.** −4 **59.** 7 **61.** −12 **63.** −4 **65.** 0 **67.** −12 **69.** 14 **71.** −22 **73.** 20 **75.** 262°F **77.** −16 **79.** −12°C **81.** false **83.** false **85.** 652 ft **87.** 144 ft **89.** answers will vary **91.** 0 **93.** 8 **95.** 4539

Exercise Set 2.4

1. 6 **3.** −36 **5.** −64 **7.** 0 **9.** −48 **11.** −8 **13.** 80 **15.** 0 **17.** −15 **19.** 4 **21.** −27 **23.** 25 **25.** −8 **27.** −28 **29.** −6 **31.** 25 **33.** 0 **35.** −4 **37.** −5 **39.** 8 **41.** 0 **43.** −13 **45.** −1 **47.** undefined **49.** 6 **51.** −10 **53.** 0 **55.** −12 **57.** −54 **59.** 42 **61.** −24 **63.** 16 **65.** −2 **67.** −7 **69.** −4 **71.** 48 **73.** −1080 **75.** 0 **77.** −5 **79.** −6 **81.** 3 **83.** 1 **85.** −243 **87.** 180 **89.** 1 **91.** −8 **93.** −966 **95.** −2050 **97.** 8 and 2 **99.** 0 and 0 **101.** $(-4)(3) = -12$, and a loss of 12 yards **103.** $(-20)(5) = -100$, a descent of 100 ft **105.** true **107.** true **109.** answers will vary **111.** 225 **113.** 109 **115.** 8

Calculator Explorations, Sec. 2.5

1. −159 **3.** −136 **5.** −559 **7.** 8 **9.** −258

Exercise Set 2.5

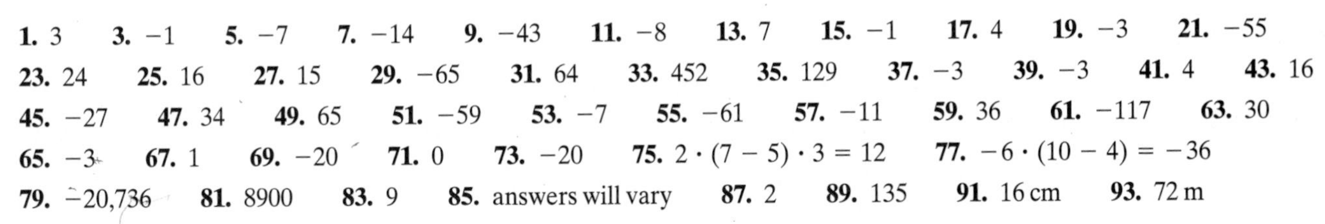

1. 3 **3.** −1 **5.** −7 **7.** −14 **9.** −43 **11.** −8 **13.** 7 **15.** −1 **17.** 4 **19.** −3 **21.** −55 **23.** 24 **25.** 16 **27.** 15 **29.** −65 **31.** 64 **33.** 452 **35.** 129 **37.** −3 **39.** −3 **41.** 4 **43.** 16 **45.** −27 **47.** 34 **49.** 65 **51.** −59 **53.** −7 **55.** −61 **57.** −11 **59.** 36 **61.** −117 **63.** 30 **65.** −3 **67.** 1 **69.** −20 **71.** 0 **73.** −20 **75.** $2 \cdot (7 - 5) \cdot 3 = 12$ **77.** $-6 \cdot (10 - 4) = -36$ **79.** −20,736 **81.** 8900 **83.** 9 **85.** answers will vary **87.** 2 **89.** 135 **91.** 16 cm **93.** 72 m

Chapter 2 Review

1. −1435 **3.** −7 −6 −5 −4 −3 −2 −1 0 1 2 3 4 5 6 7 **5.** > **7.** > **9.** 0 **11.** 12 **13.** false

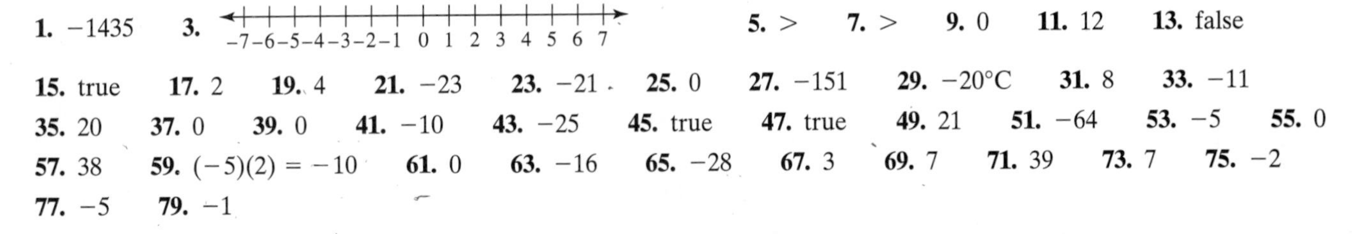

15. true **17.** 2 **19.** 4 **21.** −23 **23.** −21 **25.** 0 **27.** −151 **29.** −20°C **31.** 8 **33.** −11 **35.** 20 **37.** 0 **39.** 0 **41.** −10 **43.** −25 **45.** true **47.** true **49.** 21 **51.** −64 **53.** −5 **55.** 0 **57.** 38 **59.** $(-5)(2) = -10$ **61.** 0 **63.** −16 **65.** −28 **67.** 3 **69.** 7 **71.** 39 **73.** 7 **75.** −2 **77.** −5 **79.** −1

Chapter 2 Test

1. 3 **2.** −6 **3.** −100 **4.** 4 **5.** −30 **6.** 12 **7.** 65 **8.** 5 **9.** 12 **10.** −6 **11.** 50 **12.** −2 **13.** −11 **14.** −46 **15.** −117 **16.** 3456 **17.** 28 **18.** −213 **19.** −1 **20.** −2 **21.** 2 **22.** −5 **23.** −3 **24.** 5 **25.** −1 **26.** 8 **27.** 14,893 + (−147) = 14,746; 14,746 ft **28.** +41

Chapter 2 Cumulative Review

1. *(Sec. 1.4, Ex. 1)* 74,000 **2.** *(Sec. 1.5, Ex. 7)* 15,540 thousand bytes **3.** *(Sec. 1.6, Ex. 8)* 401 R2 **4.** *(Sec. 1.7, Ex. 2)* **(a)** 64; **(b)** 7; **(c)** 32; **(d)** 180 **5.** *(Sec. 1.7, Ex. 3)* 7 **6.** *(Sec. 1.8, Ex. 3)* 2 **7.** *(Sec. 2.1, Ex. 3)* **(a)** <; **(b)** >; **(c)** > **8.** *(Sec. 2.2, Ex. 1)* **(a)** −7; **(b)** −4 **9.** *(Sec. 2.2, Ex. 6)* −2 **10.** *(Sec. 2.3, Ex. 1)* **(a)** −15; **(b)** −7; **(c)** 1 **11.** *(Sec. 2.3, Ex. 4)* 0 **12.** *(Sec. 2.4, Ex. 3)* 25 **13.** *(Sec. 2.4, Ex. 6)* **(a)** 0; **(b)** undefined **14.** *(Sec. 2.5, Ex. 2)* 4 **15.** *(Sec. 2.5, Ex. 5)* −8 **16.** *(Sec. 2.5, Ex. 9)* −1

CHAPTER 3

Solving Equations and Problem Solving

Exercise Set 3.1

1. 5 **3.** 1 **5.** 11 **7.** $8x$ **9.** $-4n$ **11.** $-2c$ **13.** $-4x$ **15.** $13a - 8$ **17.** $30x$ **19.** $-22y$ **21.** $72a$ **23.** $2y + 4$ **25.** $5a - 40$ **27.** $12x + 28$ **29.** $2x + 15$ **31.** $27n - 20$ **33.** $15c + 3$ **35.** $7w + 15$ **37.** $11x - 8$ **39.** $-2y$ **41.** $-7z$ **43.** $8d - 3c$ **45.** $-5x + 4y$ **47.** $-q$ **49.** $2x + 22$ **51.** $-3x - 35$ **53.** $-3z - 15$ **55.** $-6x + 6$ **57.** $3x - 30$ **59.** $-r + 8$ **61.** $-7n + 3$ **63.** $9z - 14$ **65.** −6 **67.** $-12x + 20$ **69.** $2xy - 20$ **71.** $7a + 12$ **73.** $3y + 5$ **75.** $(-25x + 55)$ in. **77.** $(26y - 12)$ m **79.** $(25a - 12)$ ft **81.** $16y^2$ sq. cm **83.** $(20x + 16)$ sq. mi **85.** $4824q + 12{,}274$ **87.** $20{,}368x + 42{,}032$ **89.** answers will vary **91.** −8 **93.** −5 **95.** −8 **97.** 0

Exercise Set 3.2

1. yes **3.** yes **5.** yes **7.** yes **9.** yes **11.** yes **13.** 18 **15.** 26 **17.** 9 **19.** −16 **21.** 6 **23.** 0 **25.** 8 **27.** −1 **29.** 2 **31.** 0 **33.** −28 **35.** 73 **37.** 0 **39.** −4 **41.** −65 **43.** 13 **45.** −3 **47.** −22 **49.** −28 **51.** 3 **53.** 162,964 **55.** −28 **57.** 590 **59.** 1000 **61.** 3000 **63.** \$322 **65.** \$62

Exercise Set 3.3

1. 4 **3.** −4 **5.** 0 **7.** −17 **9.** 5 **11.** −4 **13.** 8 **15.** −1 **17.** 2 **19.** −8 **21.** −1 **23.** −7 **25.** 0 **27.** 6 **29.** −9 **31.** 9 **33.** 4 **35.** −6 **37.** 1 **39.** −1 **41.** 5 **43.** −2 **45.** −2 **47.** 0 **49.** 4 **51.** 7 **53.** 3 **55.** −3 **57.** 1 **59.** $4x + 7$ **61.** $2x - 17$ **63.** $-6(x + 15)$ **65.** $\frac{45}{-5x}$ **67.** answers will vary **69.** 25 **71.** 1 **73.** 10

Calculator Explorations, Sec. 3.4

1. yes **3.** no **5.** yes

Exercise Set 3.4

1. 3 **3.** −2 **5.** −4 **7.** −3 **9.** −12 **11.** 5 **13.** −5 **15.** 8 **17.** 5 **19.** 1 **21.** −4 **23.** 2 **25.** 5 **27.** −3 **29.** 2 **31.** −2 **33.** 3 **35.** 6 **37.** −10 **39.** −4 **41.** −1 **43.** 4 **45.** −4

47. 3 **49.** −1 **51.** 0 **53.** −22 **55.** 4 **57.** 1 **59.** −30 **61.** $-42 + 16 = -26$ **63.** $-5(-29) = 145$
65. $3(-14 - 2) = -48$ **67.** $\frac{100}{2(50)} = 1$ **69.** 33 **71.** −37 **73.** 49 **75.** −6

Exercise Set 3.5

1. $-5 + x = -7$ **3.** $3x = 27$ **5.** $-20 - x = 104$ **7.** $2[x + (-1)] = 50$ **9.** 8 **11.** 6 **13.** 9 **15.** 6
17. 24 **19.** 8 **21.** 5 **23.** 5 **25.** 12 **27.** 5 **29.** Bush, 168 votes; Clinton, 370 votes **31.** 80 books
33. truck, 35 mph; car, 70 mph **35.** $225 **37.** 41 points **39.** 4066 votes; yes; answers will vary **41.** 255 **43.** 1360 **45.** 2167 **47.** 255,023

Exercise Set 3.6

1. 64 ft **3.** 36 cm **5.** 21 in. **7.** 120 cm **9.** 48 ft **11.** 66 in. **13.** 21 ft **15.** 60 ft **17.** 346 yd
19. 22 ft **21.** $66 **23.** 96 m **25.** 66 ft **27.** 128 mi **29.** 42 m **31.** 7 in. **33.** 402 yd **35.** 21 km
37. perimeter **39.** area **41.** area **43.** perimeter **45.** 23 **47.** 1 **49.** 10 **51.** 216

Chapter 3 Review

1. $10y - 15$ **3.** $-6a - 7$ **5.** $-2x - 10$ **7.** $11x - 12$ **9.** $-5a + 4$ **11.** $(4x + 4)$ yd **13.** $(6x - 3)$ sq. yd
15. $600{,}822x - 9180$ **17.** yes **19.** −2 **21.** −6 **23.** −25 **25.** 7 **27.** 1 **29.** 0 **31.** 0
33. $2x + 11$ **35.** $\frac{70}{x + 6}$ **37.** 5 **39.** 17 **41.** −2 **43.** 11 **45.** 2 **47.** $20 - (-8) = 28$
49. $\frac{-75}{5 + 20} = -3$ **51.** $2x - 8 = 40$ **53.** $x - 3 = x \div 4$ **55.** 5 **57.** 2386 votes **59.** 36 m **61.** 32 ft
63. 7 m, 21 m, 21 m

Chapter 3 Test

1. $-5x + 5$ **2.** $-6y - 14$ **3.** $14z - 8$ **4.** $(15x + 17)$ in. **5.** $(19x - 3)$ sq. m **6.** −6 **7.** −6 **8.** 10
9. 8 **10.** 24 **11.** 1 **12.** −2 **13.** −2 **14.** 6 **15.** 0 **16.** −2 **17.** 7 **18.** 4 **19.** 0 **20.** −2
21. 8 free throws **22.** 7 in. **23.** 244 women **24.** 62 ft **25.** length, 17 cm; width, 8 cm

Chapter 3 Cumulative Review

1. *(Sec. 1.1, Ex. 3)* 2,564,350 **2.** *(Sec. 1.2, Ex. 1)* 159 **3.** *(Sec. 1.5, Ex. 5)* 20,296 **4.** *(Sec. 1.8, Ex. 1)* 15
5. *(Sec. 2.1, Ex. 4)* **(a)** 2; **(b)** 5; **(c)** 0 **6.** *(Sec. 2.2, Ex. 2)* **(a)** −23; **(b)** −6; **(c)** 8 **7.** *(Sec. 2.3, Ex. 6)* 14
8. *(Sec. 2.4, Ex. 2)* **(a)** 84; **(b)** −24 **9.** *(Sec. 2.4, Ex. 7)* −4 **10.** *(Sec. 2.5, Ex. 4)* 82
11. *(Sec. 3.1, Ex. 2)* **(a)** $5x$; **(b)** $-6y$; **(c)** $8x^2 - 2$ **12.** *(Sec. 3.1, Ex. 5)* $-8x$ **13.** *(Sec. 3.1, Ex. 9)* $10x - 2$
14. *(Sec. 3.2, Ex. 1)* yes **15.** *(Sec. 3.2, Ex. 6)* −4 **16.** *(Sec. 3.3, Ex. 5)* −3 **17.** *(Sec. 3.4, Ex. 4)* −4
18. *(Sec. 3.5, Ex. 2)* −9 **19.** *(Sec. 3.5, Ex. 3)* 4066 votes **20.** *(Sec. 3.1, Ex. 10)* $(10z + 1)$ ft

CHAPTER 4
Fractions

Exercise Set 4.1

1. $\frac{1}{3}$ **3.** $\frac{7}{12}$ **5.** $\frac{3}{8}$ **7.** $\frac{8}{5}$ **9.** $\frac{3}{2}$ **11.** $\frac{5}{3}$

13. 0 $\frac{1}{4}$ 1 **15.** 0 $\frac{4}{7}$ 1 **17.** 0 1 $\frac{8}{5}$ 2

19. 0 1 2 $\frac{7}{3}$ 3 **21.** 0 $\frac{3}{8}$ 1 **23.** $\frac{20}{35}$ **25.** $\frac{14}{21}$ **27.** $\frac{10y}{25}$ **29.** $\frac{15}{30}$ **31.** $\frac{30}{21x}$

33. $\frac{10}{5}$ **35.** 1 **37.** -5 **39.** 12 **41.** 1 **43.** -7 **45.** $\frac{9}{12}$ **47.** $\frac{8y}{12}$ **49.** $\frac{6}{12}$ **51.** $\frac{48x}{36x}$ **53.** $\frac{20x}{36x}$

55. $\frac{36x}{36x}$ **57.** $\frac{5}{12}$ **59.** $\frac{11}{24}$ **61.** $\frac{7}{11}$ **63. (a)** $\frac{32}{50}$; **(b)** 18; **(c)** $\frac{18}{50}$ **65.** $\frac{464}{2088}$ **67.** 4084 **69.** 22,852

71. 42 **73.** 162 **75.** -20 **77.** -1

Mental Math, Sec. 4.2

1. yes, yes, yes **3.** $2 \cdot 5$ **5.** $3 \cdot 7$ **7.** 3^2

Exercise Set 4.2

1. $2^2 \cdot 5$ **3.** $2^4 \cdot 3$ **5.** 2^6 **7.** $2^4 \cdot 3 \cdot 5$ **9.** $\frac{1}{4}$ **11.** $\frac{x}{5}$ **13.** $\frac{7}{8}$ **15.** $\frac{4}{5}$ **17.** $\frac{5}{6}$ **19.** $\frac{5x}{6}$ **21.** $\frac{2}{3}$

23. $\frac{3x}{4}$ **25.** $\frac{2a}{3b}$ **27.** $\frac{7}{8}$ **29.** $\frac{3}{7}$ **31.** $\frac{3y}{5}$ **33.** $\frac{4}{7}$ **35.** $\frac{4x^2y}{5}$ **37.** $\frac{4}{5}$ **39.** $\frac{5x}{8}$ **41.** $\frac{3x}{10}$ **43.** $\frac{3a^2}{2b^3}$

45. $\frac{5}{8z}$ **47.** $\frac{1}{2}$ **49.** $\frac{3}{4}$ **51.** $\frac{2}{7}$ **53.** $\frac{9}{20}$ **55.** $\frac{1}{5}$ **57.** answers will vary **59.** d **61.** false **63.** true

65. $\frac{3}{5}$ **67.** -27 **69.** -14 **71.** 0 **73.** 29

Mental Math, Sec. 4.3

1. $\frac{2}{15}$ **3.** $\frac{6}{35}$ **5.** $\frac{9}{8}$

Exercise Set 4.3

1. $\frac{7}{12}$ **3.** $-\frac{5}{28}$ **5.** $\frac{1}{15}$ **7.** $\frac{18x}{55}$ **9.** $\frac{3a^2}{4}$ **11.** $\frac{x^2}{y}$ **13.** $\frac{1}{125}$ **15.** $\frac{4}{9}$ **17.** $-\frac{4}{27}$ **19.** $\frac{4}{5}$ **21.** $-\frac{1}{6}$

23. $\frac{16}{9x}$ **25.** $\frac{121y}{60}$ **27.** $-\frac{1}{6}$ **29.** $\frac{x}{25}$ **31.** $\frac{2}{5}$ **33.** $-\frac{5}{3}$ **35. (a)** $\frac{1}{3}$; **(b)** $\frac{12}{25}$ **37. (a)** $-\frac{36}{55}$; **(b)** $-\frac{44}{45}$

39. yes **41.** no **43.** $\frac{45}{32}$ **45.** $\frac{18x^2}{35}$ **47.** $-\frac{1}{4}$ **49.** $\frac{3}{4}$ **51.** $\frac{9}{16}$ **53.** xy^2 **55.** $\frac{77}{2}$ **57.** $-\frac{36}{x}$

59. $\frac{3}{49}$ **61.** $-\frac{19y}{7}$ **63.** $\frac{4}{11}$ **65.** $\frac{8}{3}$ **67.** $\frac{15x}{4}$ **69.** $\frac{8}{9}$ **71.** $\frac{1}{60}$ **73.** b **75. (a)** 3; **(b)** $\frac{49}{3}$

77. (a) $-\frac{5}{2}$; **(b)** -10 **79.** 30 gal **81.** $\frac{3}{2}$ in. **83.** 30 lb **85.** 36 in. **87.** 3840 mi **89.** 2400 mi

91. 3 ft **93.** 7,210,000 households **95.** answers will vary **97.** $2 \cdot 3 \cdot 3 \cdot 5$ **99.** $5 \cdot 13$
101. $2 \cdot 3 \cdot 3 \cdot 7$

Exercise Set 4.4

1. $\frac{5}{7}$ **3.** 0 **5.** $\frac{2}{3x}$ **7.** $-\frac{1}{13}$ **9.** $\frac{2}{3}$ **11.** $\frac{6}{11}$ **13.** $-\frac{3}{y}$ **15.** $\frac{7a-3}{4}$ **17.** $-\frac{3}{4}$ **19.** $-\frac{1}{3}$ **21.** $\frac{5}{4}$

23. $-\frac{4}{5}$ **25.** $-\frac{3}{4}$ **27.** yes **29.** no **31.** yes **33.** 12 **35.** 45 **37.** 36 **39.** $24x$ **41.** 150 **43.** $\frac{7}{11}$

45. $\frac{8}{15}$ **47.** $\frac{2x}{3}$ **49.** $-\frac{x}{2}$ **51.** $\frac{3}{4z}$ **53.** $-\frac{3}{10}$ **55.** 2 **57.** $-\frac{2}{3}$ **59.** $\frac{3x}{4}$ **61.** 126 **63.** 75 **65.** 24

67. 50 **69.** $12a$ **71.** 168 **73.** 363 **75.** $-\frac{2}{3}$ **77.** $\frac{1}{13}$ **79.** $\frac{4}{5}$ **81.** 1 in. **83.** 2 m **85.** $\frac{1}{36}$ sq. cm

87. 2 cups **89.** 22 cups **91.** $\frac{1}{4}$ mi **93.** $\frac{3}{4}$ **95.** answers will vary **97.** $\frac{12}{35}$ **99.** 8 **101.** $\frac{4}{5}$ **103.** -4

Mental Math, Sec. 4.5

1. 6 **3.** 12 **5.** 56 **7.** 12

Exercise Set 4.5

1. $\frac{5}{6}$ **3.** $\frac{1}{6}$ **5.** $-\frac{4}{33}$ **7.** $\frac{3x-6}{14}$ **9.** $\frac{3}{5}$ **11.** $\frac{24y-5}{12}$ **13.** 1 **15.** $-\frac{11}{30}$ **17.** $\frac{7x}{8}$ **19.** $-\frac{13}{16}$

21. $\frac{13}{12}$ **23.** $\frac{1}{4}$ **25.** $\frac{11}{6}$ **27.** $\frac{11}{36}$ **29.** $\frac{12}{7}$ **31.** $\frac{89a}{99}$ **33.** $\frac{4y-1}{6}$ **35.** $\frac{x+6}{2x}$ **37.** $-\frac{8}{33}$ **39.** $\frac{3}{14}$

41. $\frac{11y-10}{35}$ **43.** $-\frac{11}{36}$ **45.** $\frac{1}{20}$ **47.** $\frac{33}{56}$ **49.** $\frac{17}{16}$ **51.** $\frac{8}{9}$ **53.** $\frac{15+11y}{33}$ **55.** $-\frac{53}{42}$ **57.** $\frac{11}{18}$

59. $\frac{44a}{39}$ **61.** $-\frac{11}{60}$ **63.** $\frac{5y+9}{9y}$ **65.** $\frac{56}{45}$ **67.** $\frac{40+9x}{72x}$ **69.** $\frac{19}{20}$ **71.** $-\frac{5}{24}$ **73.** $\frac{57+56x}{144}$

75. $\frac{37x-20}{56}$ **77.** $\frac{11}{12}$ **79.** $-\frac{7}{10}$ **81.** $\frac{11}{16}$ **83.** $\frac{11}{18}$ **85.** $\frac{34}{15}$ cm **87.** $\frac{17}{10}$ m **89.** $\frac{14}{15}$ cu. yd; yes **91.** $\frac{1}{6}$ hr

93. $\frac{3}{8}$ **95.** $\frac{2}{15}$ **97.** $\frac{49}{44}$ **99.** $-\frac{97}{846}$ **101.** answers will vary **103.** $\frac{1}{4}$ **105.** $\frac{1}{4}$ **107.** 600

109. 2330

Exercise Set 4.6

1. $\frac{1}{6}$ **3.** $\frac{3}{7}$ **5.** $\frac{x}{6}$ **7.** $\frac{23}{22}$ **9.** $\frac{2x}{13}$ **11.** $\frac{35}{9}$ **13.** $-\frac{17}{45}$ **15.** $\frac{11}{8}$ **17.** $\frac{7}{6}$ **19.** $-\frac{2}{5}$ **21.** $-\frac{2}{9}$ **23.** $\frac{5}{2}$

25. $\frac{7}{2}$ **27.** $\frac{4}{9}$ **29.** $-\frac{13}{2}$ **31.** $\frac{9}{25}$ **33.** $-\frac{5}{32}$ **35.** 1 **37.** $\frac{1}{10}$ **39.** $-\frac{11}{40}$ **41.** $\frac{x+6}{16}$ **43.** $-\frac{77}{16}$

45. $-\frac{55}{16}$ **47.** $\frac{5}{8}$ **49.** $\frac{11}{56}$ **51.** halfway between a and b **53.** false **55.** false **57.** true **59.** 9 **61.** 32

63. y **65.** m

Exercise Set 4.7

1. $\frac{2}{7}$ **3.** 12 **5.** -27 **7.** $\frac{27}{8}$ **9.** $\frac{1}{21}$ **11.** $\frac{2}{11}$ **13.** 1 **15.** $\frac{15}{2}$ **17.** -1 **19.** -15 **21.** $\frac{3x-28}{21}$

23. $\frac{y+10}{2}$ **25.** $\frac{7x}{15}$ **27.** $\frac{4}{3}$ **29.** 2 **31.** $\frac{21}{10}$ **33.** $-\frac{1}{14}$ **35.** -3 **37.** 50 **39.** $-\frac{1}{9}$ **41.** -6

43. 4 **45.** $\frac{3}{5}$ **47.** $-\frac{1}{24}$ **49.** $-\frac{5}{14}$ **51.** 4 **53.** -36 **55.** answers will vary **57.** $\frac{8}{3}$ **59.** $\frac{15}{8}$ **61.** $\frac{19}{5}$

Exercise Set 4.8

1. $\frac{7}{3}$ **3.** $\frac{27}{8}$ **5.** $\frac{83}{7}$ **7.** $1\frac{6}{7}$ **9.** $3\frac{2}{15}$ **11.** $4\frac{5}{8}$ **13.** $\frac{8}{21}$ **15.** $4\frac{2}{3}$ **17.** $5\frac{1}{2}$ **19.** $18\frac{2}{3}$ **21.** $6\frac{4}{5}$

23. $25\frac{5}{14}$ **25.** $13\frac{13}{24}$ **27.** $2\frac{3}{5}$ **29.** $7\frac{5}{14}$ **31.** $\frac{24}{25}$ **33.** 4 **35.** $5\frac{11}{14}$ **37.** $6\frac{2}{9}$ **39.** $\frac{25}{33}$ **41.** $35\frac{13}{18}$

43. $2\frac{1}{2}$ **45.** $72\frac{19}{30}$ **47.** $\frac{11}{14}$ **49.** $5\frac{4}{7}$ **51.** $13\frac{16}{33}$ **53.** $4\frac{7}{40}$ lb **55.** $2\frac{3}{8}$ hr **57.** $1\frac{5}{12}$ hr **59.** $7\frac{13}{20}$ in.

61. $10\frac{1}{4}$ hr **63.** 130 pages **65.** $11\frac{5}{12}$ **67.** 55 mi **69.** $22\frac{1}{2}$ g **71.** $147\frac{1}{2}$ mi **73.** 20 ft **75.** 111 hamburgers

77. 24 cans **79.** $6-3y$ **81.** $16y+4$ **83.** 39 **85.** 0

Chapter 4 Review

1. $\frac{3}{4}$ **3.** 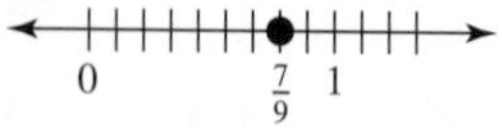

5. 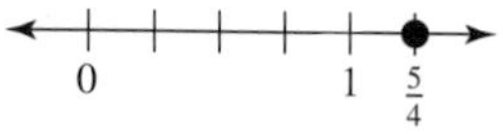

7. 20 **9.** $49a$ **11.** 40 **13.** $\frac{3}{7}$ **15.** $\frac{1}{3x}$ **17.** $\frac{29}{32c}$ **19.** $\frac{5x}{3y^2}$ **21.** $\frac{2}{3}$ **23.** $\frac{3}{10}$ **25.** $-\frac{7}{12x}$ **27.** $\frac{9}{x^2}$

29. $-\frac{1}{27}$ **31.** x^2y^2 **33.** $\frac{2}{15}$ **35.** -2 **37.** $9x^2$ **39.** $-\frac{5}{6y}$ **41.** $\frac{12}{7}$ **43.** $\frac{10}{11}$ **45.** $-\frac{1}{3}$ **47.** $\frac{4x}{5}$

49. $\frac{1}{3}$ **51.** $3x$ **53.** yes **55.** $\frac{1}{2}$ hr **57.** $\frac{3}{4}$ **59.** $\frac{7}{26}$ **61.** $-\frac{5}{12}$ **63.** $\frac{4+3b}{15}$ **65.** $\frac{11x}{18}$ **67.** $-\frac{15}{14}$

69. $\frac{5}{18}$ **71.** $-\frac{9}{10}$ **73.** $\frac{3}{2}$ ft **75.** yes **77.** $\frac{3}{10}$ **79.** $\frac{3y}{11}$ **81.** $\frac{17}{16}$ **83.** $-7y$ **85.** $\frac{15}{4}$ **87.** $\frac{19}{30}$

89. $-\frac{3}{10}$ **91.** -6 **93.** 1 **95.** -4 **97.** 3 **99.** $31\frac{1}{4}$ **101.** $\frac{5}{1}$ **103.** $\frac{3}{1}$ **105.** 60 **107.** $3\frac{19}{60}$

109. $20\frac{7}{24}$ **111.** $1\frac{4}{11}$ **113.** $12\frac{3}{4}$ **115.** $2\frac{29}{46}$ **117.** $6\frac{2}{5}$ **119.** $44\frac{1}{2}$ yd **121.** $7\frac{1}{2}$ sq. ft

123. $\frac{1}{40}$ oz more in the $15\frac{5}{8}$ oz can **125.** 203 calories **127.** $2\frac{4}{5}$ hr

Chapter 4 Test

1. $\frac{4}{3}$ **2.** $-\frac{4}{3}$ **3.** $\frac{8x}{9}$ **4.** $\frac{x-21}{7x}$ **5.** y^2 **6.** $\frac{16}{45}$ **7.** $\frac{9a+4}{10}$ **8.** $-\frac{2}{3y}$ **9.** $24y^2$ **10.** 9 **11.** $14\frac{1}{40}$

12. $\frac{1}{a^2}$ **13.** $\frac{64}{3}$ **14.** $22\frac{1}{2}$ **15.** $3\frac{3}{5}$ **16.** $\frac{1}{3}$ **17.** $\frac{3}{4}$ **18.** $\frac{3}{4x}$ **19.** $\frac{76}{21}$ **20.** -2 **21.** -4 **22.** 1

23. $\frac{5}{2}$ **24.** $\frac{4}{31}$ **25.** $3\frac{3}{4}$ ft **26.** **(a)** $\frac{5}{16}$; **(b)** 125,000 backpacks

Chapter 4 Cumulative Review

1. *(Sec. 1.1, Ex. 5)* **(a)** $<$; **(b)** $>$; **(c)** $>$ **2.** *(Sec. 1.4, Ex. 4)* 2300 **3.** *(Sec. 1.6, Ex. 1)* **(a)** 6; **(b)** 9; **(c)** 6

4. *(Sec. 1.8, Ex. 6)* **(a)** $x+7$; **(b)** $15-x$; **(c)** $2 \cdot x$; **(d)** $x \div 5$ or $\frac{x}{5}$; **(e)** $x-2$ **5.** *(Sec. 2.1, Ex. 6)* **(a)** 4; **(b)** -5

6. *(Sec. 2.2, Ex. 4)* -10 **7.** *(Sec. 2.4, Ex. 4)* -14 **8.** *(Sec. 2.5, Ex. 1)* **(a)** 9; **(b)** -9 **9.** *(Sec. 3.1, Ex. 3)* $6y+2$

10. *(Sec. 3.2, Ex. 4)* -9 **11.** *(Sec. 3.4, Ex. 3)* 5 **12.** *(Sec. 3.5, Ex. 4)* software, $420; computer system, $1680

13. *(Sec. 4.1, Ex. 1)* **(a)** 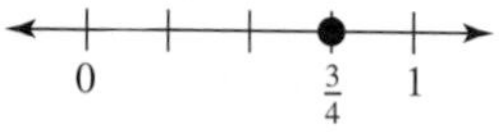

(b) 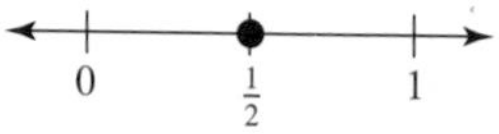

(c) 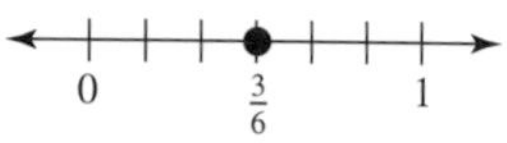

14. *(Sec. 4.1, Ex. 4)* $\frac{36x}{44}$ **15.** *(Sec. 4.2, Ex. 4)* $\frac{3}{5}$ **16.** *(Sec. 4.3, Ex. 1)* $\frac{10}{33}$ **17.** *(Sec. 4.3, Ex. 10)* $\frac{5}{12}$

18. *(Sec. 4.4, Ex. 2)* $-\frac{5}{8}$ **19.** *(Sec. 4.5, Ex. 4)* $\frac{6-x}{3}$ **20.** *(Sec. 4.7, Ex. 5)* 2 **21.** *(Sec. 4.8, Ex. 6)* $7\frac{17}{24}$

CHAPTER 5
Decimals

Mental Math, Sec. 5.1

1. tens **3.** tenths

Exercise Set 5.1

1. six and fifty-two hundredths **3.** sixteen and twenty-three hundredths **5.** three and two hundred five thousandths **7.** one hundred sixty-seven and nine thousandths **9.** 6.5 **11.** 10.6 **13.** 9.08 **15.** 60.65 **17.** 5.625 **19.** 0.0064 **21.** $\frac{3}{10}$ **23.** $\frac{27}{100}$ **25.** $5\frac{47}{100}$ **27.** $\frac{6}{125}$ **29.** $<$ **31.** $>$ **33.** $=$ **35.** $<$ **37.** $>$ **39.** 0.6 **41.** 0.23 **43.** 0.594 **45.** 98,210 **47.** 12.3 **49.** $7\frac{7}{100}$ **51.** $\frac{9}{50}$ **53.** $15\frac{401}{500}$ **55.** $\frac{601}{2000}$ **57.** $487\frac{8}{25}$ **59.** 9.67, 9.672, 9.68, 9.682 **61.** 17.67 **63.** 0.5 **65.** 0.130 **67.** 3830 **69.** 0.26499, 0.25786 **71.** \$29.62 **73.** \$0.969 **75.** 40,000 **77.** 2.42 hours **79.** answers will vary **81.** answers will vary **83.** answers will vary **85.** 68 **87.** 5766 **89.** 71 **91.** 243

Calculator Explorations, Sec. 5.2

1. 328.742 **3.** −4.03 **5.** 865.392

Exercise Set 5.2

1. 3.5 **3.** 6.83 **5.** 27.0578 **7.** 56.432 **9.** −8.57 **11.** 11.16 **13.** 6.5 **15.** 15.3 **17.** 598.23 **19.** 16.3 **21.** −6.32 **23.** −6.4 **25.** 9.982 **27.** 12.3 **29.** no **31.** yes **33.** $6.9x + 6.9$ **35.** $-10.97 + 3.47y$ **37.** 3.1 **39.** 450.738 **41.** −9.9 **43.** 25.67 **45.** −5.62 **47.** 776.89 **49.** 11,983.32 **51.** 465.56 **53.** −549.8 **55.** 861.6 **57.** 115.123 **59.** 876.6 **61.** 0.088 **63.** 465.902 **65.** 3.81 **67.** 3.39 **69.** 1.4 **71.** yes **73.** yes **75.** no **77.** \$1.69 **79.** 319.64 m **81.** \$0.06 **83.** 67.44 ft **85.** 29.614 mph **87.** 715.05 hr **89.** \$5.92—yes **91.** $22.181x - 22.984$ **93.** answers will vary **95.** 46 **97.** 3870 **99.** $\frac{4}{9}$ **101.** 8

Exercise Set 5.3

1. 0.12 **3.** 0.6 **5.** −17.595 **7.** 39.273 **9.** 65 **11.** 0.65 **13.** −7093 **15.** 0.0983 **17.** 162 **19.** 7.5 **21.** −38.1 **23.** no **25.** no **27.** yes **29.** 8π m ≈ 25.12 m **31.** 10π cm ≈ 31.4 cm **33.** 18.2π yd ≈ 57.148 yd **35.** 0.42 **37.** 22.26 **39.** 43.274 **41.** 8.23854 **43.** 2084.004 **45.** 14,790 **47.** −9.3762 **49.** 84.97593 **51.** 1.12746 **53.** −0.14444 **55.** 53.256 **57.** −0.6 **59.** 17.1 **61.** 114 **63.** 32 cm **65.** 31 in. **67.** 6π in. ≈ 18.84 in. **69.** \$9.97 **71.** \$7.47 **73.** 3,831,600 mi **75.** 64.96 in. **77.** 250π ft ≈ 785 ft **79.** **(a)** 62.8 meters, 125.6 m; **(b)** yes **81.** answers will vary **83.** $18\frac{1}{3}$ **85.** 27 **87.** 486 **89.** 54

Exercise Set 5.4

1. 0.094 **3.** −300 **5.** 5.8 **7.** −6.6 **9.** 200 **11.** 23.87 **13.** 110 **15.** 0.54982 **17.** −0.0129 **19.** 8.7 **21.** 5.65 **23.** 7.0625 **25.** 0.4 **27.** yes **29.** no **31.** yes **33.** 0.413 **35.** −7 **37.** 4.8 **39.** 2100 **41.** 30 **43.** 7000 **45.** 9.8 **47.** 9.6 **49.** 0.024 **51.** −0.69 **53.** 65 **55.** 0.0003 **57.** 120,000 **59.** 1200 kilobytes **61.** 23.7 gal **63.** 202.1 lb **65.** 5.1 m **67.** 8.6 ft **69.** 85 **71.** 70.3 **73.** answers will vary **75.** 345.22 **77.** 1001.0 **79.** $\frac{44}{21}$ **81.** $\frac{9}{5}$

Exercise Set 5.5

1. 12 **3.** 2.8 **5.** 2898.66 **7.** 4.2 **9.** 0.16 **11.** −3 **13.** 0.28 **15.** −9 **17.** 3 **19.** yes **21.** no **23.** 149.8 **25.** 22.89 **27.** 50.49 **29.** 2.3 **31.** 5.29 **33.** 7.6 **35.** 0.2025 **37.** −1.29 **39.** 5.76 **41.** 5.7 **43.** 3.6 **45.** 36 in. **47.** 39 ft **49.** 110 sq. cm **51.** 37.68 cm **53.** \$49.77 **55.** yes **57.** 65 ft **59.** 22,992 people **61.** answers will vary **63.** 14 **65.** $\frac{25}{49}$ **67.** $\frac{109}{105}$

Exercise Set 5.6

1. 0.2 **3.** 0.5 **5.** 0.75 **7.** 0.08 **9.** 0.375 **11.** 0.33 **13.** 0.44 **15.** 0.22 **17.** 0.38 **19.** $\frac{23}{100}$ **21.** $\frac{3}{50}$ **23.** $\frac{147}{500}$ **25.** $\frac{1}{100}$ **27.** $\frac{1}{5}$ **29.** $\frac{9021}{10{,}000}$ **31.** $2\frac{1}{2}$ **33.** $13\frac{9}{50}$ **35.** $6\frac{111}{500}$ **37.** $47\frac{1}{25}$ **39.** $<$ **41.** $>$ **43.** $<$ **45.** $<$ **47.** $<$ **49.** $>$ **51.** $<$ **53.** $<$ **55.** 0.32, 0.34, 0.35 **57.** 0.49, 0.491, 0.498 **59.** $0.73, \frac{3}{4}, 0.78$ **61.** $0.412, 0.453, \frac{4}{7}$ **63.** $5.23, \frac{42}{8}, 5.34$ **65.** $\frac{17}{8}, 2.37, \frac{12}{5}$ **67.** -15.4 **69.** 8 **71.** -3.7 **73.** 25.65 sq. in. **75.** 9.36 sq. cm **77.** 0.625 **79.** answers will vary **81.** 625 **83.** 324 **85.** $\frac{64}{125}$ **87.** $\frac{49}{4}$ **89.** $\frac{3}{2}$

Exercise Set 5.7

1. 5.9 **3.** 0.43 **5.** 0.45 **7.** 4.2 **9.** -4 **11.** 1.8 **13.** 10 **15.** 7.6 **17.** 60 **19.** -0.07 **21.** 0.0148 **23.** -8.13 **25.** -1 **27.** 7 **29.** 7 **31.** answers will vary **33.** 7.683 **35.** 4.683 **37.** $6x - 16$ **39.** $3x - 5$ **41.** $-2y + 6.8$

Calculator Explorations, Sec. 5.8

1. 32 **3.** 3.873 **5.** 9.849

Exercise Set 5.8

1. 2 **3.** 25 **5.** $\frac{1}{9}$ **7.** $\frac{12}{8} = \frac{3}{2}$ **9.** 72 **11.** -21 **13.** 11 **15.** 8 **17.** 1.732 **19.** 3.873 **21.** 3.742 **23.** 6.856 **25.** 5 **27.** 8 **29.** 17.205 **31.** 13 in. **33.** 6.633 cm **35.** 16 **37.** 63 **39.** -22 **41.** 28 **43.** -6 **45.** $\frac{3}{2}$ **47.** 2.828 **49.** 5.099 **51.** 8.426 **53.** 2.646 **55.** 16.125 **57.** 12 **59.** 44.822 **61.** 42.426 **63.** 1.732 **65.** 141.42 yd **67.** 25.0 ft **69.** 340 ft **71.** 16 knots; answers will vary **73.** answers will vary **75.** 1015 **77.** 7 **79.** 5 **81.** -13

Exercise Set 5.9

1. 7 sq. m **3.** $9\frac{3}{4}$ sq. yd **5.** 15 sq. yd **7.** 2.25π sq. in. ≈ 7.065 sq. in. **9.** 36.75 sq. ft. **11.** 28 sq. m **13.** 22 sq. yd **15.** $36\frac{3}{4}$ sq. ft **17.** $22\frac{1}{2}$ sq. in. **19.** 25 sq. cm **21.** 86 sq. mi **23.** 24 sq. cm **25.** 72 cu. in. **27.** 512 cu. cm **29.** $12\frac{4}{7}$ cu. yd **31.** $56\frac{4}{7}$ cu. in. **33.** 75 cu. cm **35.** 168 sq. ft **37.** 31.25 sq. yd **39.** 128 sq. in.; $\frac{8}{9}$ sq. ft **41.** 510 sq. in. **43.** 168 sq. ft **45.** 4320 sq. ft **47.** 240 sq. ft **49.** 33.49 cu. cm **51.** $10\frac{5}{6}$ cu. in. **53.** 3080 cu. ft **55.** $2095\frac{5}{21}$ cu. in. **57.** 98π cu. ft **59.** $7\frac{7}{8}$ cu. ft **61.** 74,747,700 cu. ft **63.** 87,150 sq. mi **65.** 14π in. ≈ 44 in. **67.** 25 ft **69.** $12\frac{3}{4}$ ft

Chapter 5 Review

1. tenths **3.** twenty-three and forty-five hundredths **5.** one hundred nine and twenty-three hundredths **7.** 2.15 **9.** 16,025.0014 **11.** $12\frac{23}{1000}$ **13.** $\frac{231}{100{,}000}$ **15.** $>$ **17.** $>$ **19.** 0.6 **21.** 42.90 **23.** 13,500 people **25.** 9.5 **27.** -7.28 **29.** 320.312 **31.** 1.7 **33.** -1324.5 **35.** 65.02 **37.** \$0.44 **39.** $4.2x + 12.8$ **41.** 72

43. -78.246 **45.** 14π m ≈ 43.96 m **47.** 70 **49.** -4900 **51.** 8.059 **53.** 7.3 m **55.** -6.5 **57.** 50
59. 76.72 **61.** 20.91 **63.** 54.1 **65.** 0.3526 **67.** 30.4 **69.** 87.9 sq. cm **71.** $23\frac{893}{1000}$ **73.** 0.923
75. 0.217 **77.** 51.057 **79.** $<$ **81.** $<$ **83.** $0.42, \frac{3}{7}, 0.43$ **85.** $\frac{3}{4}, \frac{6}{7}, \frac{8}{9}$ **87.** 47.19 **89.** 80 **91.** 34
93. 0.28 **95.** 1 **97.** 12 **99.** 17 **101.** $\frac{1}{10}$ **103.** 29 **105.** 93 **107.** 28.28 cm **109.** 240 sq. ft
111. 600 sq. cm **113.** 49π sq. ft ≈ 153.86 sq. ft **115.** 119 sq. in. **117.** 144 sq. m **119.** 130 sq. ft **121.** 84 cu. ft
123. $346\frac{1}{2}$ cu. in. **125.** 307.72 cu. in. **127.** 0.5π cu. ft **129.** 28.728 cu. ft

Chapter 5 Test

1. forty-five and ninety-two thousandths **2.** 3000.059 **3.** 17.595 **4.** -51.20 **5.** -20.42 **6.** 40.902
7. 0.037 **8.** 34.9 **9.** 0.862 **10.** $<$ **11.** $<$ **12.** $\frac{69}{200}$ **13.** $24\frac{73}{100}$ **14.** 0.5 **15.** 0.941 **16.** 1.93
17. -6.2 **18.** $0.5x - 13.4$ **19.** 7 **20.** 12.530 **21.** -12 **22.** 5.66 cm **23.** 198.08 oz **24.** -3 **25.** 3.7
26. 2.31 sq. mi **27.** 185 sq. in. **28.** $62\frac{6}{7}$ cu. in. **29.** 30 cu. ft **30.** 18 cu. ft

Chapter 5 Cumulative Review

1. *(Sec. 1.7, Ex. 4)* 28 **2.** *(Sec. 2.1, Ex. 5)* **(a)** -11; **(b)** 2 **3.** *(Sec. 2.4, Ex. 1)* **(a)** -21; **(b)** 10; **(c)** 0
4. *(Sec. 2.5, Ex. 7)* 3 **5.** *(Sec. 3.1, Ex. 7)* $-15a + 6$ **6.** *(Sec. 3.3, Ex. 2)* -4 **7.** *(Sec. 3.4, Ex. 1)* 12
8. *(Sec. 3.5, Ex. 1)* **(a)** $x + 9 = 5$; **(b)** $2x = -10$; **(c)** $x - 6 = 168$; **(d)** $3(x + 5) = -30$; **(e)** $8 \div 2x = 2$ or $\frac{8}{2x} = 2$
9. *(Sec. 4.2, Ex. 7)* $\frac{10}{27}$ **10.** *(Sec. 4.3, Ex. 4)* $-\frac{1}{8}$ **11.** *(Sec. 4.3, Ex. 8)* **(a)** $\frac{16}{625}$; **(b)** $\frac{1}{16}$
12. *(Sec. 4.3, Ex. 13)* **(a)** $-\frac{7}{24}$; **(b)** $-\frac{21}{8}$ **13.** *(Sec. 4.4, Ex. 4)* $\frac{2}{7}$ **14.** *(Sec. 4.5, Ex. 1)* $\frac{2}{3}$
15. *(Sec. 4.6, Ex. 3)* $\frac{2y - 20}{3}$ **16.** *(Sec. 5.1, Ex. 6)* **(a)** $>$; **(b)** $<$; **(c)** $=$ **17.** *(Sec. 5.2, Ex. 4)* -1.16
18. *(Sec. 5.3, Ex. 1)* 18.408 **19.** *(Sec. 5.5, Ex. 5)* -3.7 **20.** *(Sec. 5.7, Ex. 1)* 9.5
21. *(Sec. 5.8, Ex. 1)* **(a)** 7; **(b)** 6; **(c)** 1; **(d)** 9 **22.** *(Sec. 5.9, Ex. 2)* 5.1 sq. mi

CHAPTER 6

Graphing and Introduction to Statistics

Exercise Set 6.1

1. 1995 **3.** 4000 **5.** 1991, 1992, 1996 **7.** 1990, 1993 **9.** 10 oz **11.** 1985, 1990, 1993
13. 2 ounces per week **15.** Consumption of chicken is increasing. **17.** April **19.** 12
21. March, April, May, June **23.** Tokyo, 26.5 million or 26,500,000 **25.** New York City, 16.2 million or 16,200,000
27. 12 million or 12,000,000 **29.**

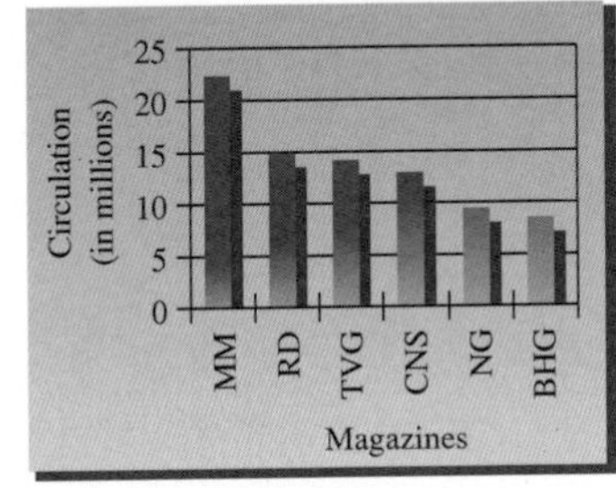

31. 54.5 **33.** 1975 **35.** Increase **37.** 83°F **39.** Sunday, 68°F **41.** Tuesday, 13°F **43.** yes **45.** no **47.** no

Exercise Set 6.2

1.

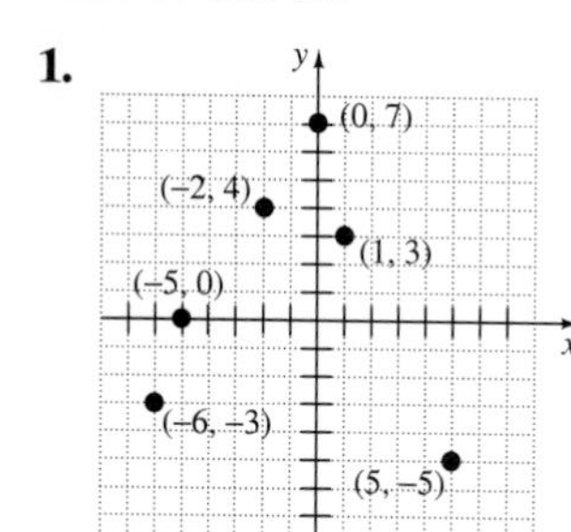

3.

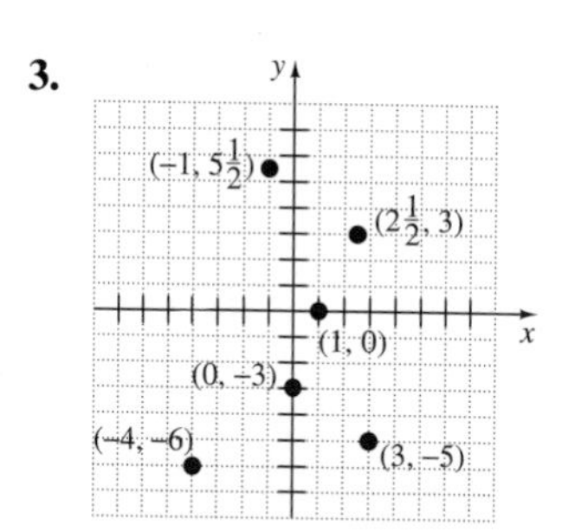

5. $A(0,0); B\left(3\frac{1}{2},0\right); C(3,2); D(-1,3); E(-2,-2); F(0,-1); G(2,-1)$ **7.** yes

9. no **11.** yes

13.

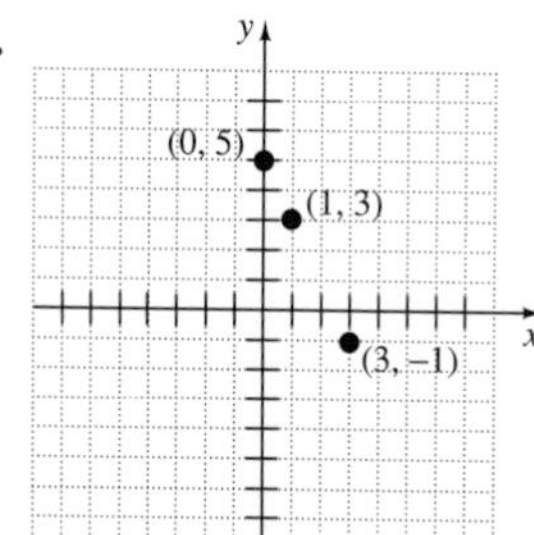

15.

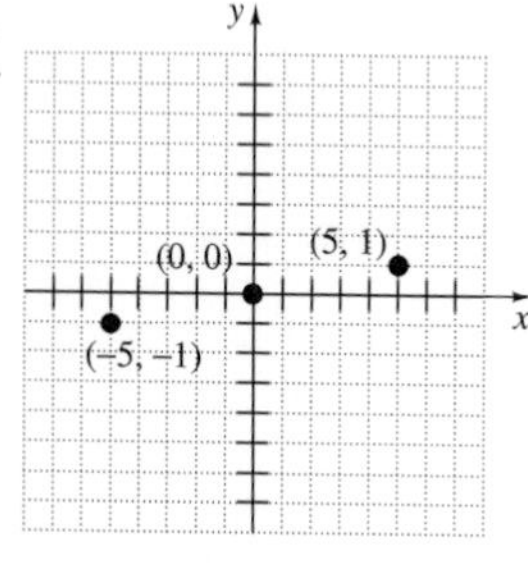

17.

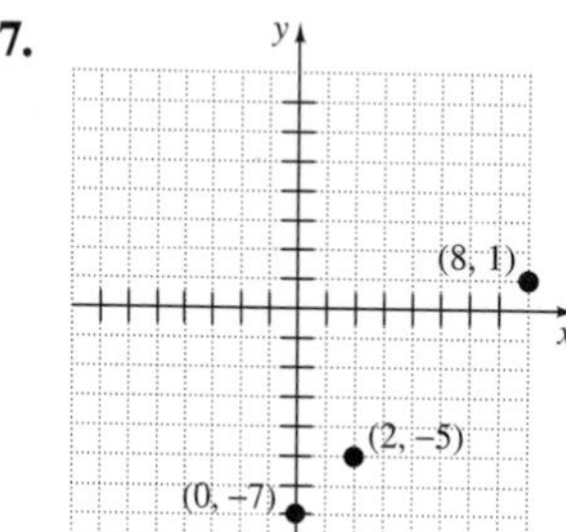

19. $(1,8), (0,0), (-2,-16)$ **21.** $(2,12), (22,-8), (0,14)$ **23.** $(1,6), (2,7), (3,8)$ **25.** yes **27.** yes **29.** no **31.** $(1,-2), (2,1), (3,4)$ **33.** $(0,0), (2,-2), (-2,2)$ **35.** $(-2,0), (1,-3), (-7,5)$ **37.** true **39.** true **41.** false **43.** false **45.** 1.7 **47.** 21.84 **49.** −23.6

Exercise Set 6.3

1.

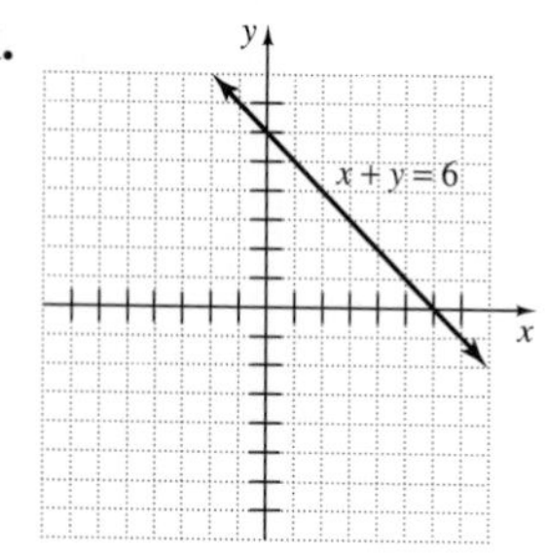

3.

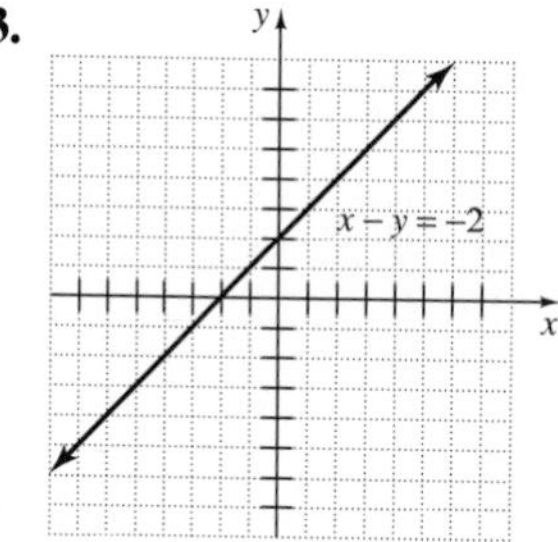

5.

$y = 4x$

7.

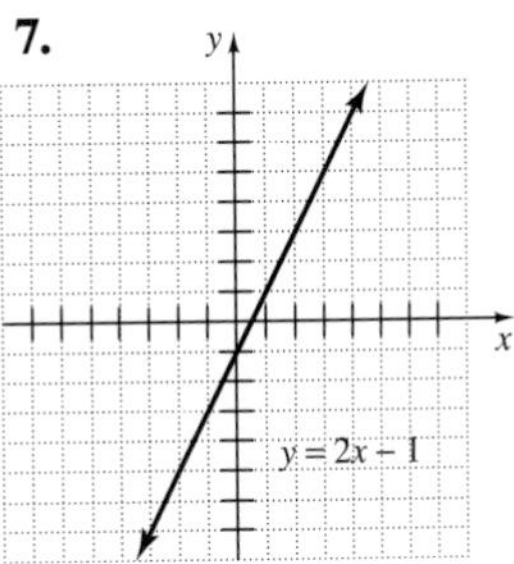

9.

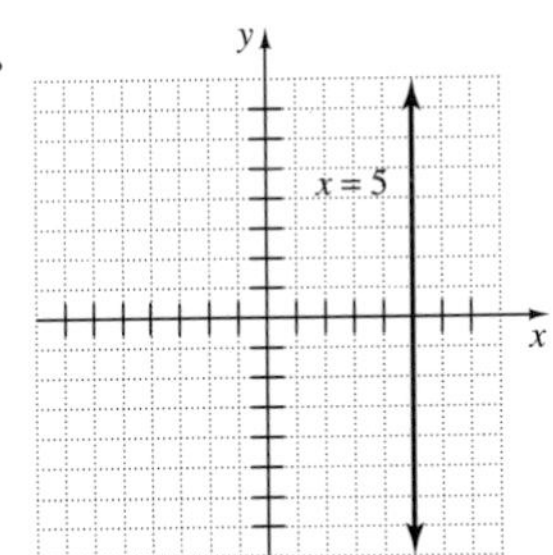

11.

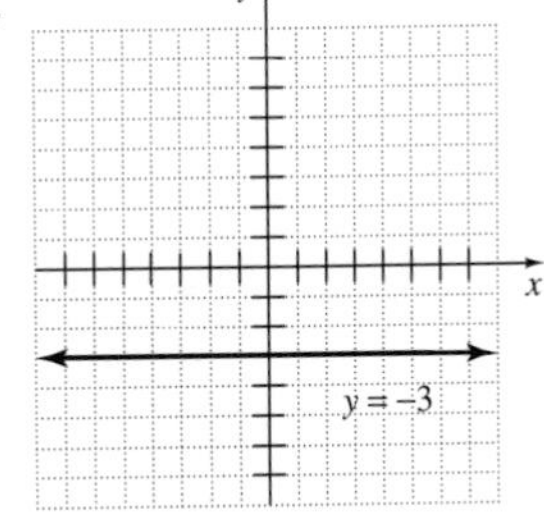

13.

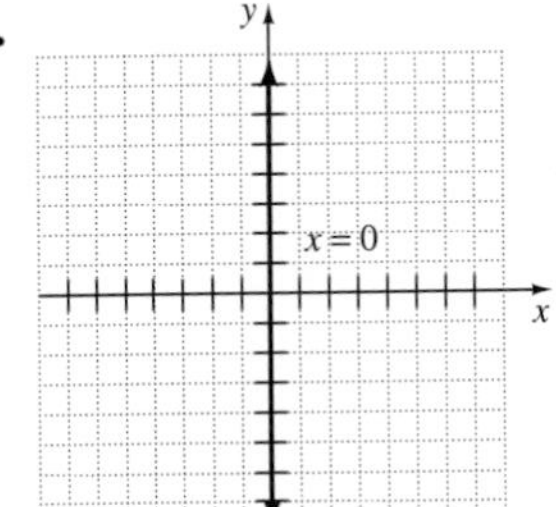

15.

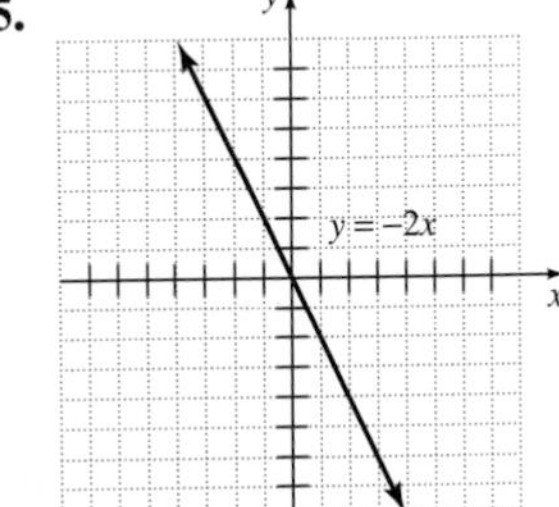

17.

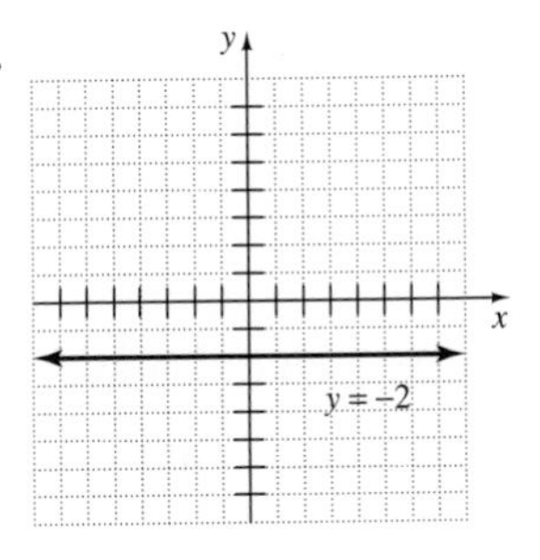

19.

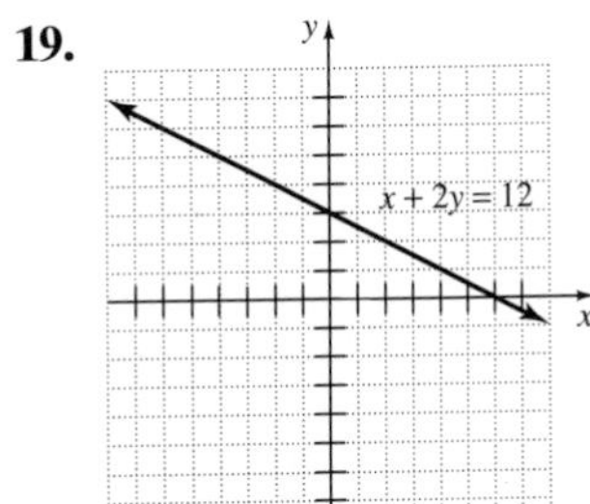

21.

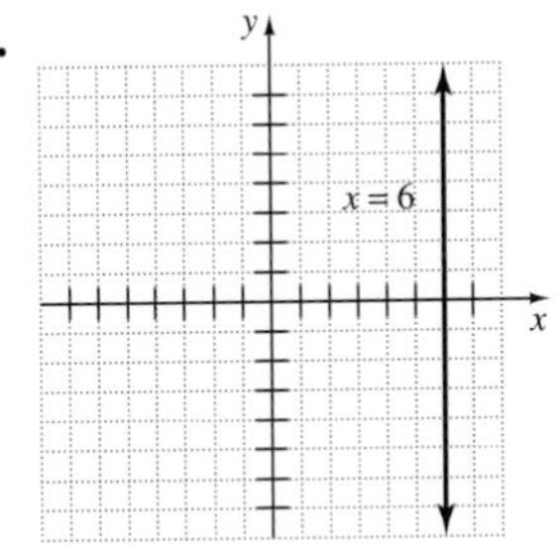

23.

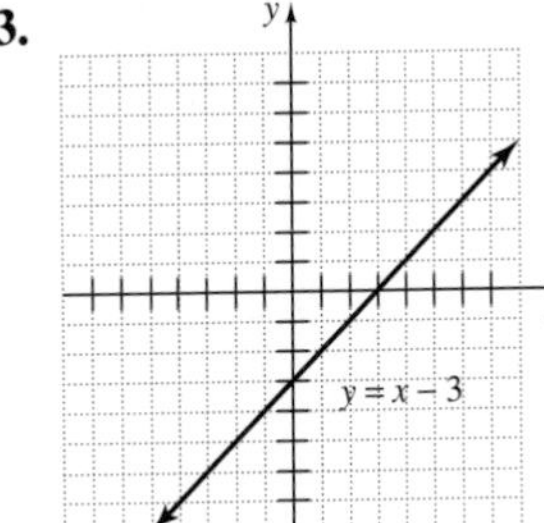

25.

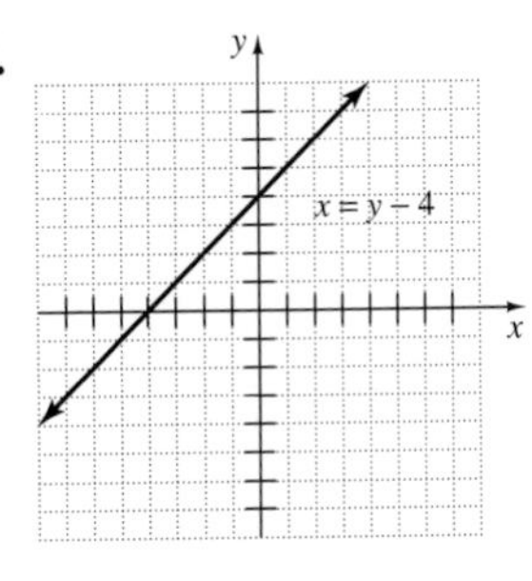

27.

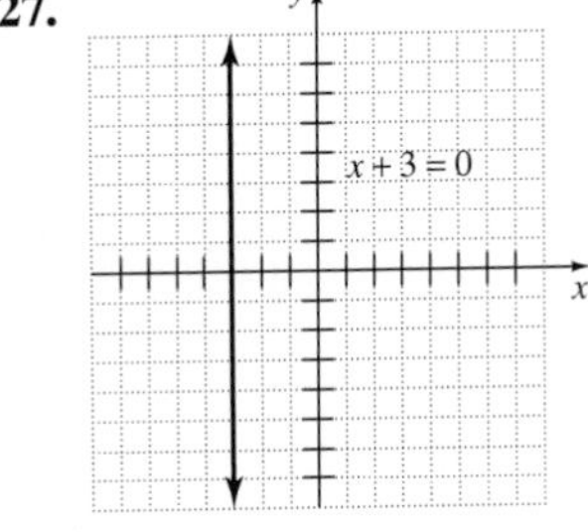

29.

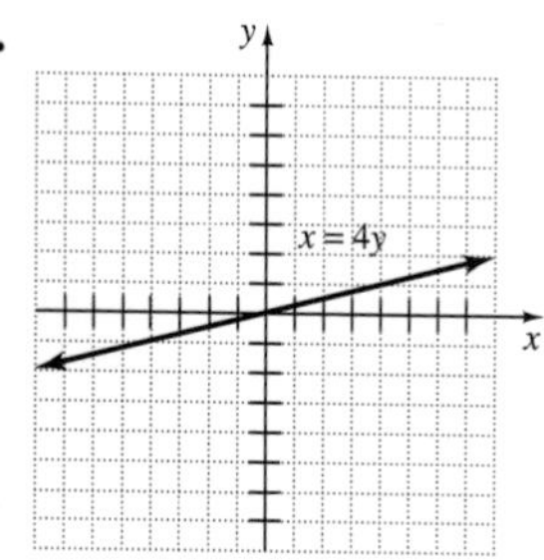

31.

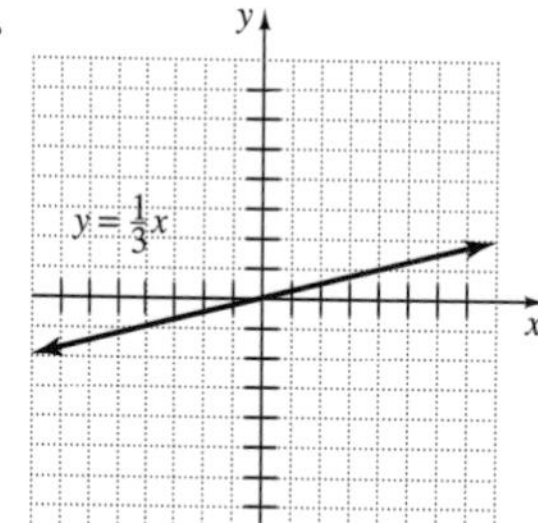

33.

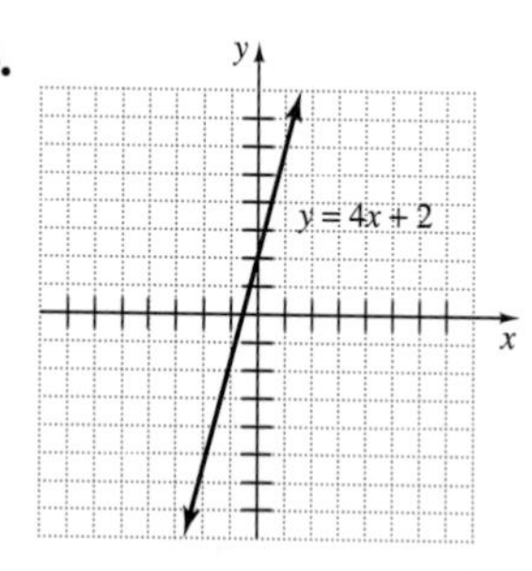

35.

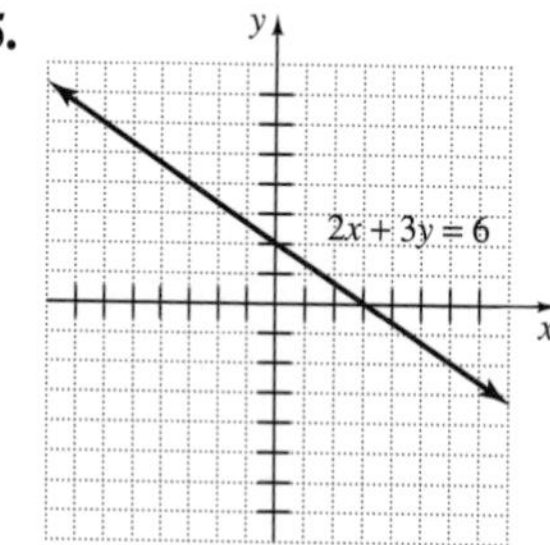

37.

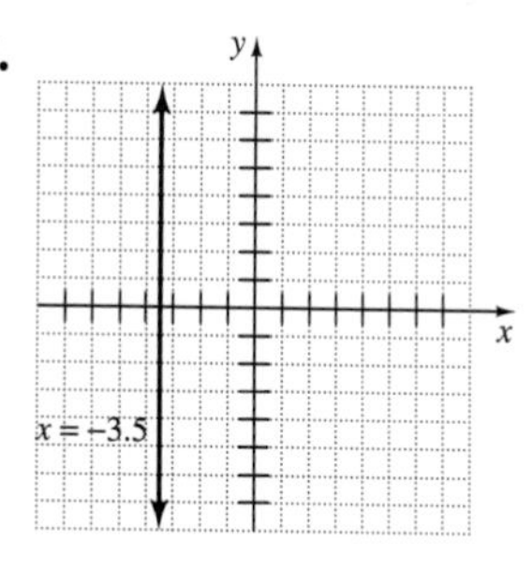

39.

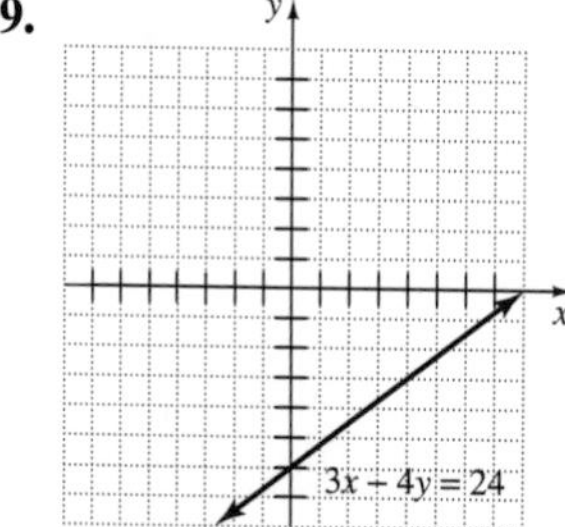

41. -29 **43.** 150 **45.** -1

Exercise Set 6.4

1. mean: 29; median: 28; no mode **3.** mean: 8.1; median: 8.2; mode: 8.2 **5.** mean: 0.6; median: 0.6; mode: 0.2 and 0.6 **7.** mean: 370.9; median: 313.5; no mode **9.** 1314 ft **11.** 1131.5 ft **13.** 2.79 **15.** 3.46 **17.** 6.8 **19.** 6.9 **21.** 85.5 **23.** 73 **25.** 70 and 71 **27.** 9 **29.** 21, 21, 20 **31.** $\frac{3}{5}$ **33.** $-\frac{1}{9}$ **35.** $\frac{6}{7}$ **37.** $\frac{x}{y}$

Chapter 6 Review

1. 12,000 **3.** Alabama **5.** Utah and Alabama **7.** 8% **9.** 1980, 1990, 1994 **11.** 13 **13.** 1979, 1984 **15.** answers will vary

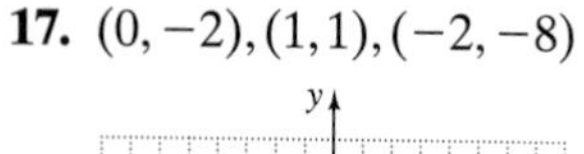

17. $(0, -2), (1, 1), (-2, -8)$

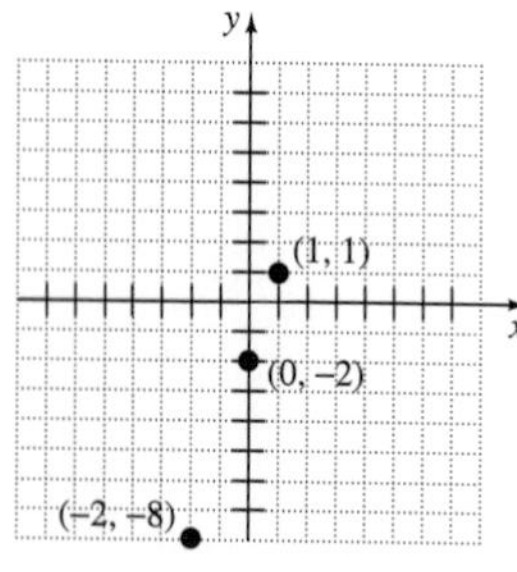

19. $(4, 1), (0, -3), (6, 3)$

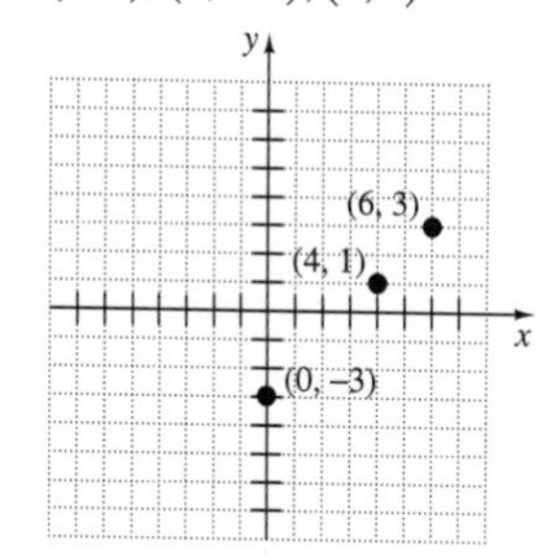

21. $(1, -5), (6, 0), (-1, -7)$

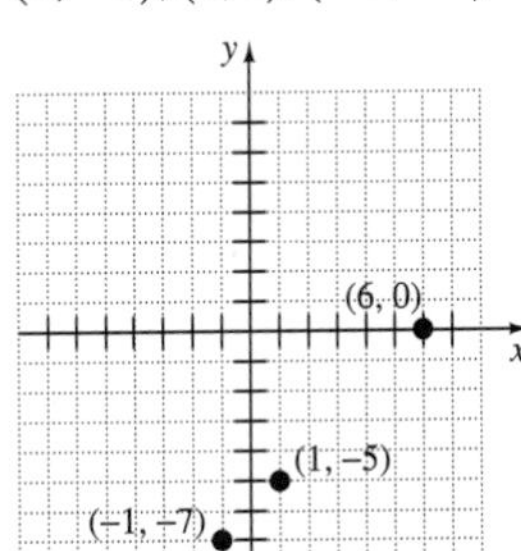

23.

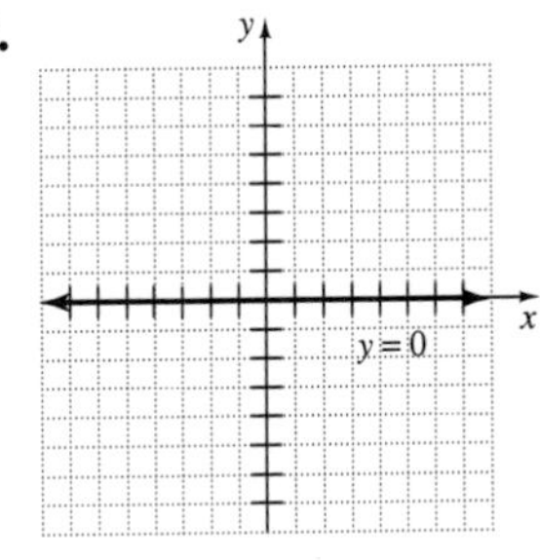

25.

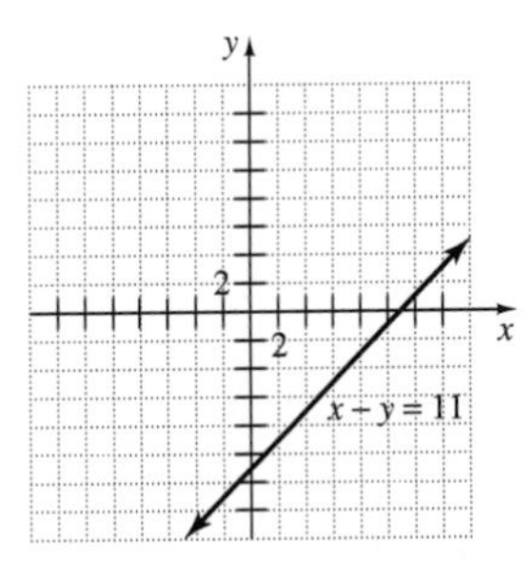

27.

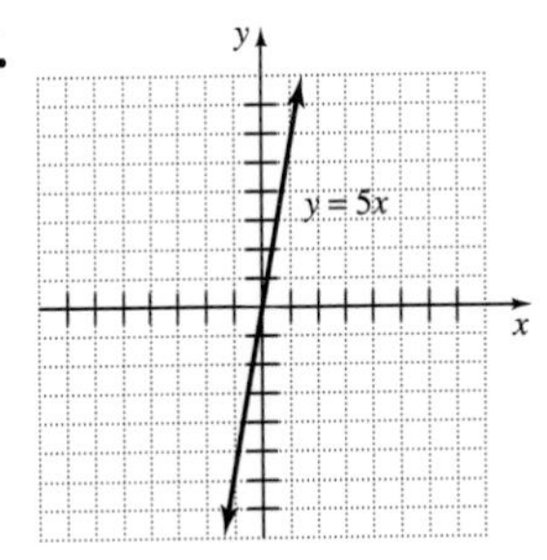

29.

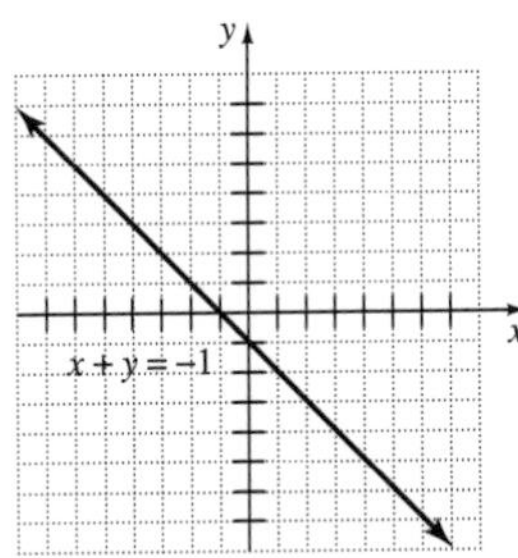

31.

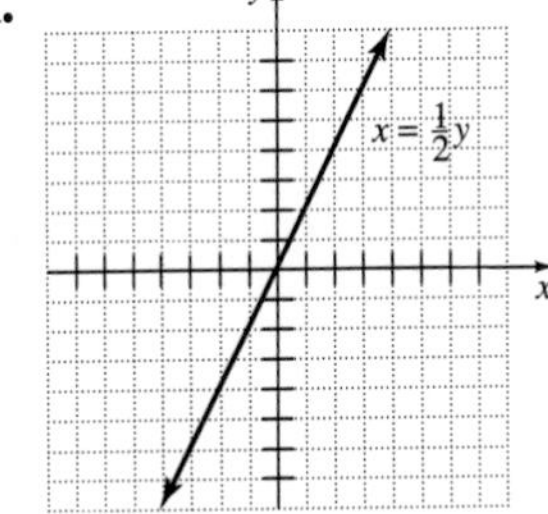

33. average: 55.2; median: 60; no mode

35. average: 447.3; median: 420; mode: 400 **37.** 2.57

Chapter 6 Test

1. \$90 **2.** Third week, \$140 **3.** \$440 **4.** June, August, September **5.** February; 3 cm

6. March and November **7.** $(4, 0)$ **8.** $(0, -3)$ **9.** $(-3, 4)$ **10.** $(-2, -1)$

11. $(0, 0), (-6, 1), (12, -2)$

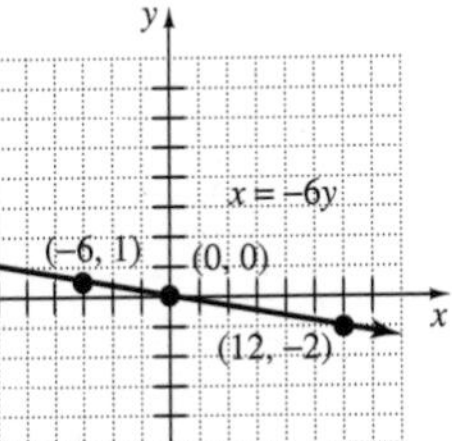

12. $(2, 10), (-1, -11), (0, -4)$

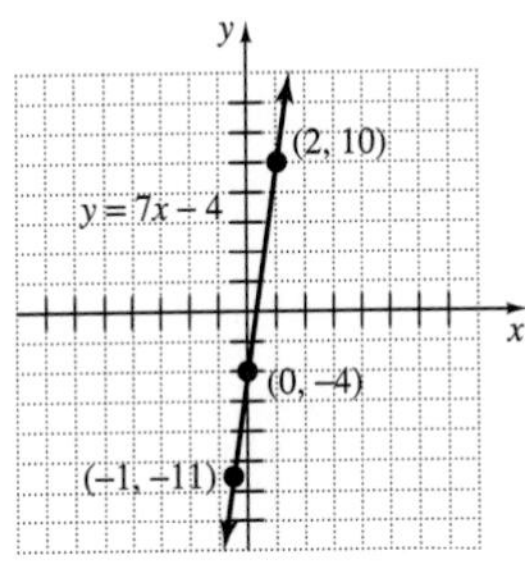

13.

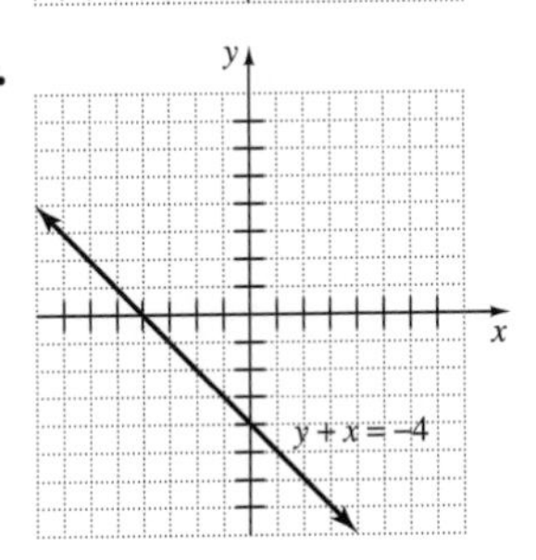

14.

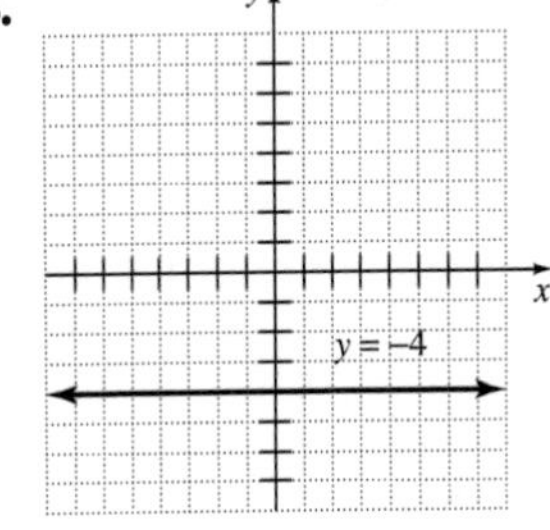

15.

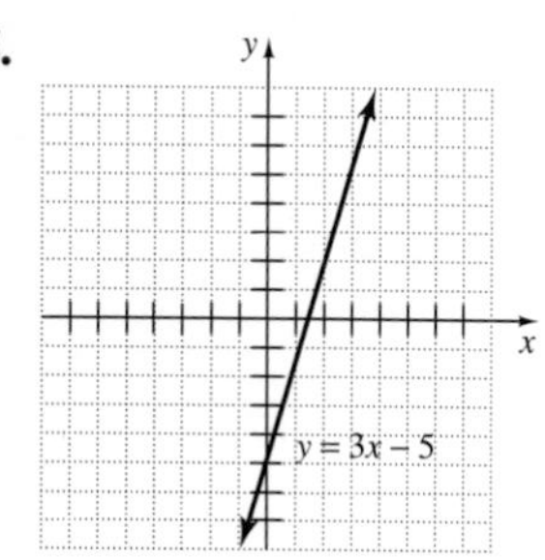

16.

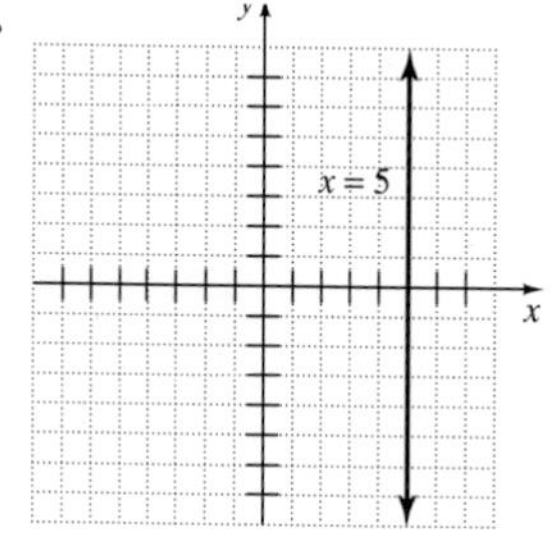

17.

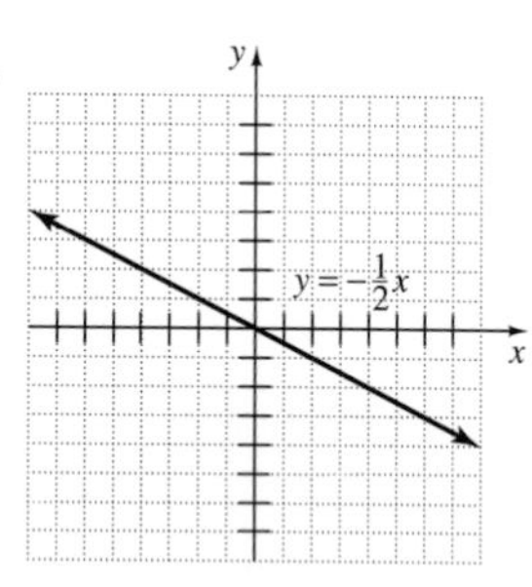

18.

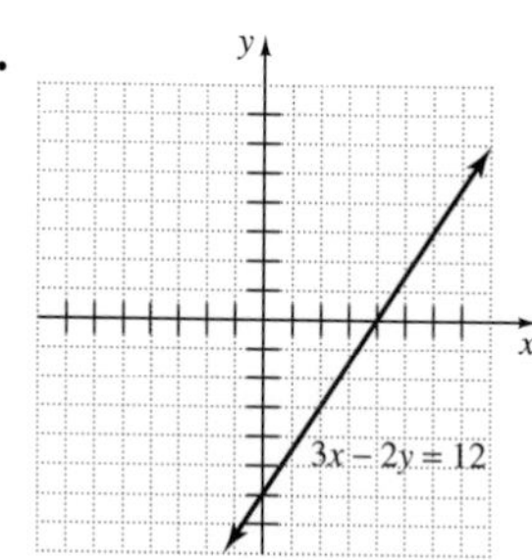

19. average: 38.4; median: 42; no mode **20.** average: 12.625; median: 12.5; mode: 12 and 16 **21.** 3.07

Chapter 6 Cumulative Review

1. *(Sec. 2.2, Ex. 3)* **(a)** 3; **(b)** -4 **2.** *(Sec. 2.5, Ex. 3)* -2 **3.** *(Sec. 3.3, Ex. 1)* -3 **4.** *(Sec. 3.6, Ex. 3)* 20 ft

5. *(Sec. 4.1, Ex. 3)* $\frac{6}{15}$ **6.** *(Sec. 4.3, Ex. 2)* $\frac{4}{9}$ **7.** *(Sec. 4.3, Ex. 11)* $\frac{2}{9x}$ **8.** *(Sec. 4.4, Ex. 6)* $-\frac{1}{2}$

9. *(Sec. 4.5, Ex. 5)* $\frac{1}{28}$ **10.** *(Sec. 4.6, Ex. 5)* $\frac{77}{72}$ **11.** *(Sec. 4.8, Ex. 1)* $\frac{11}{4}$ **12.** *(Sec. 5.2, Ex. 8)* -7.5

13. *(Sec. 5.3, Ex. 4)* -2.08 **14.** *(Sec. 5.5, Ex. 8)* -2.6 **15.** *(Sec. 5.6, Ex. 4)* $\frac{2}{25}$ **16.** *(Sec. 5.6, Ex. 8)* 8.4 sq. ft

17. *(Sec. 5.8, Ex. 5)* 25

18. *(Sec. 6.2, Ex. 1)*

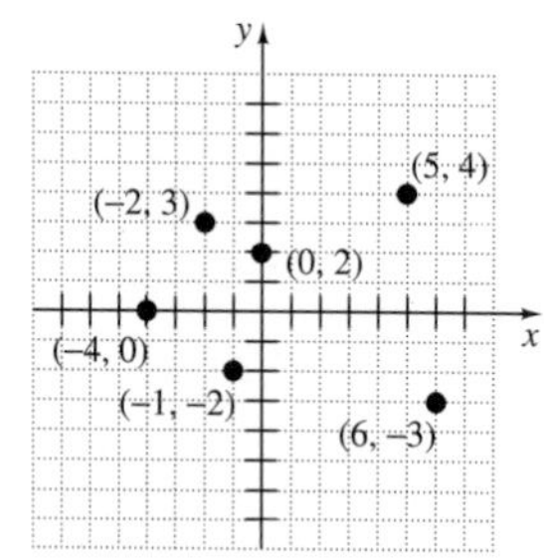

19. *(Sec. 6.2, Ex. 5)* **(a)** $(3, 6)$; **(b)** $(0, 0)$; **(c)** $(-2, -4)$

20. *(Sec. 6.3, Ex. 3)*

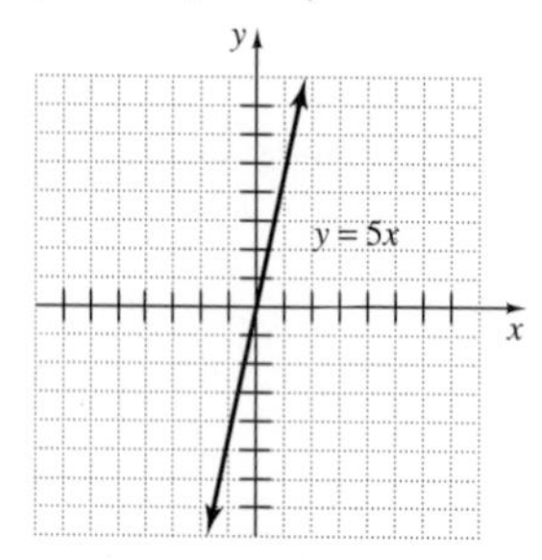

CHAPTER 7

Ratio, Proportion, and Measurement

Exercise Set 7.1

1. $\frac{2}{3}$ **3.** $\frac{11}{20}$ **5.** $\frac{3}{7}$ **7.** $\frac{1}{10}$ **9.** $\frac{9}{4}$ **11.** **(a)** $\frac{3}{14}$; **(b)** $\frac{14}{3}$; **(c)** $\frac{1}{12}$ **13.** $\frac{5}{3}$ **15.** $\frac{17}{40}$ **17.** $\frac{\text{1 shrub}}{\text{3 ft}}$

19. $\frac{\text{3 returns}}{\text{20 sales}}$ **21.** $\frac{\text{2 phone lines}}{\text{9 employees}}$ **23.** $\frac{\text{75 riders}}{\text{1 subway car}}$ **25.** $\frac{\text{110 calories}}{\text{1 ounce}}$ **27.** 6 diapers/baby **29.** $50,000/year

31. $\frac{12}{7}$ **33.** $\frac{16}{23}$ **35.** $\frac{2}{5}$ **37.** $\frac{1}{12}$ **39.** $\frac{9}{2}$ **41.** $\frac{2}{1}$ **43.** $\frac{14}{33}$ **45.** $\frac{\text{2 in.}}{\text{9 hr}}$ **47.** $\frac{\text{3 attendants}}{\text{100 passengers}}$

49. $6\frac{2}{3}$ km/min **51.** 1,225,000 voters/senator **53.** 300 good/defective **55.** $11.50/disk **57.** $\frac{\text{5 women}}{\text{4 men}}$

59. $\frac{\text{3 single}}{\text{1 married}}$ **61.** 12-oz size **63.** 16-oz size **65.** **(a)** $31\frac{1}{4}$ boards/hr; **(b)** $33\frac{1}{3}$ boards/hr; **(c)** Lamont

67. 2/1 **69.** 3/17 **71.** $\frac{1}{17}$ **73.** yes **75.** 17.5 bricks/minute **77.** answers will vary **79.** answers will vary

81. 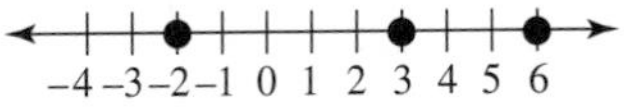 −4 −3 −2 −1 0 1 2 3 4 5 6 **83.** 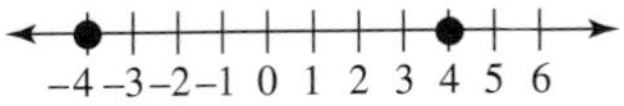 −4 −3 −2 −1 0 1 2 3 4 5 6

85. 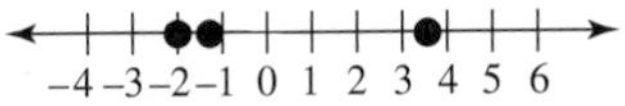 −4 −3 −2 −1 0 1 2 3 4 5 6 **87.** 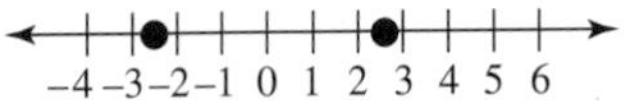 −4 −3 −2 −1 0 1 2 3 4 5 6

Exercise Set 7.2

1. $\frac{\text{10 diamonds}}{\text{6 opals}} = \frac{\text{5 diamonds}}{\text{3 opals}}$ **3.** $\frac{\text{3 printers}}{\text{12 computers}} = \frac{\text{1 printer}}{\text{4 computers}}$ **5.** $\frac{\text{1 raisin}}{\text{5 cornflakes}} = \frac{\text{8 raisins}}{\text{40 cornflakes}}$

7. $\frac{\text{6 eagles}}{\text{58 sparrows}} = \frac{\text{3 eagles}}{\text{29 sparrows}}$ **9.** $\frac{2\frac{1}{4}\text{ cups flour}}{\text{24 cookies}} = \frac{6\frac{3}{4}\text{ cups flour}}{\text{72 cookies}}$ **11.** true **13.** false **15.** true **17.** true

19. false **21.** true **23.** false **25.** true **27.** true **29.** false **31.** 3 **33.** 5 **35.** 4 **37.** 3.2

39. 19.2 **41.** $\frac{3}{4}$ **43.** 3 **45.** 12 **47.** 0.0025 **49.** 25 **51.** 14.9 **53.** 0.07 **55.** 3.3 **57.** 3.163

59. 0 **61.** 1400 **63.** 252.5 **65.** answers will vary **67.** $<$ **69.** $>$ **71.** $<$ **73.** $>$

Exercise Set 7.3

1. 12 passes **3.** 165 min **5.** 180 students **7.** 23 ft **9.** 270 sq. ft **11.** 56 mi **13.** 450 km **15.** 24 oz

17. 16 bags **19.** $162,000 **21.** 15 hits **23.** 27 people **25.** 86 weeks **27.** 2:1 **29.** 3:2 **31.** 6 **33.** 5

35. 13.5 **37.** 17.5 **39.** 8 **41.** 21.25 **43.** 50 ft **45.** $x = 4.4$ ft, $y = 5.6$ ft **47.** 14.4 ft **49.** $4\frac{2}{3}$ ft

51. 8 ft **53.** answers will vary **55.** 23 **57.** 2 **59.** 0 **61.** $5\frac{1}{2}$

Calculator Explorations, Sec. 7.4

1. ≈ 22.96 ft **3.** ≈ 21.59 cm **5.** ≈ 3.1 mi

Mental Math, Sec. 7.4

1. 1 ft **3.** 2 ft **5.** 1 yd **7.** no **9.** no **11.** no

Exercise Set 7.4

1. 5 ft **3.** 36 ft **5.** 8 mi **7.** $8\frac{1}{2}$ ft **9.** $3\frac{1}{3}$ yd **11.** 33,792 ft **13.** 13 yd 1 ft **15.** 3 ft 5 in.

17. 1 mi 4720 ft **19.** 62 in. **21.** 17 ft **23.** 84 in. **25.** 12 ft 3 in. **27.** 22 yd 1 ft **29.** 8 ft 5 in. **31.** 5 ft 6 in. **33.** 13 ft 4 in. **35.** 50 yd 2 ft **37.** 4000 cm **39.** 4.0 cm **41.** 0.3 km **43.** 1.4 m **45.** 15 m **47.** 83 mm **49.** 0.201 dm **51.** 40 mm **53.** 8.94 m **55.** 2.94 m **57.** 1.29 cm **59.** 12.640 km **61.** 54.9 m **63.** 1.55 km **65.** 10 ft 6 in. **67.** 224 yd 2 ft **69.** 10.75 sq. m **71.** 9.12 m **73.** 15 ft 9 in. **75.** 3 ft 1 in. **77.** $105\frac{1}{3}$ yd **79.** 26.7 mm **81.** 41.25 m **83.** 86 ft 6 in. **85.** 3.35 m **87.** 6.009 km **89.** 15 tiles **91.** answers will vary **93.** 9 **95.** 12 **97.** 7 **99.** $\frac{3}{4}$

Calculator Explorations, Sec. 7.5

1. ≈ 425.25 g **3.** ≈ 15.4 lb **5.** ≈ 0.175 oz

Mental Math, Sec. 7.5

1. 1 lb **3.** 2000 lb **5.** 16 oz **7.** 1 ton **9.** no **11.** yes **13.** no

Exercise Set 7.5

1. 32 oz **3.** 10,000 lb **5.** 6 tons **7.** $3\frac{3}{4}$ lb **9.** $1\frac{3}{4}$ tons **11.** 260 oz **13.** 9800 lb **15.** 76 oz **17.** 1.5 tons **19.** 53 lb 10 oz **21.** 9 tons 390 lb **23.** 3 tons 175 lb **25.** 8 lb 11 oz **27.** 3 lb 2 oz **29.** 1 ton 700 lb **31.** 0.5 kg **33.** 4000 mg **35.** 25,000 g **37.** 0.048 g **39.** 0.0063 kg **41.** 15,140 mg **43.** 4010 g **45.** 13.5 mg **47.** 5.815 g **49.** 1850 mg **51.** 1360 g **53.** 13.52 kg **55.** 2.125 kg **57.** 8.064 kg **59.** 30 mg **61.** 5 lb 8 oz **63.** 35 lb 14 oz **65.** 850 g **67.** 2.38 kg **69.** 5 lb 1 oz **71.** 6.12 kg **73.** 130 lb **75.** 211 lb **77.** 250 mg **79.** 144 mg **81.** answers will vary **83.** 0.25 **85.** 0.16 **87.** 0.875 **89.** 0.833

Calculator Explorations, Sec. 7.6

1. ≈ 4.73 L **3.** ≈ 4.62 gal **5.** ≈ 3.785 L

Mental Math, Sec 7.6

1. 1 pt **3.** 1 gal **5.** 1 qt **7.** 1 c **9.** 2 c **11.** 4 qt **13.** no **15.** no

Exercise Set 7.6

1. 4 c **3.** 16 pt **5.** $2\frac{1}{2}$ gal **7.** 5 pt **9.** 8 c **11.** $3\frac{3}{4}$ qt **13.** 768 fl oz **15.** 9 c **17.** 22 pt **19.** 10 gal 1 qt **21.** 4 c 4 fl oz **23.** 1 gal 1 qt **25.** 2 gal 3 qt 1 pt **27.** 2 qt 1 c **29.** 17 gal **31.** 4 gal 3 qt **33.** 5000 ml **35.** 4.5 L **37.** 0.41 kl **39.** 0.064 L **41.** 160 L **43.** 3600 ml **45.** 0.00016 kl **47.** 22.5 L **49.** 4.5 L **51.** 8410 ml **53.** 10,600 ml **55.** 3840 ml **57.** 162.4 L **59.** 48 fl oz **61.** 2 qt **63.** 1.59 L **65.** yes **67.** 2 qt 1 c **69.** 18.954 L **71.** $0.316 **73.** 9 qt **75.** 474 ml **77.** no **79.** answers will vary **81.** answers will vary **83.** $\frac{7}{10}$ **85.** $\frac{3}{100}$ **87.** $\frac{3}{500}$ **89.** $\frac{7}{20}$

Mental Math, Sec. 7.7

1. yes **3.** no **5.** no **7.** yes

Exercise Set 7.7

1. 5°C **3.** 40°C **5.** 140°F **7.** 239°F **9.** 16.7°C **11.** 61.2°C **13.** 197.6°F **15.** 61.3°F **17.** 56.7°C **19.** 80.6°F **21.** 21.1°C **23.** 37.9°C **25.** 244.4°F **27.** 260°C **29.** 462.2°C **31.** 12 in. **33.** 12 cm **35.** 8 ft 4 in.

Chapter 7 Review

1. $\frac{5}{4}$ **3.** $\frac{9}{40}$ **5.** $\frac{4}{15}$ **7.** $\frac{1}{2}$ **9.** $\frac{\text{1 stillborn birth}}{\text{125 live births}}$ **11.** $\frac{\text{5 pages}}{\text{2 min}}$ **13.** 52 mph **15.** \$0.31 per pear

17. 65 km/hr **19.** \$36.80 per course **21.** 8-oz size **23.** 1-gallon size **25.** $\frac{\text{20 men}}{\text{14 women}} = \frac{\text{10 men}}{\text{7 women}}$

27. $\frac{\text{16 sandwiches}}{\text{8 players}} = \frac{\text{2 sandwiches}}{\text{1 player}}$ **29.** no **31.** no **33.** 5 **35.** 32.5 **37.** 32 **39.** 60 **41.** 0.63

43. 14 **45.** 8 bags **47.** no **49.** \$54,600 **51.** $40\frac{1}{2}$ ft **53.** 37.5 **55.** 17.4 **57.** 33 ft **59.** 9 ft

61. 13,200 ft **63.** 17 yd 1 ft **65.** 4200 cm **67.** 0.01218 m **69.** 21 yd 1 ft **71.** 41 ft 3 in. **73.** 9.5 cm

75. 9117 m **77.** 169 yd 2 ft **79.** 156.3 km **81.** 4.125 lb **83.** 3 lb 4 oz **85.** 1.4 g **87.** 21 dag

89. 3 lb 9 oz **91.** 2 tons 750 lb **93.** 4.9 g **95.** 8.1 g **97.** 4 lb 4 oz **99.** 7.85 kg **101.** 8 qt **103.** 27 qt

105. 4 qt 1 pt **107.** 3800 ml **109.** 1.4 kl **111.** 1 gal 1 qt **113.** 736 ml **115.** 2 gal 3 qt **117.** 10.88 L

119. 473°F **121.** 107.6°F **123.** 34°C **125.** 5.2°C **127.** 1.7°C

Chapter 7 Test

1. $\frac{9}{13}$ **2.** $\frac{5}{6}$ **3.** 81.25 km/hr **4.** 0.3 in./day **5.** 8-oz size **6.** 16-oz size **7.** 5 **8.** $\frac{7}{3}$ **9.** 8

10. 23 ft 4 in. **11.** 10 qt **12.** 1.875 lb **13.** 5600 lb **14.** $4\frac{3}{4}$ gal **15.** 0.04 g **16.** 2400 g **17.** 36 mm

18. 0.43 g **19.** 830 ml **20.** 1 gal 2 qt **21.** 3 lb 13 oz **22.** 8 ft 3 in. **23.** 2 gal 3 qt **24.** 66 mm

25. 2.256 km **26.** 28.9°C **27.** 54.7°F **28.** 7.5 **29.** 69 ft **30.** 5.6 m **31.** 4 gal 3 qt **32.** 105.8°F

33. 91.4 m **34.** 16 ft 6 in.

Chapter 7 Cumulative Review

1. *(Sec. 1.7, Ex. 6)* 2 **2.** *(Sec. 2.3, Ex. 3)* 3 **3.** *(Sec. 3.1, Ex. 6)* $6x + 24$ **4.** *(Sec. 3.3, Ex. 6)* -1 **5.** *(Sec. 4.3, Ex. 5)* $\frac{5}{2}$

6. *(Sec. 4.5, Ex. 2)* $-\frac{8}{33}$ **7.** *(Sec. 4.7, Ex. 6)* -45 **8.** *(Sec. 5.2, Ex. 7)* 3.98 **9.** *(Sec. 5.3, Ex. 8)* 1.012

10. *(Sec. 5.5, Ex. 6)* 1.69 **11.** *(Sec. 5.7, Ex. 2)* -3.35 **12.** *(Sec. 5.8, Ex. 3)* $\frac{2}{5}$ **13.** *(Sec. 5.9, Ex. 3)* 72 sq. ft

14. *(Sec. 6.3, Ex. 2)*

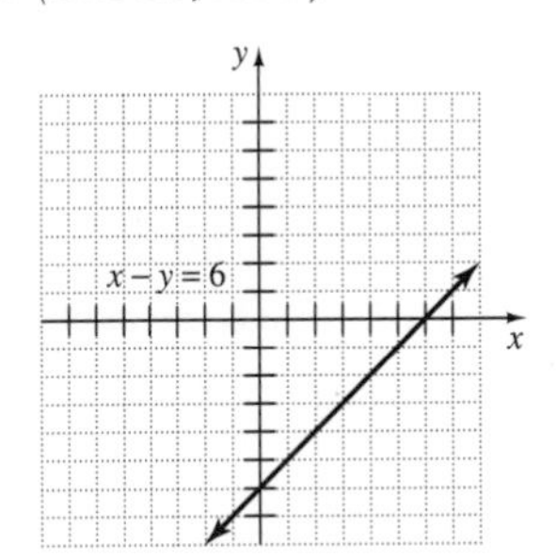

15. *(Sec. 6.2, Ex. 3)* yes **16.** *(Sec. 6.4, Ex. 4)* 80.5 **17.** *(Sec. 7.2, Ex. 5)* 1.6 **18.** *(Sec. 7.4, Ex. 4)* 43 in.

19. *(Sec. 7.4, Ex. 10)* 450 m **20.** *(Sec. 7.5, Ex. 11)* 82 kg **21.** *(Sec. 7.7, Ex. 5)* 37°C

CHAPTER 8

Percent

Exercise Set 8.1

1. 0.48 **3.** 0.06 **5.** 1.00 **7.** 0.613 **9.** 98% **11.** 310% **13.** 2900% **15.** 0.3% **17.** $\frac{3}{25}$
19. $\frac{1}{25}$ **21.** $\frac{9}{200}$ **23.** $\frac{5}{4}$ **25.** 25% **27.** 18.75% **29.** 140% **31.** 27.27% **33.** 94.62% **35.** 266.67%
37. 0.35 **39.** 0.17 **41.** 1.57 **43.** 6.35 **45.** $\frac{73}{100}$ **47.** $\frac{1}{8}$ **49.** $\frac{1}{16}$ **51.** $\frac{2}{25}$ **53.** 22% **55.** 310%
57. 2% **59.** 175% **61.** 71.43% **63.** 0.2% **65.** 53.33% **67.** 30 students **69.** 90% **71.** 10%
73. 70% **75.** 0.0625 **77.** $\frac{21}{40}$ **79.** 20% **81.** $\frac{3}{20}$ **83.** 138% **85.** 85%

87.

PERCENT	DECIMAL	FRACTION
35%	0.35	$\frac{7}{20}$
20%	0.2	$\frac{1}{5}$
50%	0.5	$\frac{1}{2}$
70%	0.7	$\frac{7}{10}$
37.5%	0.375	$\frac{3}{8}$

89. 0.266; 26.6% **91.** 1.155; 115.5% **93.** greater **95.** $\frac{1}{2}$ **97.** $\frac{5}{6}$ **99.** 4 **101.** 2.48

Exercise Set 8.2

1. $15\% \cdot 75 = x$ **3.** $30\% \cdot x = 80$ **5.** $x \cdot 90 = 20$ **7.** $19 = 40\% \cdot x$ **9.** $x = 9\% \cdot 43$ **11.** 3.5 **13.** 7.28
15. 600 **17.** 10 **19.** 7.14% **21.** 5.16% **23.** 1 **25.** 210 **27.** 45 **29.** 400% **31.** $22.50
33. $52.50 **35.** $63,272.73 **37.** 6.18% **39.** 46,080 **41.** $30,625 **43.** $1.13 **45.** 87.64% **47.** 1035
49. $81.68 **51.** 29.2% **53.** 26.3% **55.** snack cracker in Exercise 52 **57.** 9 **59.** $\frac{4}{3}$ **61.** $\frac{15}{8}$ **63.** 45

Calculator Explorations, Sec. 8.3

1. 1.56051 **3.** 8.06231 **5.** $634.49

Exercise Set 8.3

1. $32 **3.** $73.60 **5.** $750 **7.** $33.75 **9.** $700 **11.** $58.65 **13.** $15,625 **15.** $562.50 **17.** $12,580
19. $46,815.40 **21.** $2327.15 **23.** $58,163.60 **25.** $240.75 **27.** $10,881.50 **29.** $2991.07 **31.** $938.66
33. $971.90 **35.** $260.31 **37.** $637.26 **39.** answers will vary **41.** Account 2; answers will vary **43.** 5.09
45. 0.4488 **47.** 2.4983 **49.** 222

Mental Math, Sec. 8.4

1. $1000 **3.** $400.00

Exercise Set 8.4

1. \$7.50 **3.** \$858.93 **5.** \$5.40 **7.** \$703.64 **9.** \$2180 **11.** \$49,474.24 **13.** \$1888.50 **15.** \$268.38 **17.** \$1917 **19.** \$12,672.00 **21.** \$96.30 **23.** \$3941.67 **25.** \$430 **27.** \$856.98 **29.** 5 **31.** -10 **33.** -16 **35.** 9

Exercise Set 8.5

1. \$6696 **3.** 0.4% **5.** \$63,400.00 (rounded to hundreds) **7.** 21.8% **9.** \$417.87 **11.** 0.20 billion **13.** \$155,750 **15.** 23.1% **17.** 137.5% **19.** 18.3% **21.** 8.84 million **23.** \$300 **25.** \$1350 **27.** \$1290

Chapter 8 Review

1. 0.83 **3.** 0.735 **5.** 1.25 **7.** 0.006 **9.** 0.33 **11.** 260% **13.** 35% **15.** 72.5% **17.** 132% **19.** 75% **21.** 0.32 **23.** 20% **25.** $83\frac{1}{3}\%$ **27.** $166\frac{2}{3}\%$ **29.** 60% **31.** $233\frac{1}{3}\%$ **33.** $\frac{1}{100}$ **35.** $\frac{1}{4}$ **37.** $\frac{51}{500}$ **39.** $\frac{1}{6}$ **41.** $1\frac{1}{2}$ **43.** $\frac{1}{4}$ **45.** 10% **47.** 100,000 **49.** 72.4% **51.** 3000 **53.** 418 **55.** 35.1 **57.** 66% **59.** 16.2% **61.** 1.5 yr **63.** \$6.00 **65.** \$11,506.56 **67.** \$55.58 **69.** \$1.15 **71.** \$263.75 **73.** \$300.43 **75.** \$320 **77.** \$29,400 **79.** \$1791.67 **81.** \$48 **83.** 26.4% **85.** \$13.23

Chapter 8 Test

1. 5.6% **2.** 0.007 **3.** $\frac{6}{5}$ **4.** 55% **5.** 33.6 **6.** 1250 **7.** 75% **8.** $27.5 = 15\% \cdot x$ **9.** $\frac{1}{8}$ **10.** 38.4 lb **11.** \$56,750 **12.** \$14,544 **13.** \$358.43 **14.** \$27 **15.** \$647.50 **16.** \$14,598.78 **17.** \$9880 **18.** \$2.92 **19.** \$2005.63 **20.** \$120.93 **21.** \$395.01 **22.** \$18,506.28 **23.** \$588.18

Chapter 8 Cumulative Review

1. *(Sec. 1.2, Ex. 2)* 184,046 **2.** *(Sec. 1.6, Ex. 6)* 365 R2 **3.** *(Sec. 2.2, Ex. 7)* -12 **4.** *(Sec. 3.1, Ex. 1)* **(a)** 2; **(b)** -3; **(c)** 1; **d** -100 **5.** *(Sec. 3.3, Ex. 4)* -3 **6.** *(Sec. 4.2, Ex. 1)* $3 \cdot 3 \cdot 5$ **7.** *(Sec. 4.3, Ex. 6)* $\frac{6}{5}$ **8.** *(Sec. 4.5, Ex. 6)* $\frac{1}{6}$ **9.** *(Sec. 4.7, Ex. 2)* $-\frac{1}{6}$ **10.** *(Sec. 4.8, Ex. 4)* $\frac{35}{12}$ or $2\frac{11}{12}$ **11.** *(Sec. 5.2, Ex. 1)* 25.454 **12.** *(Sec. 5.4, Ex. 5)* 50 **13.** *(Sec. 5.8, Ex. 2)* $\frac{1}{6}$ **14.** *(Sec. 7.1, Ex. 8)* $\frac{5}{4}$ **15.** *(Sec. 7.3, Ex. 3)* 7 bags **16.** *(Sec. 6.3, Ex. 5)*

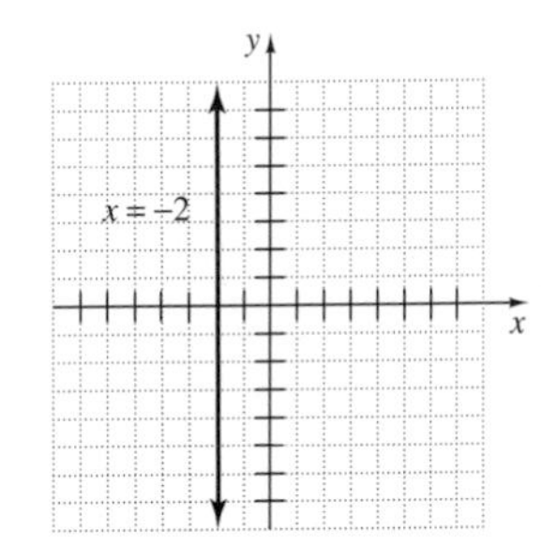

17. *(Sec. 8.1, Ex. 4)* **(a)** $\frac{2}{5}$; **(b)** $\frac{19}{1000}$; **(c)** $\frac{5}{4}$; **(d)** $\frac{1}{3}$; **(e)** 1 **18.** *(Sec. 8.2, Ex. 4)* 38 **19.** *(Sec. 8.4, Ex. 1)* sales tax, \$6.41; total \$91.91 **20.** *(Sec. 8.5, Ex. 1)* increase, \$2.50; new price, \$27.50

CHAPTER 9

Polynomials

Exercise Set 9.1

1. $-5x - 24$ **3.** $-9z^2 - 2z + 6$ **5.** $9y^2 + 25y - 40$ **7.** $-4.3a^4 - 2a^2 + 9$ **9.** $4a - 8$ **11.** $-2x^2 + 4x + 1$ **13.** $-20y^3 + 12y^2 - 4$ **15.** $-x + 16$ **17.** $12y^2 - 10y - 8$ **19.** $4x^2 + x - 16$ **21.** $-15y + 3.6$ **23.** $b^3 - b^2 + 7b - 1$ **25.** $\frac{9}{7}$ **27.** 1 **29.** -5 **31.** -8 **33.** 20 **35.** 25 **37.** 50 **39.** $(8x + 2)$ in. **41.** $(4x - 15)$ units **43.** 576 ft **45.** \$3200 **47.** 20, 6, 2 **49.** 7.2752 **51.** 81 **53.** 25 **55.** x^3 **57.** 2^2a^4

Exercise Set 9.2

1. x^{14} **3.** a^7 **5.** $15z^5$ **7.** $-40x^2$ **9.** $25x^6y^9$ **11.** $28a^5b^6$ **13.** $30x^3$ **15.** $12a^{17}$ **17.** x^{15} **19.** z^{20} **21.** b^{62} **23.** $81a^4$ **25.** $a^{33}b^{24}$ **27.** $121x^6y^{12}$ **29.** $-24y^{22}$ **31.** $256x^9y^{13}$ **33.** $16x^{12}$ sq. in. **35.** $12a^4b^5$ sq. m **37.** $18{,}003{,}384a^{45}b^{30}$ **39.** $34{,}867.84401x^{50}$ **41.** $7x - 21$ **43.** $-6a - 4b$ **45.** $9x + 18y - 27$

Exercise Set 9.3

1. $27x^3 - 9x$ **3.** $-20a^3 + 30a^2 - 5a$ **5.** $42x^4 - 35x^3 + 49x^2$ **7.** $x^2 + 13x + 30$ **9.** $2x^2 + 2x - 24$ **11.** $36a^2 + 48a + 16$ **13.** $a^3 - 33a + 18$ **15.** $8x^3 + 2x^2 - 55x + 50$ **17.** $x^5 + 4x^4 + 6x^3 + 7x^2 + 2x$ **19.** $-30r + 20r$ **21.** $-6y^3 - 2y^4 + 12y^2$ **23.** $x^2 + 14x + 24$ **25.** $4a^2 - 9$ **27.** $x^2 + 10x + 25$ **29.** $b^2 + \frac{7}{5}b + \frac{12}{25}$ **31.** $6x^3 + 25x^2 + 10x + 1$ **33.** $49x^2 + 70x + 25$ **35.** $4x^2 - 4x + 1$ **37.** $8x^5 - 8x^3 - 6x^2 - 6x + 9$ **39.** $x^5 + 2x^4 + 3x^3 + 2x^2 + x$ **41.** $10z^4 - 3z^3 + 3z - 2$ **43.** $(y^3 - 3y^2 - 16y - 12)$ sq. ft **45.** $(x^4 - 3x^2 + 1)$ sq. m **47.** $2 \cdot 5 \cdot 5$ **49.** $2 \cdot 2 \cdot 2 \cdot 3 \cdot 3$ **51.** $2 \cdot 2 \cdot 2 \cdot 5 \cdot 5$

Exercise Set 9.4

1. 3 **3.** 12 **5.** 4 **7.** 4 **9.** y^2 **11.** a^5 **13.** xy^2 **15.** x **17.** $2z^3$ **19.** $3y(y + 6)$ **21.** $5a^6(2 - a^2)$ **23.** $4x(x^2 + 3x + 5)$ **25.** $z^5(z^2 - 6)$ **27.** $-7(5 - 2y + y^2)$ or $7(-5 + 2y - y^2)$ **29.** $12a^5(1 - 3a)$ **31.** $x^2 + 2x$ **33.** 36 **35.** $\frac{4}{5}$ **37.** 37.5%

Chapter 9 Review

1. $10b - 3$ **3.** $-x + 2.8$ **5.** $4z^2 + 11z - 6$ **7.** $-11y^2 - y + \frac{3}{4}$ **9.** 45 **11.** $(26x - 28)$ ft **13.** y^7 **15.** $-15x^3y^5$ **17.** x^{28} **19.** $a^{20}b^{10}c^5$ **21.** $108x^{30}y^{25}$ **23.** $10a^3 - 12a$ **25.** $x^2 + 8x + 12$ **27.** $y^2 - 10y + 25$ **29.** $x^3 - x^2 + x + 3$ **31.** $3z^4 + 5z^3 + 6z^2 + 3z + 1$ **33.** 5 **35.** 6 **37.** x^2 **39.** xy **41.** $5a$ **43.** $2x(x + 6)$ **45.** $y^4(6 - y^2)$ **47.** $a^3(5a^4 - a + 1)$

Chapter 9 Test

1. $15x - 4$ **2.** $7x - 2$ **3.** $3.4y^2 + 2y - 3$ **4.** $-2a^2 + a + 1$ **5.** 17 **6.** y^{14} **7.** y^{33} **8.** $16x^8$ **9.** $-12a^{10}$ **10.** p^{54} **11.** $72a^{20}b^5$ **12.** $10x^3 + 6.5x$ **13.** $-2y^4 - 12y^3 + 8y$ **14.** $x^2 - x - 6$ **15.** $25x^2 + 20x + 4$ **16.** $a^3 + 8$ **17.** perimeter: $(14x - 4)$ in.; area: $(5x^2 + 33x - 14)$ sq. in. **18.** 15 **19.** $3y^3$ **20.** $3y(y + 6)$ **21.** $a^3(a^2 - 5)$ **22.** $5(x^2 + 2x - 3)$ **23.** When $t = 1$ second, height is 184 feet; when $t = 3$ seconds, height is 56 feet.

Chapter 9 Cumulative Review

1. *(Sec. 1.3, Ex. 2)* 7321 **2.** *(Sec. 1.8, Ex. 4)* 26 **3.** *(Sec. 2.5, Ex. 5)* -8 **4.** *(Sec. 3.1, Ex. 8)* $2x - 9$

5. *(Sec. 3.4, Ex. 5)* 2 **6.** *(Sec. 4.2, Ex. 8)* $\frac{36}{13x}$ **7.** *(Sec. 4.5, Ex. 3)* $\frac{1}{3}$ **8.** *(Sec. 4.6, Ex. 6)* $-\frac{8}{9}$

9. *(Sec. 5.3, Ex. 3)* 0.8496 **10.** *(Sec. 5.4, Ex. 2)* −0.052 **11.** *(Sec. 5.5, Ex. 4)* 200 mi

12. *(Sec. 5.8, Ex. 9)* approximately 20 m **13.** *(Sec. 6.1, Ex. 2)* **(a)** 55; **(b)** amphibians **14.** *(Sec. 6.3, Ex. 2)*

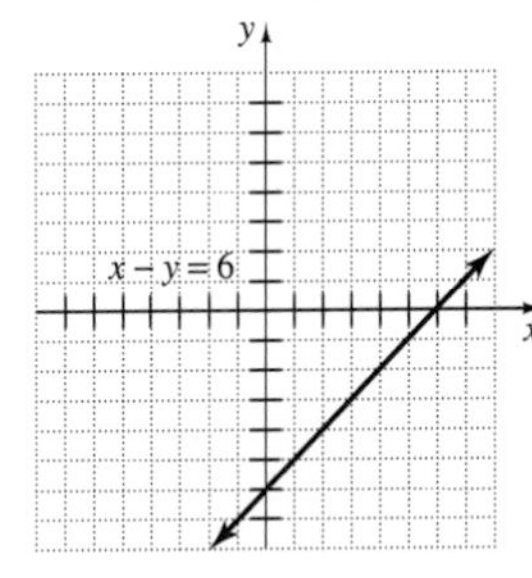

15. *(Sec. 6.4, Ex. 6)* median: 83.5; mode: 89 **16.** *(Sec. 7.2, Ex. 3)* 5 **17.** *(Sec. 7.3, Ex. 1)* $17\frac{1}{2}$ **18.** *(Sec. 8.2, Ex. 6)* 17

19. *(Sec. 8.5, Ex. 3)* decrease: \$16.25; new price: \$48.75 **20.** *(Sec. 9.1, Ex. 2)* $16y^2 + 4y + 2$ **21.** *(Sec. 9.2, Ex. 7)* $125t^3$

22. *(Sec. 9.3, Ex. 5)* $4x^2 + 4x + 1$ **23.** *(Sec. 9.4, Ex. 4)* $7x^2(x + 2)$

Appendix A One Hundred Addition Facts

1. 5 **2.** 11 **3.** 5 **4.** 15 **5.** 12 **6.** 7 **7.** 8 **8.** 6 **9.** 14 **10.** 10 **11.** 12 **12.** 5 **13.** 10
14. 2 **15.** 9 **16.** 12 **17.** 11 **18.** 8 **19.** 18 **20.** 7 **21.** 10 **22.** 0 **23.** 10 **24.** 10 **25.** 17
26. 8 **27.** 13 **28.** 3 **29.** 12 **30.** 17 **31.** 16 **32.** 8 **33.** 7 **34.** 13 **35.** 4 **36.** 3 **37.** 8
38. 7 **39.** 8 **40.** 9 **41.** 6 **42.** 9 **43.** 16 **44.** 8 **45.** 9 **46.** 15 **47.** 12 **48.** 1 **49.** 7
50. 6 **51.** 2 **52.** 6 **53.** 11 **54.** 11 **55.** 8 **56.** 10 **57.** 5 **58.** 10 **59.** 14 **60.** 6 **61.** 13
62. 4 **63.** 7 **64.** 9 **65.** 7 **66.** 13 **67.** 13 **68.** 7 **69.** 5 **70.** 9 **71.** 4 **72.** 3 **73.** 16
74. 11 **75.** 14 **76.** 13 **77.** 1 **78.** 11 **79.** 4 **80.** 4 **81.** 10 **82.** 14 **83.** 8 **84.** 15 **85.** 5
86. 15 **87.** 10 **88.** 2 **89.** 12 **90.** 6 **91.** 12 **92.** 6 **93.** 9 **94.** 11 **95.** 9 **96.** 10 **97.** 14
98. 12 **99.** 9 **100.** 3

Appendix B One Hundred Multiplication Facts

1. 1 **2.** 35 **3.** 56 **4.** 9 **5.** 32 **6.** 45 **7.** 28 **8.** 7 **9.** 4 **10.** 0 **11.** 63 **12.** 64 **13.** 6
14. 0 **15.** 30 **16.** 10 **17.** 24 **18.** 0 **19.** 18 **20.** 72 **21.** 40 **22.** 14 **23.** 32 **24.** 2 **25.** 54
26. 3 **27.** 56 **28.** 16 **29.** 54 **30.** 25 **31.** 2 **32.** 0 **33.** 36 **34.** 24 **35.** 12 **36.** 20 **37.** 36
38. 18 **39.** 12 **40.** 6 **41.** 48 **42.** 72 **43.** 8 **44.** 5 **45.** 0 **46.** 28 **47.** 27 **48.** 0 **49.** 15
50. 48 **51.** 45 **52.** 12 **53.** 0 **54.** 27 **55.** 81 **56.** 20 **57.** 0 **58.** 9 **59.** 0 **60.** 6 **61.** 18
62. 7 **63.** 3 **64.** 21 **65.** 36 **66.** 0 **67.** 63 **68.** 12 **69.** 35 **70.** 0 **71.** 42 **72.** 0 **73.** 40
74. 8 **75.** 0 **76.** 24 **77.** 9 **78.** 0 **79.** 15 **80.** 16 **81.** 6 **82.** 30 **83.** 0 **84.** 4 **85.** 21
86. 8 **87.** 0 **88.** 49 **89.** 16 **90.** 24 **91.** 0 **92.** 14 **93.** 4 **94.** 0 **95.** 6 **96.** 8 **97.** 18
98. 10 **99.** 0 **100.** 42

Appendix F Review of Angles, Lines, and Special Triangles

1. 71° **3.** 19.2° **5.** $78\frac{3}{4}°$ **7.** 30° **9.** 149.8° **11.** $100\frac{1}{2}°$

13. $m\angle 1 = m\angle 5 = m\angle 7 = 110°; m\angle 2 = m\angle 3 = m\angle 4 = m\angle 6 = 70°$ **15.** 90° **17.** 90° **19.** 90° **21.** 45°, 90°

23. 73°, 90° **25.** $50\frac{1}{4}°$, 90° **27.** 6 **29.** 4.5 **31.** 10 **33.** 12

INDEX